Lecture Notes in Civil Engineering 646

Lecture Notes in Civil Engineering (LNCE) publishes the latest developments in Civil Engineering—quickly, informally and in top quality. Though original research reported in proceedings and post-proceedings represents the core of LNCE, edited volumes of exceptionally high quality and interest may also be considered for publication. Volumes published in LNCE embrace all aspects and subfields of, as well as new challenges in, Civil Engineering. Topics in the series include:

- Construction and Structural Mechanics
- Building Materials
- Concrete, Steel and Timber Structures
- Geotechnical Engineering
- Earthquake Engineering
- Coastal Engineering
- Ocean and Offshore Engineering; Ships and Floating Structures
- Hydraulics, Hydrology and Water Resources Engineering
- Environmental Engineering and Sustainability
- Structural Health and Monitoring
- Surveying and Geographical Information Systems
- Indoor Environments
- Transportation and Traffic
- Risk Analysis
- Safety and Security

To submit a proposal or request further information, please contact the appropriate Springer Editor:

- Pierpaolo Riva at pierpaolo.riva@springer.com (Europe and Americas);
- Swati Meherishi at swati.meherishi@springer.com (Asia—except China, Australia, and New Zealand);
- Wayne Hu at wayne.hu@springer.com (China).

All books in the series now indexed by Scopus and EI Compendex database!

Nikos D. Lagaros · Savvas P. Triantafyllou ·
Rajai Z. Alrousan · Mohammad A. Alhassan ·
Khairedin M. Abdalla
Editors

Proceedings of the International Conferences on Digital Technology Driven Engineering 2024

ADDitively Manufactured OPTimized Structures by means of Machine Learning (ADDOPTML)

Springer

Editors
Nikos D. Lagaros
Department of Structural Engineering
National Technical University of Athens
Athens, Greece

Savvas P. Triantafyllou
Department of Structural Engineering
National Technical University of Athens
Athens, Greece

Rajai Z. Alrousan
Civil Engineering
Jordan University of Science and Technology
Irbid, Jordan

Mohammad A. Alhassan
Civil Engineering
Jordan University of Science and Technology
Irbid, Jordan

Khairedin M. Abdalla
Jordan University of Technology
Irbid, Jordan

ISSN 2366-2557 ISSN 2366-2565 (electronic)
Lecture Notes in Civil Engineering
ISBN 978-3-031-92028-8 ISBN 978-3-031-92029-5 (eBook)
https://doi.org/10.1007/978-3-031-92029-5

This Springer imprint is published by the registered company Springer Nature Switzerland AG
The registered company address is: Gewerbestrasse 11, 6330 Cham, Switzerland

Contents

Integration of 3D Printing and Machine Learning in Sustainable Construction: Feasibility and Challenges

Ayah Alkhawaldeh[1], Mohammad Alhassan[2,3](✉), Ansam Sawalha[1], Nour Betoush[3], Layla Amaireh[2], Khairedin Abdalla[3], and Nikos Lagaros[4]

[1] Department of Civil Engineering, American University of Madaba, Madaba, Jordan
[2] Civil Engineering Program, Al Ain University, Al Ain, UAE
mohammad.alhassan@aau.ac.ae
[3] Department of Civil Engineering, Jordan University of Science and Technology, Irbid, Jordan
[4] Institute of Structural Analysis and Antiseismic Research, National Technical University of Athens, Athens, Greece

Abstract. In recent years, three-dimensional (3D) printing and machine learning applications in sustainable construction have evolved. Recent innovations in 3D printing technology have led to its integration in a variety of industries, including buildings and constructions. It is a technology for creating 3D objects layer by layer using a computer-generated design through machine learning depending on pre-identified parameters and the desired output. This paper aims to provide a review about the current practices of the 3D printing and machine learning in sustainable construction and useful suggestions to increase its popularity and potential. Several countries started using 3D printing in construction by creating small-scale prototypes of steel or concrete that can allow for actual printing of full-scale structural systems. The advantages of this technology range from being environmentally friendly to reducing the overall weight of the individual structural elements without affecting their capacities and performance. However, the technology is still facing some major challenges that need to be resolved in order to consider the 3D printing as a competitive and effective technology in the construction industry with optimized shapes, reduced materials, and maintained sustainability criteria.

Keywords: 3D printing · construction industry · machine learning · sustainability

1 Introduction and Background

3D printing technology or additive manufacturing (AM) has gained momentum recently with innovative applications relevant to structural engineering, the construction industry, and concrete technology. These innovations in 3D printing technology have led to its integration in buildings and construction as a new competitive approach, allowing for fast-track projects entailing complex architectural geometries. It is a technology for

N. D. Lagaros et al. (Eds.): ADDOPTML 2024, LNCE 646, pp. 1–10, 2025.
https://doi.org/10.1007/978-3-031-92029-5_1

creating 3D objects layer by layer using a computer-generated design through machine learning, depending on target parameters and outputs. The advantages of this technology range from being environmentally friendly to reducing the overall weight of the individual structural elements without affecting their capacities and performance. However, the 3D printing technology is still in need for further improvements necessary for its globalization and full adoption by decision makers and design codes as a construction method resulting in optimized shapes, reduced materials, and maintained sustainability design goals (SDGs). 3D printing technology is widely used in construction in order to create prototypes, architectural models, complex structures, repair and retrofitting, topology optimization, structural health monitoring sensors, structural elements and components, and even entire buildings [1]. It can promote sustainable construction since it ensures that only the needed amount of materials is utilized, leading to waste reduction and cost savings. 3D printing technology has been used since the mid-1990s by using contour crafting and other technologies. Contour crafting can provide flexibility in creating concrete shapes, which were difficult to achieve in traditional pouring techniques; with this technique, houses can be constructed efficiently, sustainably, and environmentally friendlier [2].

In recent years, the usage of 3D printing on construction objects other than ceramic has grown. 3D printers, for example, can now print plastic and nylon components that are often used in building projects as window frame fixtures, plug fixtures, and plumbing fittings. More crucially, concrete printing has been demonstrated to be capable of printing geometrically complex concrete products [3]. However, by adopting 3D printing, the size of the concrete products was limited. For instance, Lim et al. [2] used a 3D printer with a print dimension of up to 5.4 m (height) $\times$ 4.4 m (width) $\times$ 5.4 m (length). However, this size could be enough to print basic precast concrete elements. As a result, the key challenge with using 3D printers to create construction parts is not associated with the size but whether there is enough customization and flexibility [4].

One of the newest environmentally friendly binder ingredients for 3D printing is geopolymer, which is utilized in place of more conventional cement-based materials [5–7]. Geopolymers are made by combining two parts, which are composed of aluminosilicate materials in solid form and alkaline solutions in liquid form, resulting in a durable and strong material that is also more environmentally friendly. Aluminosilicates are derived primarily from two sources: slag, a result of iron production, and fly ash, a byproduct of coal-fired power stations. It would be advantageous to use these industrial by-products as the main raw material for 3D concrete printing since it would reduce their global problems with landfill storage [8, 9]. The carbon emissions from the manufacture of geopolymers can be up to 80% lower than those from the production of OPC, according to [10]. Besides being environmentally friendly, geopolymer is also preferred because of its excellent thixotropy, ability to quickly reach high strengths, and flexible setting properties that can be achieved by replacing alkaline activators and/or precursors [11, 12]. Nematollahi et al. [13] investigated the properties of a geopolymer mortar that can be printed in three dimensions, both with and without fibers.

A recently developed "one-part" (just add water) geopolymer was examined by Bong et al. [14] for 3D concrete printing applications to improve its mechanical properties and printability performance. After determining the optimum composition for 3D concrete printing of structural elements, they discovered that the printed specimens had high mechanical performance. The 3D-printable geopolymer mixture needs 14% less energy and emits 61% less CO_2 during manufacture than a 3D-printable Portland cement-based mixture with comparable compressive strength, according to the material sustainability indicators. The creation of such "one-part" geopolymers that can be printed in 3D is expected to greatly advance the commercial feasibility and wide-ranging applications of geopolymers in the context of three-dimensional concrete printing. A large-scale staircase (Fig. 1) with intricate geometry proved that the optimum geopolymer mixture for 3D printing in this study is suitable for large-scale 3D concrete printing applications.

Fig. 1. 3D-printing of a staircase using a geopolymer mixture (Bong et al., 2021).

Moreover, the utilization of 3D printing technology in seismic design has great potential for enhancing current detailing practices with optimized seismic resisting systems that have the required strength and ductility characteristics with neat architectural features. Recent research studies emphasized the level of adaptation and sophistication that can be achieved with the aid of 3D printing technology, allowing for unconventional designs and detailing that lead to improved seismic response with cutting-edge materials. Delavar et al. [15] proposed a design methodology using 3D printed reinforced concrete lateral force-resisting walls for affordable and resilient housing in seismic regions. They determined analytically, with experimental validation, the response modification factors and strength design equations for the proposed system. Furthermore, the properties of the concretes utilized in 3D-printed seismic resisting systems were evaluated in terms of their influence on the seismic behavior and overall integrity of the structure. Zhu et al. [16] developed 3D-printable fibrous cementitious composites having high extensibility using discontinuous polyethylene fibers; they reported that the developed cementitious composites demonstrated resilient strain-hardening performance with high tensile strength and strain capacities. Mintsaev et al. [17] explored the use of 3D printing and additive manufacturing in the construction industry using local materials, focusing on the needed improvements and adaptations in the equipment to allow for the creation of various building types qualified for solving challenges, including soil collapsing and active seismicity regions. Lorenzo and Vassiliou [18] underlined the impending of additive manufacturing for confirming conventions in numerical simulations of seismic-resisting masonry structures through improving the mechanical characteristics of the 3D-printed materials, including the compressive, shear, bending, and tensile strengths. In accordance with the reviewed studies above, this paper provides a review about the current practices of the

3D printing and machine learning in sustainable construction with useful suggestions and remarks. The aim is to increase its popularity and feasibility, so that, the involved stakeholders and design organizations consider its effectiveness and suggest any needed modifications for its full adoption and completion.

2 3D Structural Printing and Machine Learning

3D structural printing, commonly referred to as additive manufacturing or 3D printing for building, is an innovative technique that builds three-dimensional structures layer by layer utilizing a wide variety of materials. This technology has captured a great deal of interest in the construction industry because of its potential for changing the way buildings and other structures (e.g., housing, infrastructure, emergency shelters, etc.) are constructed and designed. An example of the execution steps for a 3D-printed structure is displayed in Fig. 2.

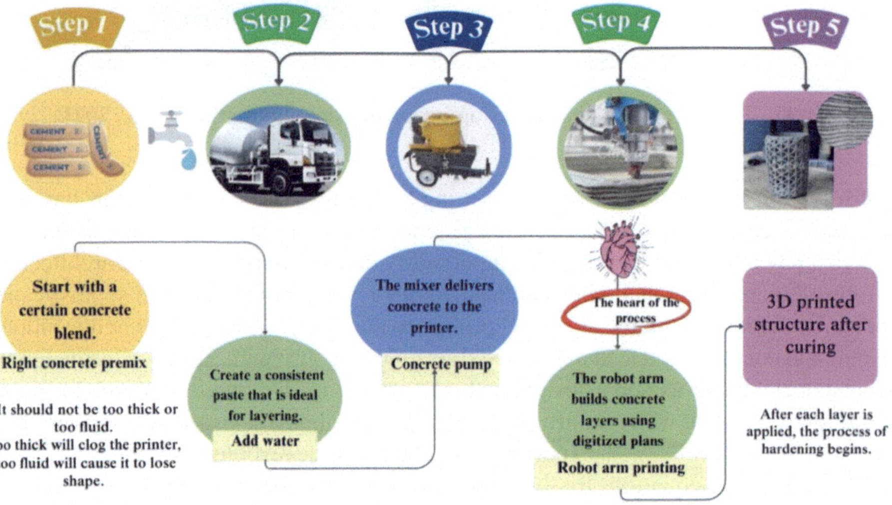

Fig. 2. Steps for creating structure using 3D-printing technology.

The integration of ML algorithms in the field of 3D printing combines several important factors that have a direct influence on the quality of the final 3D printed parts, including design for 3D printing, component quality, process optimization, printing capability, real-time adjustment, and in-situ monitoring for quality control [19]. Combining machine learning with 3D structural printing creates a strong interaction that might improve the construction sector by bringing new ideas, increased productivity, and sustainable methods. Since machine learning (ML) has the potential to allow computers to learn on their own from available data and make decisions with minimal to no human involvement, 3D printing has made it possible for users with little manufacturing experience to quickly and easily manufacture complex 3D structures using a variety of materials. The best printing settings are still difficult to determine, which lengthens the pre-printing process and wastes more material. In order to optimize printing parameters, Dabbagh et al. [20] create the first graphical user interface (GUI) that combines machine learning (ML) with 3D printing in an approachable way. In contrast to the commonly accepted orthogonal design utilized in the majority of 3D printing research, the authors used nine distinct computer-aided design (CAD) images for the first time. They also developed a self-designed method to calculate the "complexity index" of CAD designs, which allowed ML algorithms to determine the differences between designs. An illustration of the various steps involved in optimizing 3D printing parameters with ML is represented in Fig. 3.

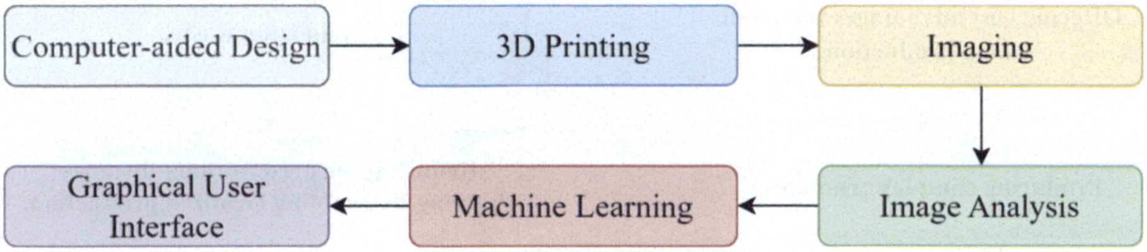

Fig. 3. Various steps involved in optimizing 3D printing parameters with ML.

3 Feasibility and Sustainability

The feasibility and sustainability of 3D printing are varied and dependent on a number of elements, such as the particular application, materials, and the full life cycle of the product, which is an important point to consider. Enhancing the sustainability of 3D printing materials and processes is a goal as technology develops. 3D printing technology has various features relating to feasibility and sustainability, as displayed in Fig. 4. The goal of ongoing research and development is to provide the 3-D printing sector with more environmentally friendly procedures and materials. This goal is explained and clarified in detail in the following subsections.

It is well acknowledged that the building sector is one of the riskiest industries in the global economy. Traditional procedures deplete natural resources, produce hard-to-recycle building waste, require significant quantities of energy, and endanger workers' occupational health and safety. As a remedy to these issues, 3D printing technology has been offered as an environmentally friendly construction approach. According to the reviewed research, it is generally acknowledged that 3D printing considerably lowers threatening negative ecological effects. Reduced greenhouse gas (GHG) emissions are achieved by the use of limited building materials, waste removal, decreased energy consumption, and increased production efficiency [21]. Time is a key element in any construction project. Cutting the time frame of construction offers customers several advantages in terms of construction productivity, including an early start to operations and income generation, lower overhead, and more resources available for other projects. 3D printing allows for faster construction. When compared to conventional approaches, the building process moves substantially more quickly [22].

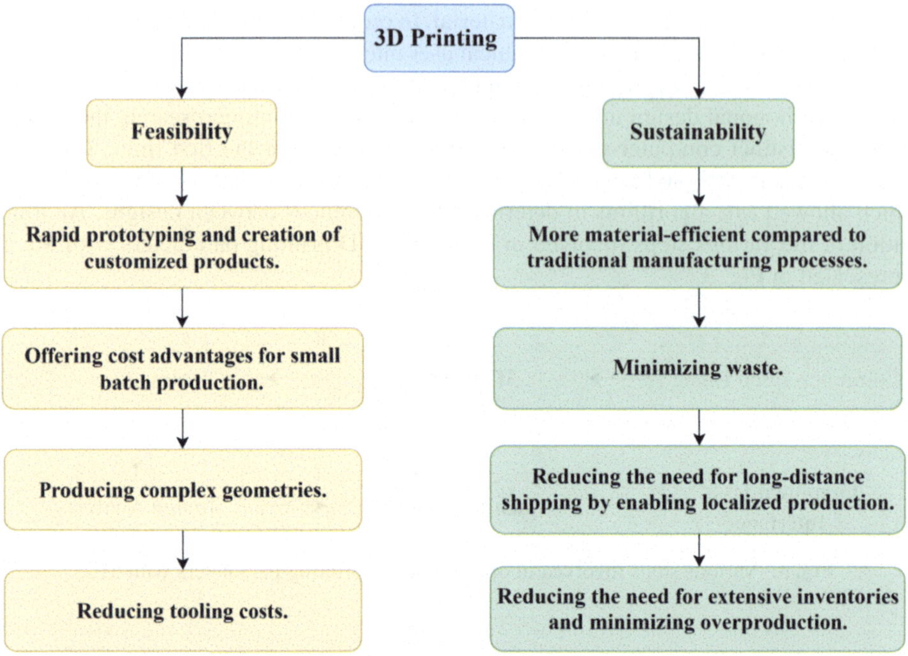

Fig. 4. Main feasibility and sustainability features of 3D printing technology

4 Challenges and Opportunities

The benefit of 3D printing technology over traditional building methods is the integration of material preparation, structural performance design, geometrical modeling, and construction, which may accomplish the unification of performance and function. At present, 3D printing technology research and application have covered all levels of materials, connections, structures, and optimization [23–25]. The potential of 3D printing to completely transform traditional methods of construction has drawn considerable interest from the structural engineering community. Despite the aforementioned benefits of 3D printing, several drawbacks are revealed in using this technique in the applications of structural engineering. Additionally, there is a danger of cybersecurity and hacking because the creation process is automated and all data is stored in a 3D model [26]. Security issues surround the 3D models' execution, transport, and storage; additive manufacturing is in danger of going over the security infrastructure required to guarantee public safety [27]. Table 1 represents the major challenges and opportunities in the field of using 3D printing in structural engineering. Figure 5 illustrates the primary advantages documented in the construction field related to the utilization of 3D printing.

Table 1. Summary of 3D printing challenges and opportunities in structural engineering.

Challenges	Opportunities
Construction materials for 3D printing are scarce. It is critical to guarantee that the printed structures fulfill legal and industrial requirements	Developments in new materials, especially those suited for 3D printing, have the potential to produce buildings that are more robust, long-lasting, and reasonably priced
When compared to traditional building methods, the time needed to 3D print large-scale constructions might be a constraint	3D printing for structural applications may become more feasible with progressions in large-scale printing technology and speed
Safety and structural integrity of 3D-printed constructions are crucial concerns. It is also essential to understand the behavior and long-term performance of printed materials	Structural integrity of 3D-printed components may be verified with the advancements in testing and simulation techniques. Collaboration with regulatory agencies, engineers, and researchers is essential
Materials and 3D printing technology associated with initial costs can be expensive	Costs can be reduced as technology progresses material optimization
Aspects of 3D-printed constructions that are unique may not be adequately covered by current building codes and regulatory systems and agencies	Creation of standards and procedures for certification of 3D-printed structures can be aided by cooperation between researchers, stakeholders, and organizations

3D printing is an innovative technology that is revolutionizing the traditional construction sector by making housing construction affordable, faster, and more sustainable. This technology allows for the creation of unique and versatile designs directly at the construction site, ensuring high precision [24, 28]. Some of the most common innovations of 3D printing in construction include:

1. Drones: drones can be utilized to monitor the process of 3D printing, evaluate the manufactured structures' quality, and deliver materials directly to the printing site. Drones can also be incorporated with 3D printers to build aerial structures that would otherwise be difficult or impossible to construct [29].
2. Connection bolts: connection bolts are 3D-printed fasteners that connect distinct components of a printed structure. They have the potential to eliminate or reduce the requirement for welding or gluing while also improving the structural integrity and endurance of printed buildings [30].
3. Hybrid structures: Hybrid constructions integrate 3D-printed components with traditional materials such as steel, concrete, or wood. They can improve the performance and appearance of printed structures while also overcoming some of the constraints of 3D printing materials [24, 28].

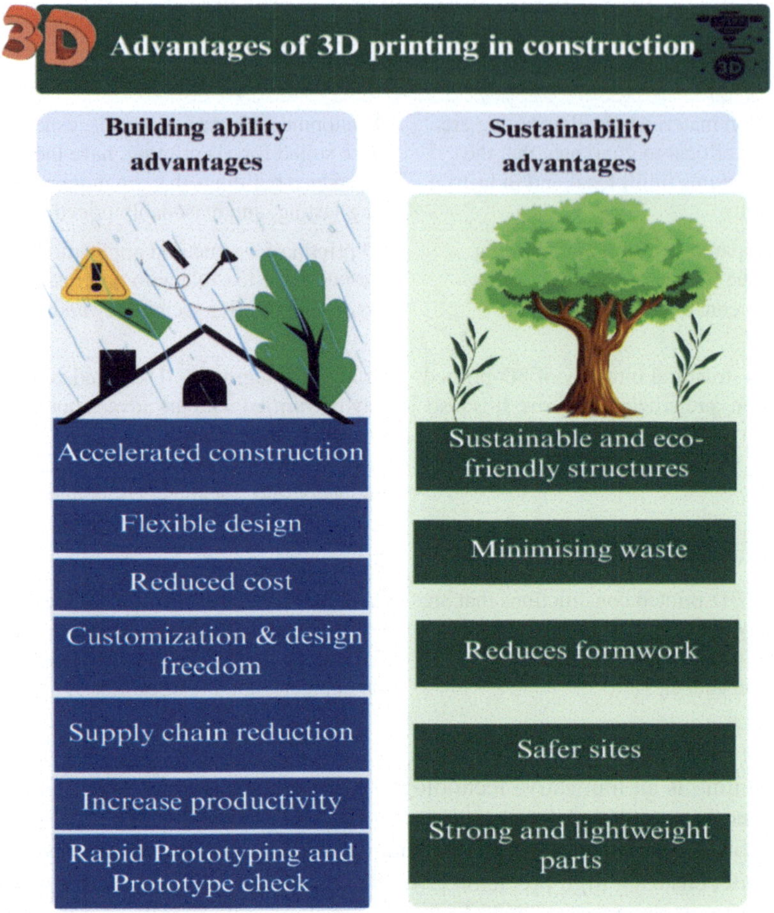

Fig. 5. Advantages documented in the construction field by using 3D printing technology.

5 Concluding Remarks

Integration of 3D printing and ML in sustainable construction gives an advantageous direction for transformational innovations. However, the feasibility of such integration provides potential for higher productivity, lower waste, higher cost-effectiveness, and innovative design opportunities such as material optimization and regulatory adaptation. The need for large data sets highlights the importance of ongoing research and collaboration. Successfully overcoming these issues might pave the path for a more sustainable and modernized future in the building sector. 3D printing in the future will be faster and more affordable. The scale of 3D printers will increase, allowing for bigger projects. Therefore, it is likely that within a year, printers will be able to create structures with up to six floors. 3D printers also have various potential applications for sustainable cities, including improving customization and resilience, fostering social innovation and inclusion, and reducing emissions and waste.

Acknowledgments. Part of this research was supported by the ADDOPTML (ADDitively Manufactured OPTimized Structures by Means of Machine Learning) project (No: 101007595) belonging to the Marie Skłodowska-Curie Actions (MSCA) Research and Innovation Staff Exchange (RISE) H2020-MSCA-RISE-2020. Their support is highly acknowledged. The authors would also like to thank the Jordan University of Science and Technology (JUST) for its support as a partner organization in ADDOPTML project.

References

1. Elhadad, A.A., Rosa-Sainz, A., Cañete, R., Peralta, E., Begines, B., Balbuena, M., Alcudia, A., Torres, Y.: Applications and multidisciplinary perspective on 3D printing techniques: recent developments and future trends. Mater. Sci. Eng. R. Rep. **156**, 100760 (2023)
2. Lim, S., Buswell, R.A., Le, T.T., Austin, S.A., Gibb, A.G.F., Thorpe, T.: Developments in construction-scale additive manufacturing processes. Autom. Constr. **21**(1), 262–268 (2012)
3. Yossef, M., Chen, A.: Applicability and limitations of 3D printing for civil structures. In: The Proceedings of the 2015 Conference on Autonomous and Robotic Construction of Infrastructure (2015)
4. Gibson, I., Kvan, T., Wai Ming, L.: Rapid prototyping for architectural models. Rapid Prototyp. J. **8**(2), 91–95 (2002)
5. Shilar, F.A., Ganachari, S.V., Patil, V.B., Bhojaraja, B.E., Yunus Khan, T.M., Almakayeel, N.: A review of 3D printing of geopolymer composites for structural and functional applications. Constr. Build. Mater. **400**, 132869 (2023)
6. Alhassan, M., Alkhawaldeh, A., Betoush, N., Alkhawaldeh, M., Huseien, G.F., Amaireh, L., Elrefae, A.: Life cycle assessment of the sustainability of alkali-activated binders. Biomimetics. **8**(1), 58 (2023)
7. Alkhawaldeh, A.A., Alhassan, M.A., Elrefae, A.: A case study of implementing life cycle cost analysis in sustainability assessment. In: 2022 International Arab Conference on Information Technology (ACIT). IEEE (2022)
8. Panda, B., Paul, S.C., Hui, L.J., Tay, Y.W.D., Tan, M.J.: Additive manufacturing of geopolymer for sustainable built environment. J. Clean. Prod. **167**, 281–288 (2017)
9. Zhao, J., Tong, L., Li, B., Chen, T., Wang, C., Yang, G., Zheng, Y.: Eco-friendly geopolymer materials: a review of performance improvement, potential application and sustainability assessment. J. Clean. Prod. **307**, 127085 (2021)
10. Duxson, P., Provis, J.L., Lukey, G.C., van Deventer, J.S.J.: The role of inorganic polymer technology in the development of 'Green Concrete'. Cem. Concr. Res. **37**(12), 1590–1597 (2007)
11. Guo, X., Yang, J., Xiong, G.: Influence of supplementary cementitious materials on rheological properties of 3D printed Fly ash based geopolymer. Cem. Concr. Compos. **114**, 103820 (2020)
12. Panda, B., Tan, M.J.: Experimental study on mix proportion and fresh properties of Fly ash based geopolymer for 3D concrete printing. Ceram. Int. **44**(9), 10258–10265 (2018)
13. Nematollahi, B., Vijay, P., Sanjayan, J., Nazari, A., Xia, M., Naidu Nerella, V., Mechtcherine, V.: Effect of polypropylene fibre addition on properties of geopolymers made by 3D printing for digital construction. Materials. **11**(12), 2352 (2018)
14. Bong, S.H., Xia, M., Nematollahi, B., Shi, C.: Ambient temperature cured 'Just-Add-Water' geopolymer for 3D concrete printing applications. Cem. Concr. Compos. **121**, 104060 (2021)
15. Delavar, M.A., Chen, H., Sideris, P.: Seismic design methodology for 3D printed concrete buildings. City. **25**(1), 177–197 (2023)

16. Zhu, B., Pan, J., Nematollahi, B., Zhou, Z., Zhang, Y., Sanjayan, J.: Development of 3D printable engineered cementitious composites with ultra-high tensile ductility for digital construction. Mater. Des. **181**, 108088 (2019)
17. Mintsaev, M.S., Bataev, D.S., Mazhiev, K.K., Mazhiev, A.K., Mazhieva, A.K., Mazhiev, A.K., Mazhiev, M.K.: Prospects for using 3D-printing technologies in construction of buildings in seismic areas. In: International Symposium "Engineering and Earth Sciences: Applied and Fundamental Research" (ISEES 2018), pp. 311–315. Atlantis Press (2018)
18. Del Giudice, L., Vassiliou, M.F.: Mechanical properties of 3D printed material with binder jet technology and potential applications of additive manufacturing in seismic testing of structures. Addit. Manuf. **36**, 101714 (2020)
19. Goh, G.D., Sing, S.L., Yeong, W.Y.: A review on machine learning in 3D printing: applications, potential, and challenges. Artif. Intell. Rev. **54**(1), 63–94 (2020)
20. Rahmani Dabbagh, S., Ozcan, O., Tasoglu, S.: Machine learning-enabled optimization of extrusion-based 3D printing. Methods. **206**, 27–40 (2022)
21. Ibrahim, I., Eltarabishi, F., Abdalla, H., Abdallah, M.: 3D printing in sustainable buildings: systematic review and applications in the United Arab Emirates. Buildings. **12**(10), 1703 (2022)
22. Hager, I., Golonka, A., Putanowicz, R.: 3D printing of buildings and building components as the future of sustainable construction? Procedia. Eng. **151**, 292–299 (2016)
23. Liu, D., Zhang, Z., Zhang, X., Chen, Z.: 3D printing concrete structures: state of the art, challenges, & opportunities. Constr. Build. Mater. **405**, 133364 (2023)
24. El-Sayegh, S., Romdhane, L., Manjikian, S.: A critical review of 3D printing in construction: benefits, challenges, and risks. Arch. Civ. Mech. Eng. **20**(2), 1–25 (2020)
25. Lowke, D., Dini, E., Perrot, A., Weger, D., Gehlen, C., Dillenburger, B.: Particle-Bed 3D printing in concrete construction—possibilities and challenges. Cem. Concr. Res. **112**, 50–65 (2018)
26. Labonnote, N., Rønnquist, A., Manum, B., Rüther, P.: Additive construction: state-of-the-art, challenges and opportunities. Autom. Constr. **72**, 347–366 (2016)
27. Bridges, S.M., Keiser, K., Sissom, N., Graves, S.J.: Cyber security for additive manufacturing. In: Proceedings of the 10th Annual Cyber and Information Security Research Conference. CISR'15: 10th Annual Cyber and Information Security Research Conference (2015)
28. Pan, Y., Zhang, Y., Zhang, D., Song, Y.: 3D printing in construction: state of the art and applications. Int. J. Adv. Manuf. Technol. **115**(5–6), 1329–1348 (2021)
29. Kazemian, A., Seylabi, E., Ekenel, M.: Concrete 3D printing: challenges and opportunities for the construction industry. In: Innovation in Construction, pp. 277–299 (2022)
30. Wi, J.-H., Lee, K.-H., Lee, C.-H.: Self-loosening of 3D printed bolted joints for engineering applications. MATEC. Web. Conf. **185**, 00029 (2018)

Form-Finding and Automated Fabrication of GFRP Panels with Double Curvate for a Canopy Structure

Andreas Kyriakou[1(✉)], George Tryfonos[2], Apostolos Michopoulos[2], Marios C. Phocas[2], and Aimilios Michael[2]

[1] DERGON, Spyrou Kyprianou, 44, 4193 Limassol, Cyprus
communications@dergon.com

[2] Department of Architecture, University of Cyprus, Kallipoleos 75, 1678 Nicosia, Cyprus

Abstract. Developments in digital design technology currently applied in the architectural and engineering fields enable the extension of design processes from the conceptual phase to detailed design stage and fabrication. Non-conventional structures with complex architectural forms are often realized through technology transfer and material advancements and characterized with enhanced geometrical and mechanical features. In this framework, architectural form, structure and material comprise an indispensable, homogenized unity in the development of the prototype. Along these lines, the current paper refers to the form-finding and automated fabrication process of a canopy structure of steel members with glass-fiber reinforced polymer (GFRP) panels of double curvature. The canopy structure serves as roof element for an electricity changing station and is located in a public parking area. The preliminary design of the structure succeeded in parametric associative design environment. The adopted final shape derived from a form-finding process of the respective adaptive system of cables actuated bending-active steel members stabilized by the prefabricated GFRP panels. Following fabrication of the steel members, the member structure was assembled and erected indoors and 3D-scannned for redefinition of the exact geometries of the panels. The latter were then 3D-printed in series by an industrial robot (6-axis KUKA KR2700). Following erection of the member structure in place, the GFRP panels were assembled. The digital design and automated fabrication process serves as case example of a nonlinear interdisciplinary approach followed in achieving a non-conventional lightweight structure with double curvature that was derived from a form-finding simulation process of the respective adaptive system.

Keywords: Adaptive Structures · Parametric Associative Design · Form-Finding · Automated Fabrication · Glass-Fiber Polymer Panels

1 Introduction

The contemporary building industry deals with issues of productivity, effectiveness, ecological and environmental awareness, reduction of energy consumption and economic costs. Related developments refer to fast assembly processes, lightweight construction

N. D. Lagaros et al. (Eds.): ADDOPTML 2024, LNCE 646, pp. 11–21, 2025.
https://doi.org/10.1007/978-3-031-92029-5_2

and new materials in achieving structural innovation and efficiency. At the same time, automated fabrication, i.e., additive and robotic manufacturing, provides potential for flexible production of complex geometry objects with minimized time and energy consumption. Automated fabricated components have already found numerous applications within the European aerospace and transport industry due to related advantages, e.g., less machine, material, and labor costs, less manufacturing waste and responsible raw material usage/management.

Bending-active structures exhibit complex curved shapes that originate from flat elements [1, 2]. They have enhanced capabilities in their kinematics in providing different configuration transitions while maintaining geometrical reversibility. The elastic deformation is handled through a form-finding procedure, in which form and force counteract. The work of Frei Otto in the early 60s introduced form-finding as a design tool and opened up a new spectrum of a material-oriented approach in Architecture [3]. Meanwhile, an integration between form, material and structure may be simulated through parametric associative design coupled with form-finding and structural analysis tools [4].

The application of bending-active members has been originally demonstrated with a pavilion prototype construction of birch plywood lamellas in 2010 [5]. The internal module structure connections used finger joints and automated wood sewing technology to connect the edges together. A pavilion of glass-fiber polymer (GFRP) material simulated in its bending-active behavior to obtain the final form was automated manufactured in real scale and presented in [6]. The load-bearing funicular structure has a modular composition. Both projects are representative of the development of lightweight natural structures and the study of form-finding.

The current paper refers to the design and automated fabrication process of a canopy structure of steel members with GFRP panels of double curvature. The structure serves as roof element for an electricity changing station of two cars and is located in a public parking area. The preliminary design of the structure was based on a form-finding process of the respective adaptive system of cables actuated bending-active steel members stabilized by the prefabricated GFRP panels. The automated fabrication of the structural elements serves as case example of a nonlinear interdisciplinary approach followed in the realization of the prototype structure.

2 Structural System

The parametric form-finding approach applied aimed at defining the geometry and form of the structure. A parametric algorithm developed provides control of the global dimensions of the structure, i.e., height, length and width. The methodology followed utilizes the parametric tools of Grasshopper plug-in for Rhinoceros software [7, 8] and Kangaroo plugin for Grasshopper [9] as standard components. The use of the parametric algorithm allows the interdisciplinary development of the steel structure based on its form-finding and static analysis, its connections design, the robotic fabrication of the GFRP panels, as well as the assembly of the canopy onsite, Fig. 1.

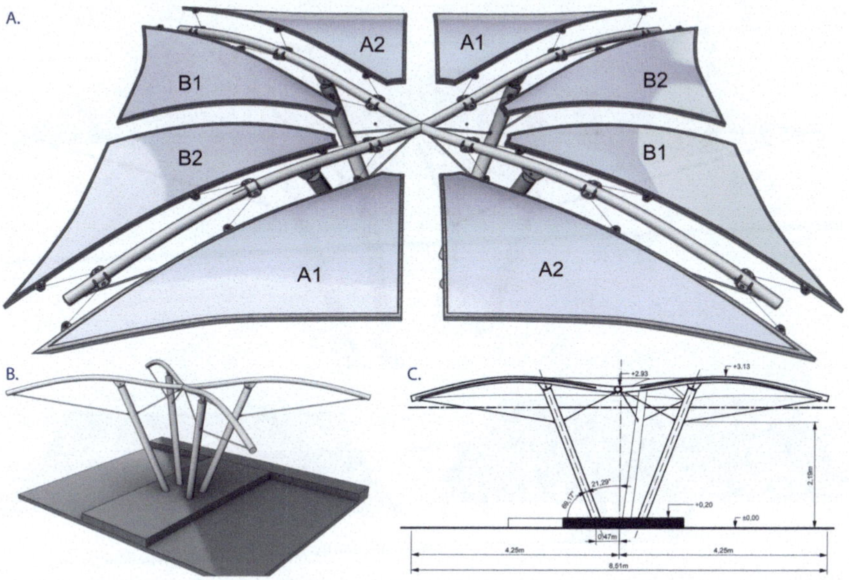

Fig. 1. Canopy structure in its components of steel columns and beams, as well as GFRP panels for the roof surface with double curvature: (**a**) Blow-up drawing of structural components; (**b**) Isometric view of structural members; (**c**) Elevation of the structure.

The four inclined columns of the structure supporting four curved beams are rigidly connected to the ground. The inclination of the columns amounts to 69.17 °C to the horizontal plane. This value was determined by architectural criteria. The rectangular roof surface of the structure has overall dimensions of 7 × 5 m. In horizontal plan, the support points of the beams are positioned at a distance of 2.5 m from the centre of the rectangular roof surface. The beams are stabilized in their position through diagonal cables that connect their end points and the central joint of the beams with the columns at a height of 2.19 m. The structure has a maximum length of 8.51 m in the longitudinal elevation. Assuming the cables to have variable length and the beams to be able to undergo large elastic deformations, the cables' activation induces bending deformations to the beams, controlled by the bending parameters in Kangaroo plug-in, Fig. 2. The roof surface follows the bending deformations of the beams in obtaining a corresponding double curvature shape.

The structural analysis considered the self-weight of the steel structure, a surface load for the GFRP panels of 0.60 kN/m^2, a non-permanent vertical load of 0.60 kN/m^2, as well as a resulting seismic point load at the end points of the columns of 3.4 kN based on Eurocode 8 and a wind velocity of 24 m/s for the definition of the wind loads based on Eurocode 1. The steel members consist of CHS (circular hollow section) with 168.3 × 5.0 and 114.3 × 3.6 mm cross-section for the columns and the beams respectively. The steel members are assigned to 210 GPa elastic modulus and 275 MPa yield strength. The cables have a diameter of 10 mm and are assigned to 195 GPa elastic modulus and 275 MPa yield strength. The cables are assigned with 2 kN prestress.

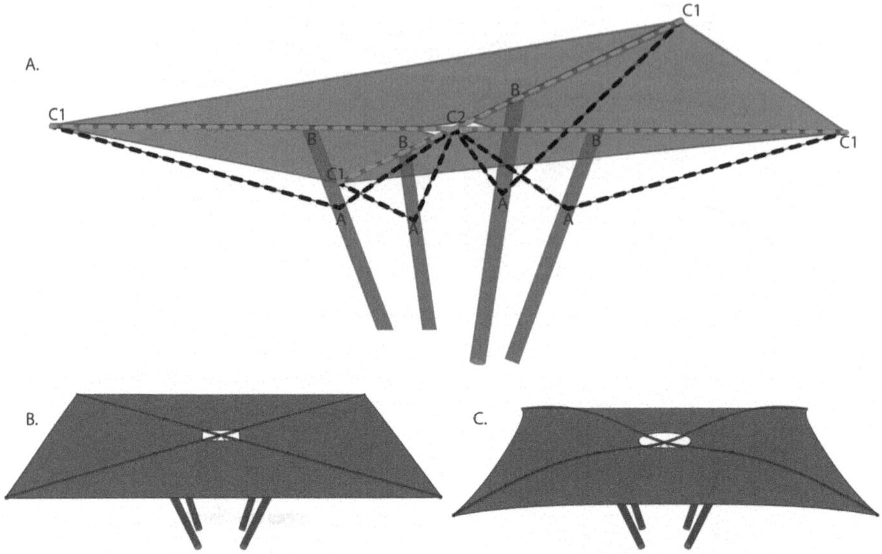

Fig. 2. Form-finding methodology of the canopy structure: (**a**) Initial state with green colored lines showing the undeformed beams and black colored lines, the activation cables with variable length; (**b**) Isometric view of undeformed structural shape; (**c**) Isometric view of deformed structural shape corresponding to the form-found state.

The material of the panels was significant for the elements' fabrication through the automated 3D printing process applied. The material comprised a combination of PP (polypropylene) and GF (glass-fiber) in percentage reaching up to 50%. The reason for using PPGF material was threefold: to achieve material endurance in the local weather conditions of Cyprus, to utilize the structural properties of the glass-fiber matrix, as well as its physical properties in reducing shrinkage effects during the printing process. Certain additives, such as UV and humidity additives, were added to ensure a longer endurance of the material to the local weather conditions. The GFRP panels have a thickness of 4 cm.

3 Components Fabrication

3.1 Steel Members

The beams bending process using a roller machine played a significant role in the fabrication process. In achieving accurate bending results of the circular beams curvature, the profile of the 3D geometry of the beams was extracted and fabricated on a water jet cut using a PVC plastic sheet of the target curvature of the members. The latter acted as reference throughout the manual bending process of the beams, Fig. 3.

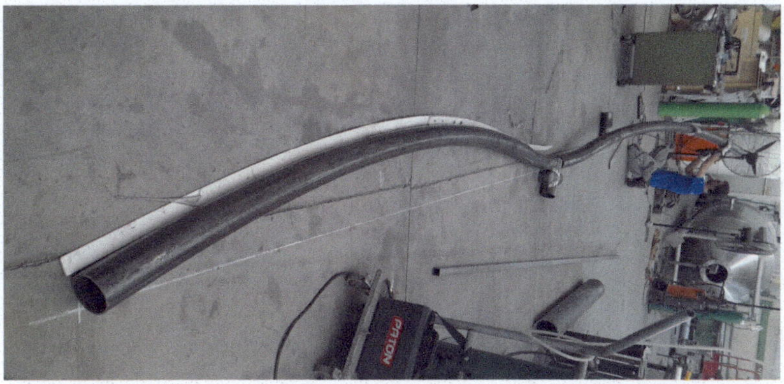

Fig. 3. Beams bending as to the target curvature of the members.

The folding of the beams at their connection point was digitally simulated to create the joint connection of the members. The beams were joined individually to a solid central element through use of a lathe. An axis was created in each direction upon which the beams interlocked with each other. Once in place, all four tubes were welded in position, Fig. 4.

Fig. 4. Beams connection to the central joint.

Crucial point of the four main columns construction was the correct welding inclination of each steel section to the base plate. Due to the rectangular section of the base plate, an alternation in the degrees of inclination of each of the four columns was necessary, Fig. 5.

Fig. 5. Columns connection to the base plate.

3.2 Off-Site Assembly

The base plate was fixed to a floor concrete plate with spot welded threaded rods, in order to temporarily erect the steel structure. This process was expected to clarify eventual inaccuracies compared to the digital design. Once the columns were aligned, the beams were placed in place using a crane. The distances between the beams and the columns were evenly distributed to match the drawing dimensions. The central beams joint was spot welded with the columns accordingly, Fig. 6.

Fig. 6. Temporary indoors erection of the steel structure.

An initial alignment aimed at adjusting the beams openings in closest possible approximation to the actual drawings. The beam alignment was significant for the definition of the GFRP panels geometries and their fabrication. A 3D scanning of the structure succeeded. The geometry of the structure was measured with a Leica RTC360 3D Laser Scanner to indicate any inaccuracies compared to the digital 3D structure, Fig. 7.

Fig. 7. 3D scan of the steel structure.

3.3 GFRP Panels

In providing an accurate measurement of the GFRP panels in accordance with the 3D scan of the structure, a surface geometry of the panels was created with Grasshopper using a four-member polyline method, Fig. 8. To create a solid piece based on the granular deposition modeling method, a double layer surface geometry was created that allowed to work in a closed loop avoiding travelling (i.e., jumping from one point of the print to another without material dripping) of the robotic arm. In this way, constant plastic pressure coming out of the nozzle of the robotic arm is ensured. The granular deposition modelling method applied has a higher processing performance compared to FDM using filament printing and the ability to print complex polymers such as PPGF used for the fabrication of the panels.

The motion path of the robotic arm was reconsidered according to the printing speed and flow rates to print the polypropylene and glass-fiber material. Having a material of PP and GF reaching up to 50% of the overall mixture, an adjustment had to be made regarding the flow rate, since GF raises exponentially extrusion pressure of the overall extrusion system. A key element when printing with granular deposition is the pressure created over the extrusion screw, which can only be bypassed by raising the temperature and lowering the print speed. This calibration process was conducted through a code used in Grasshopper and KUKA PRC to balance printing, pressure, temperature and speed.

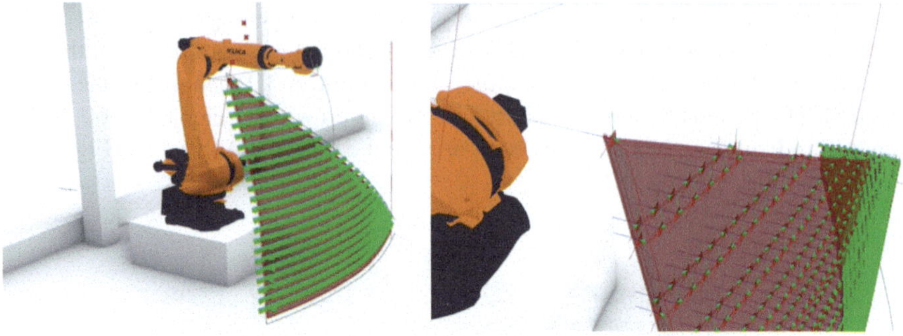

Fig. 8. Surface geometry definition of a GFRP panel.

Eight panels were printed, with the largest one having a height of 3.2 m. Using a KUKA KR 120 R2700-2 robotic arm, which only has a 2.7 m reach and a 0.5 m base, i.e., leaving no excess space for movement, inclination printing was firstly considered for the fabrication of the panels. In this way, the issue of height could be bypassed based on the parallel motion of the 6-axis robotic arm. However, when higher points in printing were reached, the element tilted due to its self-weight and flexibility resulting in geometrical inaccuracies. For this. Reason, an alternative direction of the 6-axis robotic arm was implemented. The 6-axis direction was vertically defined as opposed to a parallel arrangement to the floor, Fig. 9.

Fig. 9. 3D printing simulation of a GFRP panel based on the placement of the print material and the robotic arm.

To avoid further shrinkage issues that occurred at the printing points of the panels over 1.5 m in height, with constant thickness of 65 mm of each of the two layers, that created frizzy panel overlapping, specific modifications were made in the geometry of the panels. At the lower points of the panels, the distance between the two layers was set to 11.5 mm and at the highest points of the print above 1.5 m in height, to 13.5 mm. This created a smooth printing process and a lot more quality finish, Fig. 10. A system of brackets was used at the base of the panels to hold the printed elements in place and avoid warping especially at the edges.

Fig. 10. 3D printed panel.

Once all panels were fabricated, a steel frame was produced for the connection of GFRP panels in pairs, while also accounting to transporting issues and the structural integrity of each element, Fig. 11. The frames were fabricated using waterjet pieces of steel sheet of 8 mm in thickness of the scanned geometry. The production of the steel members and assembly of the elements succeeded offsite to avoid any inaccuracies.

Fig. 11. GFRP panels assembly in pairs with steel frame.

4 Onsite Assembly and Erection

The structural base was fabricated using a water jet cut process of a 15 mm steel plate. The base consisted of two plates, a base plate of 15 mm in thickness that was casted in situ and an identical plate used as a guide, which connected the threaded rods upon which the four main columns were placed. For achieving accuracy in erection, the plates were mechanically interconnected, Fig. 12.

Fig. 12. Structural base construction onsite.

The framed 3D printed panels assemblies were transported to the site individually following the onsite placement of the columns and the connection of the beams, Fig. 13. The panel elements were individually placed in pairs of two to maintain a correct weight distribution to the structure.

Fig. 13. Canopy structure in erected state.

5 Conclusions

The preliminary design to automated fabrication of a prototype structure has been presented in the current paper. In the preliminary design phase, the simulated structure followed a form-finding process through actuation of integrated cables and reversible elastic deformations of its bending-active members. The final form derived defined the geometry of the beams and GFRP panels. The automated fabrication followed required a redefinition of the actual geometry and the fabrication mode of the components. The project development proved that collaboration, cross disciplinary communication, experimentation and research at all stages of the design and fabrication process were necessary. Furthermore, interactivity within the decisions making processes may provide promising modes of operation for a bottom-up approach, enabling an iterative feedback loop of integrated computational design, physical realization and automated fabrication to achieve efficiency, sustainability and technological innovation.

Acknowledgements. This research was conducted within the framework of the cooperation project "Cooperative Intelligent education & electromobility Zero Energy Buildings" with the acronym "C-IZEBs". The project is co-funded by the European Union (ERDF) and national funds of Greece and Cyprus through the Interreg V-A Greece-Cyprus Cooperation Program 2014–2020.

References

1. Phocas, M.C., Alexandrou, K.: On decoding the structural behaviour of hybrid cable bending-active units in fastening, prestress and load-bearing state. Int. J. Space. Struct. **33**(2), 98–111 (2018)
2. Phocas, M.C., Alexandrou, K.: Numerical analysis and cable activation in hybrid bending-active structures with multiple cables. Eng. Struct. **174**, 561–572 (2018)
3. Liddell, I.: Frei Otto and the development of gridshells. Case. Stud. Struct. Eng. **4**, 39–49 (2015)
4. Schleicher, S., Rastetter, A., La Magna, R., Schönbrunner, A., Haberbosch, N., Knippers, J.: Form-finding and design potentials of bending-active plate structures. In: Thomsen, M.R., Tamke, M., Gengnagel, C., Faircloth, B., Scheurer, F. (eds.) Modelling Behaviour: Design Modelling Symposium, pp. 53–63. Springer International Publishing, Cham (2015)
5. Fleischmann, M., Knippers, J., Lienhard, J., Menges, A., Schleicher, S.: Material behaviour. Architect. Des. **216**, 44–51 (2012)
6. Liu, B., Faisal, T.R.: Computational design and fabrication of a bending-active structure using fiberglass: a bioinspired pavilion mimicking marine microorganism radiolaria. J. Bionic Eng. **19**(2), 471–482 (2022)
7. Grasshopper. http://www.grasshopper3d.com/. Accessed 15 Mar 2024
8. Rhino 3D. https://www.rhino3d.com. Accessed 15 Mar 2024
9. Kangaroo Physics. http://www.food4rhino.com/app/kangaroo-physics. Accessed 15 Mar 2024

Enhancing the Mix Design in 3D Concrete Printing Through Systematic Optimization Process

Mohammad R. Irshidat[1(✉)], Umar Amjad[1], Kishor Kumar[1], John-John[2], Naeem Albeitjali[1], and Khadeejath Rizmin[1]

[1] Center for Advanced Materials (CAM), Qatar University, Doha, Qatar
{mirshidat,umaramjad,Kishorkumars,nalbeitjali, khadeejath.r}@qu.edu.qa

[2] Mechanical and Industrial Engineering Department, Qatar University, Doha, Qatar
john.cabibihan@qu.edu.qa

Abstract. 3D printing of concrete has emerged as a promising technology in the construction industry, offering faster and more efficient building processes. However, the mix design heavily influences the success of 3D-printed concrete. The mix design of the concrete has a direct impact on the printed structure's printability, mechanical properties, and overall quality. A comprehensive understanding of different mix designs is crucial for optimizing 3D printing processes and achieving high-performance concrete structures. The major challenge in concrete printing is to identify and maintain the mixture characteristics suitable for printing and stacking up in layers. The mix design includes the selection of suitable materials, proportions, and additives to achieve the desired rheological properties that facilitate extrusion and layering during the printing process. Different mix designs may vary in terms of their composition, particle size distribution, water-to-cement ratio, admixtures, and curing methods. A comprehensive search in the literature was conducted to identify various mix designs that could be employed when printing cementitious materials. The literature review revealed that mix designs for 3D printing of concrete can vary significantly in terms of their composition and properties. Some mix designs use traditional materials such as Portland cement, sand, and aggregates, while others incorporate innovative materials like silica fume or fly ash to improve strength and durability. Furthermore, the use of admixtures such as superplasticizers can enhance the workability of the concrete mixture, allowing for easier extrusion and improved rheological properties. In addition to the composition and materials used, the proportions of each component also play a crucial role in the mix design. The water-to-cement ratio is an important parameter that affects the workability and strength of the printed concrete. Different mix designs may have varying water-to-cement ratios to achieve the desired consistency for printing. Moreover, the particle size distribution of the aggregates and fillers can also impact the flowability and packing density of the concrete mixture. Comparing different mix designs, it can be observed that mix designs incorporating silica fume and superplasticizer have been found to exhibit favorable rheological properties for 3D printing of concrete. These mix designs have shown improved workability, reduced viscosity, and enhanced printability compared to traditional mix designs.

N. D. Lagaros et al. (Eds.): ADDOPTML 2024, LNCE 646, pp. 22–33, 2025.
https://doi.org/10.1007/978-3-031-92029-5_3

Keywords: 3DPC · Fast Setting Cement · Fly ash · OPC · Silica Fume · Flowability · 3D ready mixture

1 Introduction

The success of 3D printed concrete (3DPC) structures depends upon a precisely developed concrete mix that keeps a delicate balance between printability and structural integrity. Achieving printability necessitates a mixture that exhibits smooth flow characteristics through the nozzle, avoiding issues such as clogging or segregation, while also maintaining cohesiveness to retain its shape post-deposition. This balance is attained through a profound understanding and modification of the rheological properties of the mix [1–3], often supplemented by the utilization of superplasticizers and viscosity-modifying agents to precisely adjust flow attributes without compromising the structural reliability [4]. The concept of buildability, analogous to the assembly of a layered cake, demands a mix that swiftly gains strength to support subsequent layers effectively. The incorporation of accelerators can aid in this process; however, their application must be accurately controlled to prevent premature setting, which could potentially disrupt the printing procedure [5]. In the field of 3DPC, structural integrity encompasses more than just compressive strength; factors such as tensile strength, ductility, and toughness hold equal significance. The integration of fiber reinforcement [6–8], utilizing materials like polypropylene or steel fibers [9–11], can substantially enhance these properties, addressing the inherent lower tensile strength and ductility of 3DPC when compared to conventional concrete. While a higher cement content typically leads to increased strength, it may also elevate shrinkage rates and adversely impact flow characteristics. To address these challenges, supplementary cementitious materials (SCMs) [12–17] such as silica fume, fly ash, or slag are used to reinforce strength, enhance flowability, and reduce the environmental impact of the concrete mixture. The inclusion of SCMs contributes to a denser microstructure and augments the overall durability of the material in the long term. The water-to-binder (w/b) ratio stands as a critical factor; lower ratios generally result in heightened strength but can potentially reduce workability. Keeping a balance where the mix exhibits adequate fluidity for extrusion yet possesses sufficient stiffness to maintain its shape is important for both the construction process and the ultimate performance of the 3DPC structure [18]. Aggregates play a pivotal role, with their size and distribution directly influencing the strength and extrusion quality of the concrete. Optimal mechanical properties are achieved through a well-graded aggregate blend with a suitable maximum size that ensures optimal packing density. Larger aggregates may present challenges during extrusion, whereas excessively fine aggregates can escalate water demand and shrinkage issues [19]. Furthermore, the mix design must account for the anisotropic nature of 3DPC, where the orientation of loading concerning the printing direction can impact strength attributes. For example, specimens produced via printing often showcase heightened compressive strength in the longitudinal direction compared to the downward and lateral orientations. Additionally, the microstructure of 3DPC, molded by the mix design and printing methodology, exerts a substantial influence on its macroscopic mechanical properties. Enhancing the understanding of the correlation between microstructure and mechanical behavior through multiscale modeling and testing can refine the design of 3D printed structures. The complex relationship between the ability to print, construct, and maintain structural integrity in 3D-printed concrete (3DPC), mix design illustrates the ongoing advancements in construction technology. This symbolizes a future marked by improved efficiency, sustainability, and unique building methods (Fig. 1).

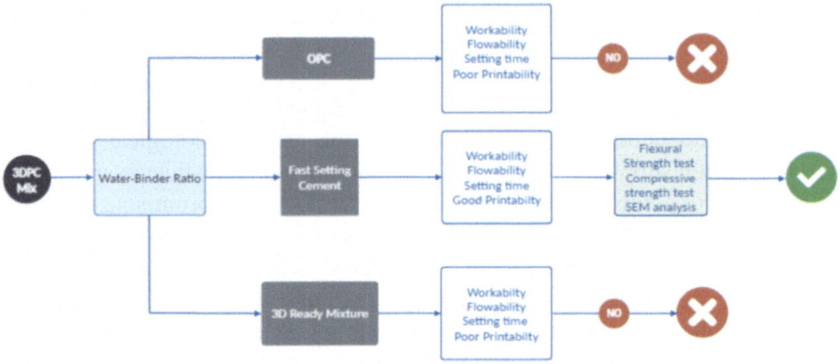

Fig. 1. Visual representation of the research study

2 Materials and Mix Design

Material selection and mix design are crucial in 3DPC printing for attaining successful production and the required qualities. It is crucial for the mixture to have a balance between printability (extrusion, shape retention), buildability (layer stability, adhesion), and hardened performance (strength, durability). Tables 1 and 2 present the mortar mix designs investigated, incorporating two distinct cement types: Ordinary Portland Cement and a fast-setting cement. The binder component, comprising 70% of the total mix by weight, is supplemented with silica fume at 10 wt% and fly ash at 20 wt% to enhance both strength and workability. Fine sand serves as the primary aggregate, while polypropylene fibers (4 mm in length) provide additional reinforcement. For the 3D printing application, a commercially available ready-mix concrete, with its composition detailed in Table 3, was sourced from DCP. A water-to-binder ratio of 0.34 was selected to balance flowability and strength requirements.

To optimize the fresh properties and setting time of the mixtures, two chemical admixtures were incorporated. A polycarboxylate-based superplasticizer, Hyperplast PC240L, was added at a dosage of 1% by weight of total binder content (8.27 g). This admixture enhances flowability, improves mechanical properties, and reduces water demand. To control the hydration process and extend workability, a retarding admixture, Hydrocon R3, was utilized. In the OPC cement trial, the retarder dosage was set at one-third the total binder content by weight (2.75 g).

2.1 Mixing Water and Water-Binder Ratio

The water-to-binder ratio (w/b), which is the ratio of the mass of water to the mass of cementitious materials in the mixture, has a significant effect on how cement-based mortars flow and how easy they are to work with. The quantity of water introduced during mixing is critical in determining the balance between flowability, which governs printability, and the mechanical properties of the hardened mortar. An optimized w/b ratio is essential for achieving adequate workability while ensuring sufficient strength and durability. Therefore, a series of experimental trials were conducted to determine the optimal w/b ratio for the mortar compositions under investigation. After several experimental trials, a water-to-binder ratio of 0.34 showed good flow properties and improved flexural and compressive strength. Tables 1, 2, and 3 provide the mix designs used to compare fast-setting cement, ordinary Portland cement, and ready-mix concrete (Fig. 2).

Table 1. Mix Design Used for Fast-Setting Cement

W/P ratio	Binder (g)			Fine Sand (g)	Polypropylene Fiber 4 mm (g)	Superplasticizer (Hyperplast PC240L) (g)	Water (g)
	Fast Setting Cement	Fly Ash	Silica Fume				
0.34	579	165	83	1241	1.2	8.27	281

Table 2. Mix Design Used for OPC Cement

W/P ratio	Binder (g)			Fine Sand (g)	Polypropylene Fiber 4 mm (g)	Superplasticizer (Hyperplast PC240L) (g)	Retarder (g)	Water (g)
	Fast Setting Cement	Fly Ash	Silica Fume					
0.34	579	165	83	1241	1.2	8.27	2.75	281

Table 3. 3D Ready Mixture Design.

W/P ratio	Binder (g)	Water (g)
0.34	2000	680

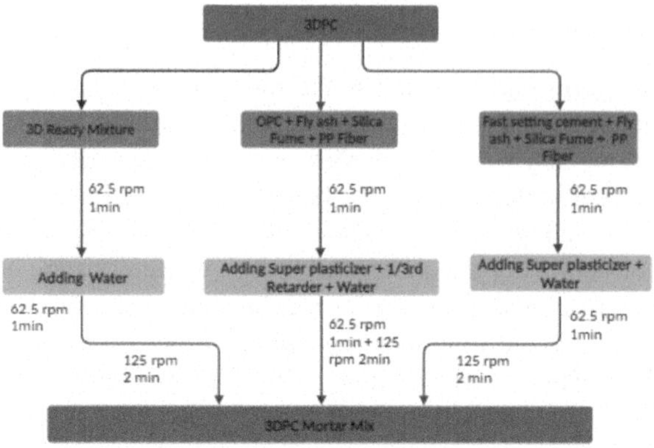

Fig. 2. Preparation of 3DPC Mixture

3 Results and Discussion

3.1 Tests Result

Flow Table Test

To assess the concrete rheology, specifically its flowability, flow tests were conducted on the prepared mixtures. Achieving adequate flowability is crucial for efficient pumping from the mixer to the nozzle, ensuring smooth extrusion, and guaranteeing sufficient interlayer adhesion. The flow test quantifies workability by measuring the average diameter of the mortar spread after a predetermined number of rotations. The results of these flowability measurements are summarized in Table 4.

Table 4. Flow Table Test Results

Mixes	Flow Diameter (cm)	Flow Percentage
Fast-Setting	23.5	140
OPC	24	140
3D ready Mixture	25	150

Flow Value and Workability

Several factors influence the flow value obtained from this test, with the water-to-cement ratio being a primary determinant; higher ratios generally result in increased flow. After removing the truncated cone, the mortar sample was dropped continuously 25 times. The flow percentage was then calculated using the formula below. The test results for flow value are shown in Table 4 for fast-setting, OPC, and 3D ready-mix mortars. From Table 4, it is observed that the fast-setting and OPC mortars exhibit similar workability with a flow value of 140, while the ready-mix mortar shows higher flowability with a value of 150.

$$\boldsymbol{Flow} = \left[\left(\boldsymbol{D_{avg}} - \boldsymbol{D_o}\right)/\boldsymbol{D_o}\right] * \mathbf{100}$$

where,

$\mathbf{D_{avg}}$ = Average base diameter
$\mathbf{D_o}$ = Original base diameter (Fig. 3)

Fig. 3. Flowability before and after drop

Setting Time

To determine the setting characteristics of the mortar, we employed the Vicat apparatus test, a standardized method for assessing the transition of cementitious materials from a plastic state to a hardened state. This test involves measuring the penetration resistance of a standardized needle into a mortar sample at specific time intervals. The initial setting time, a critical parameter for construction operations, was recorded as the point at which the Vicat needle could no longer penetrate the mortar surface to a predefined depth, typically 1 mm. This stage signifies a significant increase in the mortar's viscosity and marks the beginning of the setting process. The final setting time, indicating the completion of the setting process, was determined as the moment the mortar surface could withstand a specified load applied by the Vicat apparatus without exhibiting any noticeable indentation. At this stage, the mortar has developed sufficient strength to resist deformation under light loads. From our experimental data, we observed that the final setting time for the 3D ready-mix was 10 min. For the OPC mix, with a retarder dosage of one-third of the binder, the final setting time was 2 h and 47 min. With fast-setting cement, the final setting time was 2 h and 7 min (Fig. 4).

Fig. 4. Initial & Final Setting Time

Compressive Strength

In this study, compression tests were performed on concrete cube samples measuring 50 × 50 mm. This size was chosen to follow standard practices for assessing the compressive strength of concrete. The cubes were placed under a uniaxial compressive load until they failed, using a universal testing machine (UTM). This test helps determine how much load the concrete can bear, which is important for understanding its strength and structural performance.

0 (3 h of curing), 1 and 7-day Compressive strength tests were undertaken for the fast-setting mortar samples listed above. The results are presented in Tables 5. Figure 5 shows a sample used to measure compressive strength and Fig. 6 shows the compressive strength data for 0 days (3 h of curing), 1 day, and 7 days.

Fig. 5. 50 × 50 mm sample used to measure compressive strength using a universal testing machine (UTM)

Table 5. Compressive strength

Curing days	Average Compressive Strength (MPa)
0 (3 h curing)	26.70
1	27.13
7	32.32

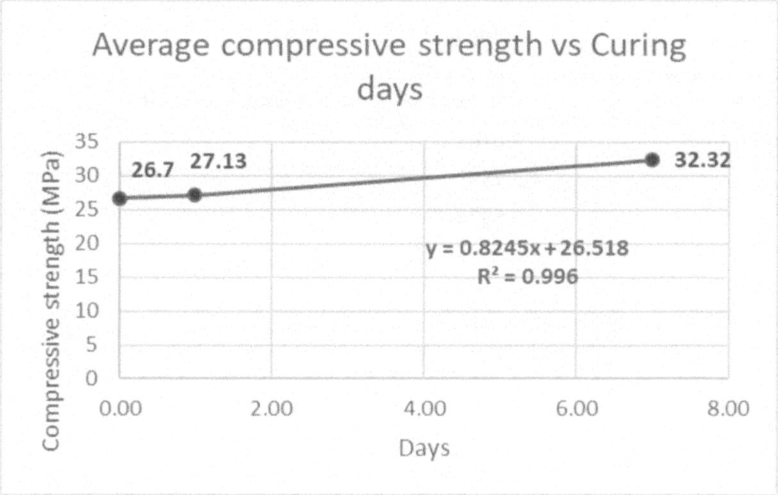

Fig. 6. Average compressive strength graph

Flexural Strength

Flexural strength was tested on prism specimens measuring 40 mm × 40 mm × 120 mm. The test followed a three-point bending method, with a span of 120 mm between supports and a load applied at the center. The universal testing machine applied the load at a constant rate until the specimens failed.

0 (3 h of curing), 1 and 7-day flexural strength tests were undertaken for the fast-setting mortar samples listed above. The results are presented in Tables 6 below. The results are presented in Table 6 below. Figure 7 shows a sample used to measure flexural and Fig. 8 shows the flexural strength data for 0 days (3 h of curing), 1 day, and 7 days. From our experimental data, we observe that the 0-day sample (3 h of curing) shows significant strength even without proper curing.

Fig. 7. 40 mm × 40 mm × 120 mm sample used to measure flexural strength using a universal testing machine (UTM)

Table 6. Flexural strength

Curing days	Average Maximum Bending Stress (MPa)
0 (3 h curing)	3.6168
1	3.7893
7	3.8974

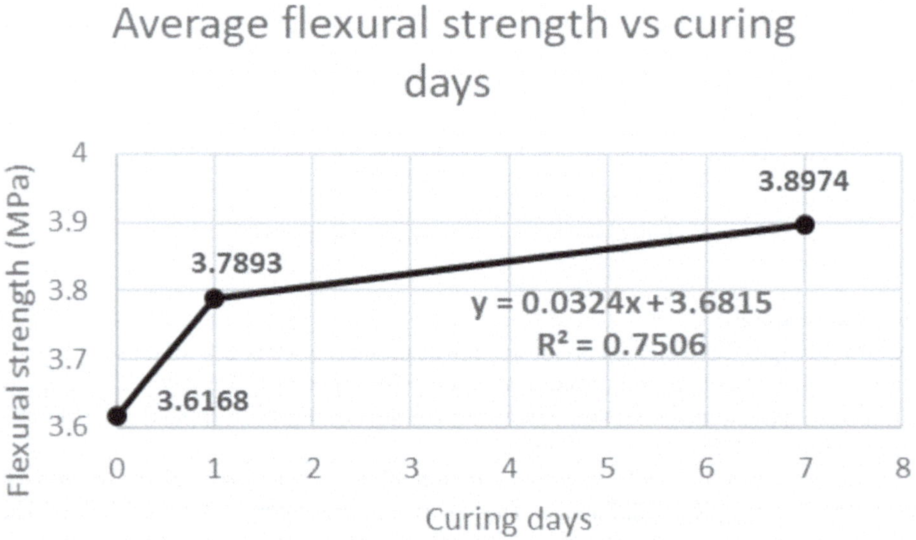

Fig. 8. Average flexural strength graph

SEM Analysis

SEM analysis of 3DPC mortar shows that, despite the layer-by-layer construction technique, the material retains structural integrity at the interlayer level and behaves as a single structure. SEM also revealed variations in pore size distribution.

The addition of polypropylene (PP) fibers to 3DCP mortar appears to help delay the onset of cracks, as seen in the SEM images of samples shown in Fig. 9. By seven days, the fibers begin to fill up gaps and increase the durability of the mortar. The fibers successfully stop cracks from getting bigger after 1 day, which increases the material's overall durability. The period between the 1 and 7th day of curing exhibits the greatest variations in fracture strength and resistance.

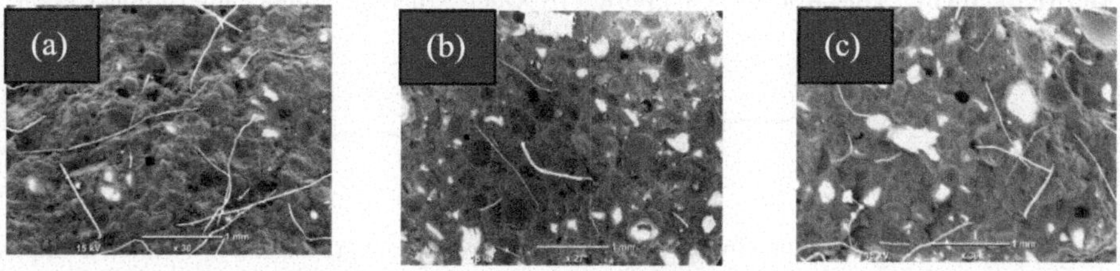

Fig. 9. SEM Images of PP Fibers in Mortar Specimens at 0 (3 h of curing), 1 and 7 Days of curing: (a) 0-Day (3 h of curing), (b) 1-Day and (c) 7-Day

4 Conclusion

This study investigated the printability of various 3D printing concrete (3DPC) mortars, encompassing Ordinary Portland Cement (OPC), fast-setting cement, and a commercially available 3D ready-mix formulation. Results indicated notable variations in flow behavior, with the 3D ready-mix exhibiting superior flowability desirable for extrusion-based 3D printing. A water-to-binder ratio of 0.34 was maintained throughout the study, maintaining a balance between flowability and mechanical properties. It is important to note that increasing the w/b ratio, while potentially improving flow, would likely have a detrimental effect on the mortar's strength. While both the OPC and fast-setting cement mortars exhibited suitable flowability, workability, and adjustable setting times. Their extrusion and buildability properties were sufficient for successful 3D printing, whether with higher pressure or using a pump. The addition of Hydrocon R3 retarder effectively delayed setting time; however, a direct correlation between retarder dosage and setting time was observed in the OPC mix design. The compressive and flexural strength tests of fast-setting mortar samples at 0 (3 h of curing), 1, and 7 days revealed that the incorporation of Hyperplast PC240L superplasticizer enhanced the workability of the mortar without compromising early-age strength development. The addition of polypropylene (PP) fibers to 3DCP mortar appears to help delay the onset of cracks, as seen in the SEM images of the fast-setting mortar samples. By the seventh day, the fibers begin to fill gaps and increase the mortar's durability. The 3D ready-mix shows a final setting time of 10 min, which is much faster than the fast-setting cement without adding any accelerator and retarder (2 h and 7 min) and the OPC mix with a retarder dosage as one-third of binder (2 h and 47 min), according to the trial results. This suggests that the 3D ready-mix is a good fit for uses that need for rapid setting. The substantial 0-day strength (3 h of curing) found in both compressive and flexural strength tests indicates that the fast-setting mortar has notable strength even with minimum curing time. According to the results, the fast-setting mortar develops significant strength rapidly and keeps becoming stronger over time, which makes it useful in situations where early strength acquisition is crucial. Additionally, the fast-setting cement showed proper rheological properties without the need for any accelerator or retarder and successfully printed the prototype using a 3 cm diameter nozzle.

Acknowledgement. This work was supported by Qatar University under high impact grant # QUHI-CAM-24/25-518.

References

1. Bhattacherjee, S., et al.: Sustainable materials for 3D concrete printing. Cem. Concr. Compos. **122**, 104156 (2021)
2. Niall, H., Khan, M., McNally, C.: The role of different clay types in achieving low-carbon 3D printed concretes. Buildings (2024). https://doi.org/10.3390/buildings14072194
3. Barve, P., Bahrami, A., Shah, S.: A comprehensive review on effects of material composition, mix design, and mixing regimes on rheology of 3D-printed geopolymer concrete. Open Constr. Build. Technol. J. (2024). https://doi.org/10.2174/0118748368292859240313061706
4. Rashid, A.A., et al.: Additive manufacturing: technology, applications, markets, and opportunities for the built environment. Autom. Constr. **118**, 103268 (2020). https://doi.org/10.1016/j.autcon.2020.103268
5. Colyn, M., van Zijl, G., Babafemi, A.J.: Fresh and strength properties of 3D printable concrete mixtures utilising a high volume of sustainable alternative binders. Constr. Build. Mater. **419**, 135474 (2024). https://doi.org/10.1016/j.conbuildmat.2024.135474
6. Zhou, Yi., Althoey, F., Alotaibi, B.S., Gamil, Y., Iftikhar, B.: An overview of recent advancements in fibre-reinforced 3D printing concrete. Front. Mater. **10**, 1289340 (2023). https://doi.org/10.3389/fmats.2023.1289340
7. Irshidat, M.R., Al-Nuaimi, N., Salim, S., Rabie, M.: Carbon nanotubes dosage optimization for strength enhancement of cementitious composites. Procedia Manuf. **44**, 366–370 (2020). https://doi.org/10.1016/j.promfg.2020.02.282
8. Bani-Hani, K.A., et al.: Strength optimisation of mortar with CNTs and nanoclays. Proc. Inst. Civil Eng. Struct. Build. **169**(5), 340–356 (2016). https://doi.org/10.1680/jstbu.14.00106
9. Mohamed, O., Zuaiter, H.: Fresh properties, strength, and durability of fiber-reinforced geopolymer and conventional concrete: a review. Polymers **16**(1), 141 (2024). https://doi.org/10.3390/polym16010141
10. Irshidat, M.R., Al-Nuaimi, N., Rabie, M.: The role of polypropylene microfibers in thermal properties and post-heating behavior of cementitious composites. Materials **13**(12), 2676 (2020). https://doi.org/10.3390/ma13122676
11. Nodehi, M., Ozbakkaloglu, T., Gholampour, A.: Effect of supplementary cementitious materials on properties of 3D printed conventional and alkali-activated concrete: a review. Autom. Constr. **138**, 104215 (2022)
12. Dey, D., et al.: Use of industrial waste materials for 3D printing of sustainable concrete: a review. J. Clean. Prod. **340**, 130749 (2022). https://doi.org/10.1016/j.jclepro.2022.130749
13. Wang, C., et al.: Mechanical anisotropy, rheology and carbon footprint of 3D printable concrete: a review. J. Build. Eng. **76**, 107309 (2023). https://doi.org/10.1016/j.jobe.2023.107309
14. Irshidat, M.R., Al-Nuaimi, N., Rabie, M.: Potential utilization of municipal solid waste incineration ashes as sand replacement for developing sustainable cementitious binder. Constr. Build. Mater. **312**, 125488 (2021). https://doi.org/10.1016/j.conbuildmat.2021.125488

15. Albeer, S.J., Hassan, M.S.: Silica fume modified cement-based mortar exposed to high temperatures: residual strengths and microstructure. J. Eng. **29**(7), 18–33 (2023)
16. Irshidat, M.R., Al-Nuaimi, N.: Industrial waste utilization of carbon dust in sustainable cementitious composites production. Materials **13**(15), 3295 (2020). https://doi.org/10.3390/ma13153295
17. Irshidat, M.R., Abdel-Jawad, Y.A., Al-Sughayer, R.: Feasibility of producing sustainable geopolymer composites made of locally available natural pozzolan. J. Mater. Cycles Waste Manag. **20**(3), 1751–1760 (2018). https://doi.org/10.1007/s10163-018-0742-5
18. Wang, X., et al.: Optimization of 3D printing concrete with coarse aggregate via proper mix design and printing process. J. Build. Eng. **56**, 104745 (2022). https://doi.org/10.1016/j.jobe.2022.104745
19. Chen, Y., et al.: Influence of gradation on extrusion-based 3D printing concrete with coarse aggregate. Constr. Build. Mater. **403**, 133135 (2023)

The Flexural Behavior of Engineered Cementitious Composites (ECC) One-Way 3D Printed Slabs Made of Solid and Hollow Sections

Rajai Z. Al-Rousan[1(✉)], Bara'a R. Alnemrawi[1], Khairedin M. Abdalla[1], Mohammad Alhassan[1,2], and Nikos Lagaros[3]

[1] Department of Civil Engineering, Engineering, Jordan University of Science and Technology, P.O. Box 3030, Irbid 22110, Jordan
rzalrousan@just.edu.jo
[2] Civil Engineering Program, Al Ain University, Al Ain, UAE
[3] School of Civil Engineering, National Technical University of Athens, Athens, Greece

Abstract. In light of the recent development and the reduction in the available natural resources, the interest in searching for sustainable and efficient construction resources has increased. Therefore, the method of 3D concrete printing takes place as a type of construction where forms are not required to shape the required structural element. In contrast, the 3D printing method suffers from a major deficiency related to the reinforcement method of the printed members where challenges exist in finding a practical and efficient method. Consequently, scientists work on the printed concrete properties where types of self-reinforced mixes were utilized such as the Engineered Cementitious Composites (ECC). In addition, the production of lightweight sections is also a major concern which could be achieved using a combination of hollow and solid sections. This study adopts the method of nonlinear finite analysis to investigate the flexural performance of one-way ECC slabs composed of a combination of solid and hollow rectangular sections. Different overall slab thicknesses and solid-to-hollow section areas were examined in terms of the flexural performance measures, such as the ultimate load, deflection, serviceability limit, stiffness, toughness, and flexural strength. Mathematical relationships were introduced based on the slab depth and the hollow volume percentage with high correlation values.

Keywords: 3D concrete printing · Sustainability · Finite element modeling · Lightweight sections · Hollow slabs · Engineered cementitious composites

1 Introduction

The formation of innovative construction methods is a challenge where strength, serviceability, and durability are of major concern. However, conventional construction methods adopt the forming of structural members of solid sections to ensure the rigidity, strength, and resilience of the composite elements which also requires a large amount of materials and formworks resulting in an infeasible and unsustainable construction method [1–4].

N. D. Lagaros et al. (Eds.): ADDOPTML 2024, LNCE 646, pp. 34–45, 2025.
https://doi.org/10.1007/978-3-031-92029-5_4

Consequently, the utilization of hollow cross-sections along with lightweight material characteristics forms two innovative solutions for the previously mentioned problems associated with the conventional construction process [5]. Lightweight construction was achieved previously in the literature using lightweight components such as concrete or aggregates while few others used hollow cross-sections [6]. Generally, the conventional construction method included the consumption of large amounts of formworks that could be used many times for regular shapes while requiring customized formworks for hollow cross-sections with complex geometries resulting in wasted materials and higher cost [7–10].

The adoption of innovative and efficient technology helps overcome the associated conventional casting methods limitations, such as the method of 3D concrete printing (3DCP) [11, 12]. The 3DCP method has several advantages such as feasible construction, sustainable construction, reduced carbon dioxide emissions, and improved productivity [13]. In addition, the construction of complex shapes is enabled using the free-formwork method (3DCP) such as the hollow sections [14, 15] while the reinforcement method of this type is still a major concern [16]. However, the existing reinforcement methods such as the printed [17], pre-, and post-reinforcements [18] limit the ease, simplicity, and efficiency of the 3DCP method.

The utilization of high-performance concrete plays a major role in increasing the efficiency of the 3DCP technology and limiting its associated limitations, such as the Engineered cementitious composites (ECC) concrete [19, 20]. Short fibers are included in the ECC concrete components, providing strength and ductile properties into the original mix and enabling the strain hardening stage and multiple cracking behaviors [21, 22]. Previous research show that ECC has a good tensile strain capacity of up to 11% [23] where studies revealed that beams and columns with ECC concrete material exhibited similar load-carrying capacity to an RC beam or column with 1.51% and 0.8% steel reinforcement ratios, respectively [24]. Overall, the several advantages of the ECC concrete make it a suitable choice for 3D printing technology [25]. It is believed that the combination of the 3DCP and ECC technologies enables the construction of lightweight structural members with hollow sections of complex geometry.

Slabs have the highest portion of the entire weight in a structure and their own weight has a direct effect on the applied loading of the other components, including beams, walls, and columns which could be overcome using lightweight construction [26–29]. However, few research exist on the behavior of 3DPC slabs made of hollow cross sections while others exist for beams, bridges, and other components. This study adopts the nonlinear finite element analysis method (NLFEA) to investigate the flexural behavior of 3DCP slabs made of self-reinforced ECC concrete material of hollow sections tested under four-point loading. The effect of the slab thickness was examined using different numbers of printed layers of 15 mm thickness along with the hollow-to-solid areas percentages achieved using a combination of solid and hollow rectangular sections with a reduced self-weight of up to 35% compared to a solid slab. The flexural performance was examined using the ultimate load capacity, ultimate deflection, serviceability limits, flexural strength, initial stiffness, toughness, and yield load.

2 Experimental Work Overview

The NLFEA models were first validated using the experimental work done by Ye et al. [30] where six slabs were printed using the 3DPC technology made of ECC self-reinforced concrete. Slabs were constructed with 1410 mm length, 595 mm width, and 120 mm overall depth made of eight printed layers of 15 mm each. Three different cross-sectional types were tested including one solid and two other hollow sections of rectangular and honeycomb shapes. The geometry, cross-section details, and test setup are illustrated in Fig. 1. The ECC concrete mix contains polypropylene fibers (PE) with 12 mm length and 24 mm diameter forming the self-reinforcement of the mix. The experimental testing was done using the four-point loading with a 1200 mm clear span and 400 mm loading distance.

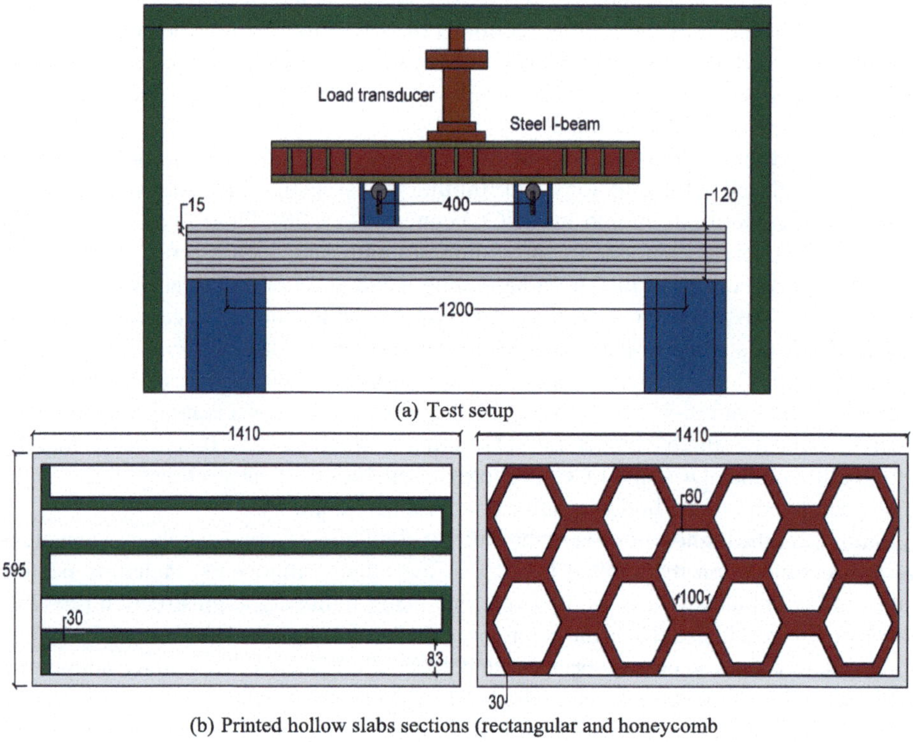

Fig. 1. Geometry and test setup

3 Nonlinear Finite Element Analysis (NLFEA) Description

The NLFEA simulation was done in this study using ABAQUS software [31] where slabs were supported on steel plates and loaded using two reference points attached to two loading steel plates that are tied to the ECC slab surfaces. In addition, the ECC and steel plate components were modeled as three-dimensional eight nodded elements with reduced integration (C3D8R). Slabs were simulated as layers with 15 mm thickness with a traction-separation cohesive interaction between them. The ECC material was modeled using the concrete damage plasticity definition available in ABAQUS software including the compression and tension behaviors.

The compressive stress-strain behavior of the ECC concrete was simulated using the Zhou et al. [32] model (Fig. 2a) while the tensile behavior was modeled using Han et al. [33] model (Fig. 2b). Steel plates were simulated as linear elastic material with 200 GPa elastic modulus and 0.3 Poisson's ratio. The interaction between the printed ECC layers was simulated using the cohesive surface with traction-separation interaction [29]. Finally, the slab was meshed with a 15 mm constant mesh size with the assembly of the validated models are illustrated in Fig. 3. The validation of the load-deflection curves was done on all of the experimentally tested slabs as illustrated in Fig. 4 with error percentage less than 5%.

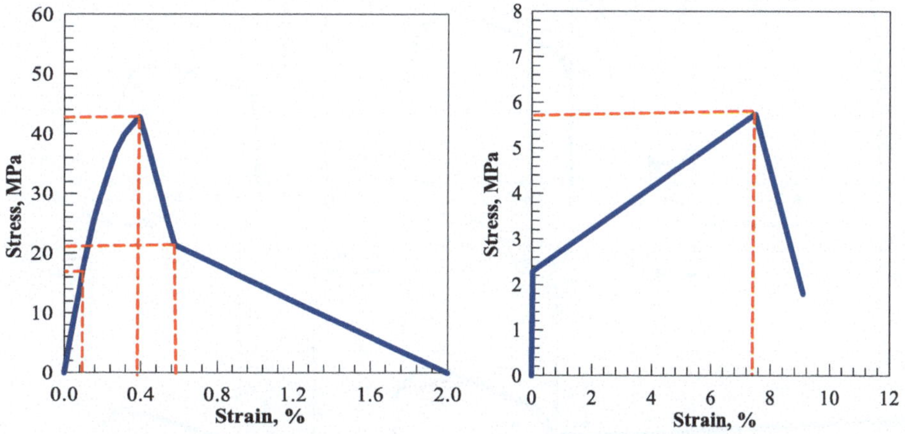

(a) Compressive behavior [1] (b) Tensile behavior [2]

Fig. 2. ECC stress-strain relationships [1, 2]

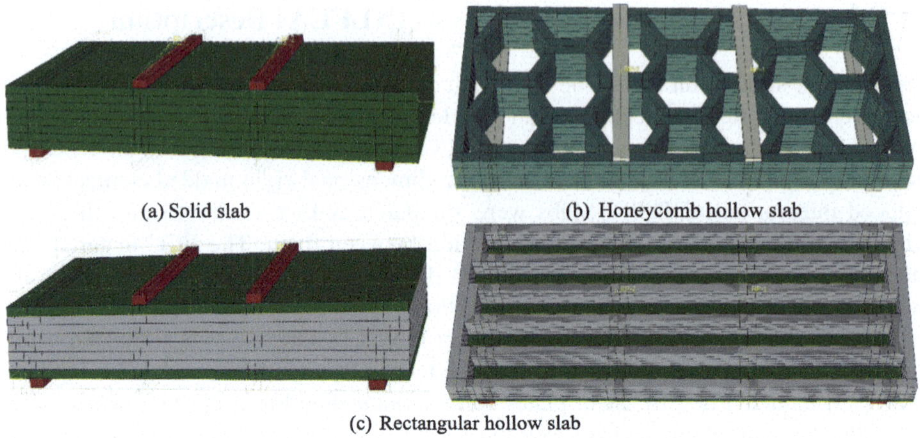

Fig. 3. Modeling assembly

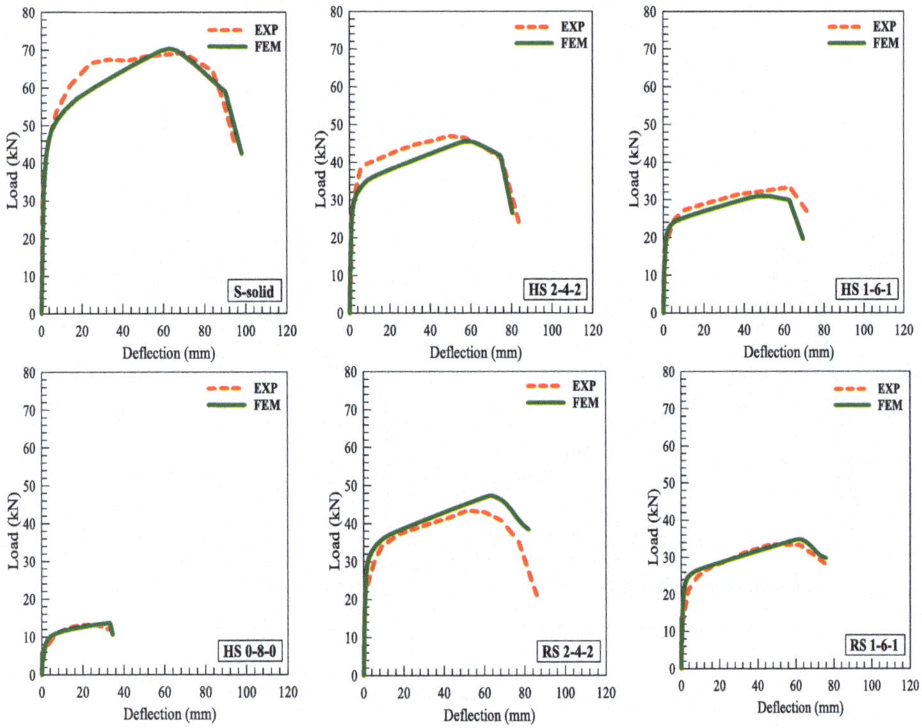

Fig. 4. Load-deflection curves validation

4 Results and Discussion

4.1 Studied Parameters

The effect of different parameters was investigated after the validation was done against Ye et al. [30] work including the effect of overall slab thickness (120 and 150 mm) and the different number of solid and rectangular hollow layers with the model's designation illustrated in Fig. 5.

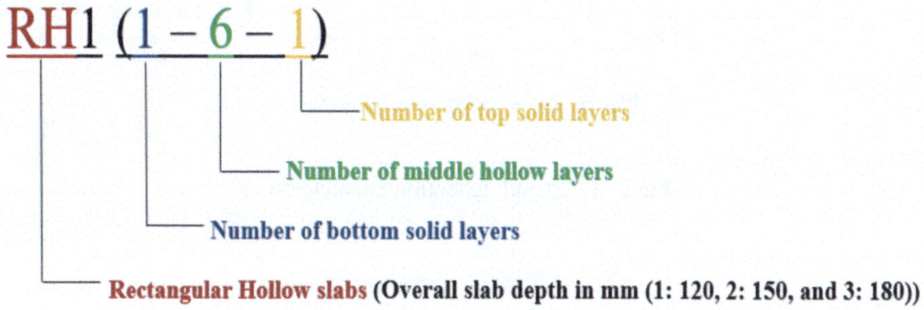

Fig. 5. NLFEA models designation

4.2 Load-Deflection Curves

The load-deflection curves of the simulated specimens are all presented in Fig. 6 including nine specimens categorized in two main groups depending on their total depth. Observing Fig. 6 reveals that increasing the number of hollow layers reduces the overall slab's performance in terms of load, ductility, and rigidity. However, it could be stated that increasing the depth of the printed slab from 120 mm to 150 mm using 8 and 10 layers of 15 mm thickness, respectively, has a significant effect on the slab's flexural performance where the capacity is increased with their ultimate deflection being decreased. In addition, increasing the number of utilized hollow layers has a marginal effect on the deflection behavior when the thickness of the slab is increased compared to the larger effect for thinner slabs. Consequently, the flexural performance, initial stiffness, and toughness have direct proportional relationships with the percentage of the solid area and the overall depth of the printed slab.

4.3 Load and Deflection Capacities

The load-carrying capacity of the 3D printed slabs was evaluated at three main stages, including the serviceability, yield, and ultimate points. The serviceability limit is defined as the point where cracking occurs at the force corresponding to 1/200 times the supported length, as given by GB 50010-2010 [34]. The load capacities (serviceability, yield, and ultimate) and deflection values (yield and ultimate) of the simulated slabs are presented in Table 1.

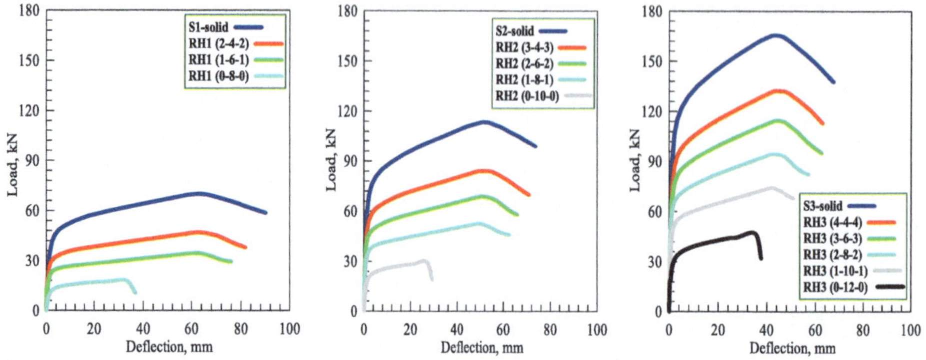

Fig. 6. Load-deflection behavior

Table 1. Load and deflection characteristics

Designation	(Solid/hollow) area, %	Load, kN			Deflection, mm	
		F1/200	Yield	Ultimate	Yield	Ultimate
S1-solid	100.00	49.55	54.09	70.25	11.91	90.2
RH1 (2-4-2)	82.65	34.19	36.12	46.96	9.75	81.93
RH1 (1-6-1)	73.97	25.89	26.35	34.84	7.42	76.09
RH1 (0-8-0)	65.30	14.58	15.22	17.95	8.53	33.34
S2-solid	100.00	82.22	89.47	113.93	11.76	73.94
RH2 (3-4-3)	86.12	62.79	65.50	84.62	9.05	71.13
RH2 (2-6-2)	79.18	51.39	54.01	69.42	9.68	66.22
RH2 (1-8-1)	72.24	40.21	41.00	52.83	7.67	62.83
RH2 (0-10-0)	65.30	24.13	24.89	29.56	7.42	26.93
S3-solid	100.00	124.44	135.53	165.88	11.67	67.53
RH3 (4-4-4)	88.43	99.83	104.27	132.87	8.95	63.01
RH3 (3-6-3)	82.65	86.93	90.46	115.03	8.97	62.52
RH3 (2-8-2)	76.86	72.37	74.70	94.75	8.37	57.00
RH3 (1-10-1)	71.08	57.67	58.54	73.99	7.18	50.75
RH3 (0-12-0)	65.30	36.66	37.01	46.76	6.37	35.37

Observing Table 1 results reveals that the serviceability limit of all slabs varies between 70.53% to 81.63% times the slab's ultimate capacity compared to 76.53% to 84.79% for the yield load. In addition, it was found that the behavior was similar for all of the performance characteristics regarding the reduction percentages associated with the different combinations of solid and hollow sections and mainly based on the ratio between the solid-to-hollow area. Moreover, it was observed that the yield deflection values vary between 9.75% to 27.55% times the ultimate deflection values. Finally, similar behavior of the load capacities behavior exists for all slab depth values with minor differences.

4.4 Stiffness, Flexural Strength, Toughness, and Ductility

The flexural performance of the 3D printed slabs was also examined based on the initial stiffness, flexural strength, toughness, and ductility were all presented in Table 2. The flexural strength was calculated as the ultimate load-carrying capacity multiplied by the loading span length and divided by the geometrical cross-section properties (bh^2). In addition, the ductility is the ratio between the ultimate and yield deflection values while the initial stiffness and toughness are measured as the slope of the load-deflection linear part and the total area under the curve, respectively. Observing Table 2 results reveals that the flexural strength has similar behavior to the ultimate capacity and the deflection values were also similar to the ductility one. Generally, it was found that increasing the percentage of the hollow sections or increasing the solid volume improves the overall structural system performance.

Table 2. Strength and serviceability measurements

Designation	Initial stiffness, (kN/mm)	Toughness, (kN mm/mm^3)	Flexural strength, MPa	Ductility
S1-solid	8.26	5550.47	9.84	7.57
RH1 (2-4-2)	5.70	3373.69	6.58	8.40
RH1 (1-6-1)	4.32	2310.71	4.88	10.25
RH1 (0-8-0)	2.43	578.60	2.51	3.91
S2-solid	13.70	7413.22	15.96	6.29
RH2 (3-4-3)	10.47	5294.05	11.85	7.86
RH2 (2-6-2)	8.57	4011.70	9.72	6.84
RH2 (1-8-1)	6.70	2907.57	7.40	8.19
RH2 (0-10-0)	4.02	741.67	4.14	3.63
S3-solid	20.74	9905.42	23.23	5.79
RH3 (4-4-4)	16.64	7386.03	18.61	7.04
RH3 (3-6-3)	14.49	6309.06	16.11	6.97
RH3 (2-8-2)	12.06	4766.41	13.27	6.81
RH3 (1-10-1)	9.61	3315.70	10.36	7.07
RH3 (0-12-0)	6.11	1525.43	6.55	5.55

4.5 General Relationships

Based on the results and the observations in the previous sections, it was concluded that solid 3D printed slabs behave similarly depending on the percentage volume increase with respect to the (120 mm) depth slabs. Therefore, their flexural performance characteristics were formulated and plotted in Fig. 7 with exact mathematical polynomial expressions.

In addition, it was found that the load capacities (serviceability, yield, and ultimate) have similar behavior based on the volume percentage between the solid and hollow

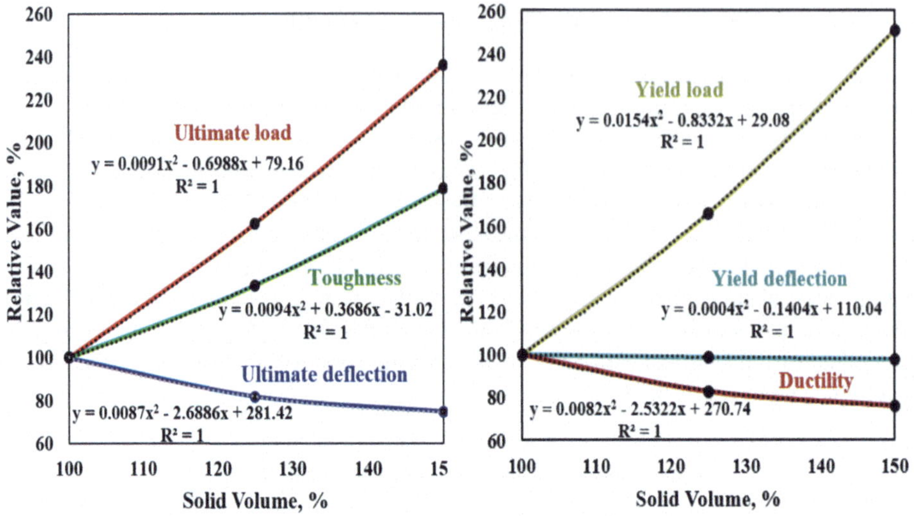

Fig. 7. The flexural behavior of solid 3DPC slabs with different thicknesses

sections, with the results being presented in Fig. 8a with an accurate correlation relationship. Finally, the toughness behavior also depends on the solid-to-hollow volume percentage, as presented in Fig. 8b.

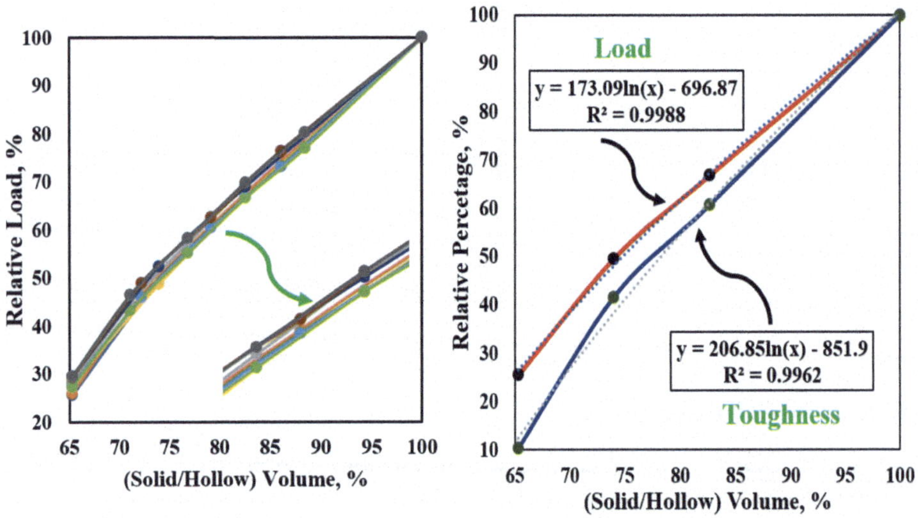

(a) General load behavior (b) Mathematical relationship

Fig. 8. The general load and toughness behavior of 3DPC hollow slabs

5 Conclusions

Based on the results of this study, the following could be concluded:

1. The flexural performance of the 3DPC hollow lightweight slabs with self-reinforced ECC material was comparable to conventional slabs with high reinforcement ratios and depths.
2. The performance of the 3DPC slabs with solid section layers depends on their overall depth and the reduction or increasing percentages are directly related using polynomial functions based on their relative solid volume percentages.
3. The combination of hollow and solid slab layers efficiently reduces the self-weight of the structural members with their ultimate load capacity and toughness basically affected by the solid-to-hollow volume percentage with logarithmic relationships.

Acknowledgments. Part of this research was supported by the ADDOPTML (ADDitively Manufactured OPTimized Structures by Means of Machine Learning) project (No: 101007595) belonging to the Marie Skłodowska-Curie Actions (MSCA) Research and Innovation Staff Exchange (RISE) H2020-MSCA-RISE-2020. Their support is highly acknowledged. The authors would also like to thank the Jordan University of Science and Technology (JUST) for its support as a partner organization in the ADDOPTML project.

References

1. Al-Rousan, R.Z., Alnemrawi, B.R.: Influence of overlay strength degradation on bond stresses of bridge deck system. Heliyon. **10**(10), e31037 (2024). https://doi.org/10.1016/j.heliyon.2024.e31037
2. Alhassan, M.A., Al-Rousan, R.Z., Taha, H.M.: Precise finite element modelling of the bond-slip contact behavior between CFRP composites and concrete. Constr. Build. Mater. **240**, 117943 (2020). https://doi.org/10.1016/j.conbuildmat.2019.117943
3. Alhassan, M., Al-Rousan, R., Ababneh, A.: Anchoring of the main CFRP sheets with transverse CFRP strips for optimum upgrade of RC Beams: parametric experimental study. Constr. Build. Mater. **293**, 123525 (2021). https://doi.org/10.1016/j.conbuildmat.2021.123525
4. Al-Rousan, R.Z., Barfed, M.H.: Impact of curvature type on the behavior of slender reinforced concrete rectangular column confined with CFRP composite. Compos. Part B Eng. **173**, 106939 (2019). https://doi.org/10.1016/j.compositesb.2019.106939
5. Zhang, Z., Yuvaraj, A., Di, J., Qian, S.: Matrix design of light weight, high strength, high ductility ECC. Constr. Build. Mater. **210**, 188–197 (2019). https://doi.org/10.1016/j.conbuildmat.2019.03.159
6. Sifan, M., Nagaratnam, B., Thamboo, J., Poologanathan, K., Corradi, M.: Development and prospectives of lightweight high strength concrete using lightweight aggregates. Constr. Build. Mater. **362**, 129628 (2023). https://doi.org/10.1016/j.conbuildmat.2022.129628
7. Ling, Y.Y., Leo, K.C.: Reusing timber formwork: importance of workmen's efficiency and attitude. Build. Environ. **35**(2), 135–143 (2000). https://doi.org/10.1016/S0360-1323(99)00008-6
8. Li, W., Lin, X., Bao, D.W., Min, X.Y.: A review of formwork systems for modern concrete construction. Structure. **38**, 52–63 (2022). https://doi.org/10.1016/j.istruc.2022.01.089

9. Al-Rousan, R.Z., Abdalla, K.M., Alnemrawi, B.R.: The behavior of heat-damaged RC beams reinforced internally with CFRP strips. In: Ungureanu, V., Bragança, L., Baniotopoulos, C., Abdalla, K.M. (eds.) 4th International Conference "Coordinating Engineering for Sustainability and Resilience" & Midterm Conference of Circular B "Implementation of Circular Economy in the Built Environment", pp. 165–174. Springer Nature Switzerland, Cham (2024). https://doi.org/10.1007/978-3-031-57800-7_15
10. Al-Rousan, R.Z., Alnemrawi, B.R.: Behaviour of thermally shocked RC columns internally confined by auxetic steel wire mesh. Proc. Inst. Civil. Eng. Struct. Build. **1**(1), 1–14 (2023). https://doi.org/10.1680/jstbu.22.00226
11. Wangler, T., Roussel, N., Bos, F.P., Salet, T.A.M., Flatt, R.J.: Digital concrete: a review. Cem. Concr. Res. **123**, 105780 (2019). https://doi.org/10.1016/j.cemconres.2019.105780
12. Khajavi, S.H., Tetik, M., Mohite, A., Peltokorpi, A., Li, M., Weng, Y., Holmström, J.: Additive manufacturing in the construction industry: the comparative competitiveness of 3D concrete printing. Appl. Sci. **11**(9), 3865 (2021). https://doi.org/10.3390/app11093865
13. Weng, Y., Li, M., Ruan, S., Wong, T.N., Tan, M.J., Ow Yeong, K.L., Qian, S.: Comparative economic, environmental and productivity assessment of a concrete bathroom unit fabricated through 3D printing and a precast approach. J. Clean. Prod. **261**, 121245 (2020). https://doi.org/10.1016/j.jclepro.2020.121245
14. Wang, L., Jiang, H., Li, Z., Ma, G.: Mechanical behaviors of 3D printed lightweight concrete structure with hollow section. Archiv. Civ. Mech. Eng. **20**(1), 16 (2020). https://doi.org/10.1007/s43452-020-00017-1
15. Asprone, D., Auricchio, F., Menna, C., Mercuri, V.: 3D printing of reinforced concrete elements: technology and design approach. Constr. Build. Mater. **165**, 218–231 (2018). https://doi.org/10.1016/j.conbuildmat.2018.01.018
16. Borg Costanzi, C., Ahmed, Z.Y., Schipper, H.R., Bos, F.P., Knaack, U., Wolfs, R.J.M.: 3D printing concrete on temporary surfaces: the design and fabrication of a concrete shell structure. Autom. Constr. **94**, 395–404 (2018). https://doi.org/10.1016/j.autcon.2018.06.013
17. Mechtcherine, V., Grafe, J., Nerella, V.N., Spaniol, E., Hertel, M., Füssel, U.: 3D-printed steel reinforcement for digital concrete construction—manufacture, mechanical properties and bond behaviour. Constr. Build. Mater. **179**, 125–137 (2018). https://doi.org/10.1016/j.conbuildmat.2018.05.202
18. Mechtcherine, V., Buswell, R., Kloft, H., Bos, F.P., Hack, N., Wolfs, R., Sanjayan, J., Nematollahi, B., Ivaniuk, E., Neef, T.: Integrating reinforcement in digital fabrication with concrete: a review and classification framework. Cem. Concr. Compos. **119**, 103964 (2021). https://doi.org/10.1016/j.cemconcomp.2021.103964
19. Ding, Y., Yu, J.-t., Yu, K.-Q., Xu, S.-l.: Basic mechanical properties of ultra-high ductility cementitious composites: from 40 MPa to 120 MPa. Compos. Struct. **185**, 634–645 (2018). https://doi.org/10.1016/j.compstruct.2017.11.034
20. Yu, J., Weng, Y., Yu, J., Chen, W., Lu, S., Yu, K.: Generative AI for performance-based design of engineered cementitious composite. Compos. Part B Eng. **266**, 110993 (2023). https://doi.org/10.1016/j.compositesb.2023.110993
21. Yu, K., Wang, Y., Yu, J., Xu, S.: A strain-hardening cementitious composites with the tensile capacity up to 8%. Constr. Build. Mater. **137**, 410–419 (2017). https://doi.org/10.1016/j.conbuildmat.2017.01.060
22. Hao, Z., Lu, C., Li, Z.: Highly accurate and automatic semantic segmentation of multiple cracks in engineered cementitious composites (ECC) under dual pre-modification deep-learning strategy. Cem. Concr. Res. **165**, 107066 (2023). https://doi.org/10.1016/j.cemconres.2022.107066

23. Li, L., Cai, Z., Yu, K., Zhang, Y.X., Ding, Y.: Performance-based design of all-grade strain hardening cementitious composites with compressive strengths from 40 MPa to 120 MPa. Cem. Concr. Compos. **97**, 202–217 (2019). https://doi.org/10.1016/j.cemconcomp.2019.01.001
24. Yu, K., Li, L., Yu, J., Xiao, J., Ye, J., Wang, Y.: Feasibility of using ultra-high ductility cementitious composites for concrete structures without steel rebar. Eng. Struct. **170**, 11–20 (2018). https://doi.org/10.1016/j.engstruct.2018.05.037
25. Soltan, D.G., Li, V.C.: A self-reinforced cementitious composite for building-scale 3D printing. Cem. Concr. Compos. **90**, 1–13 (2018). https://doi.org/10.1016/j.cemconcomp.2018.03.017
26. Al-Rousan, R.Z., Alnemrawi, B.R.: The influence of prestress level on the behavior of prefabricated precast concrete bridge deck panel systems with different overlay thicknesses. Structure. **58**, 105478 (2023). https://doi.org/10.1016/j.istruc.2023.105478
27. Al-Rousan, R., Alnemrawi, B.R.: NLFEA of the behavior of polypropylene-fiber-reinforced concrete slabs with square opening. Buidings. **14**(2), 480 (2024). https://doi.org/10.3390/buildings14020480
28. Alnemrawi, B.R., Al-Rousan, R.Z., Ababneh, A.N.: The role of CFRP strengthening in improving the punching shear behavior of heat-damaged flat slabs with openings of different sizes and locations. Eng. Fail. Anal. **160**, 108208 (2024). https://doi.org/10.1016/j.engfailanal.2024.108208
29. Al-Rousan, R.Z., Abu-Elhija, A.M.: Predicting the bond-slip relationship between concrete and CFRP using anchoring holes technique. Case. Stud. Constr. Mater. **13**, e00462 (2020). https://doi.org/10.1016/j.cscm.2020.e00462
30. Ye, J., Zhang, J., Yu, J., Yu, J., Yu, K.: Flexural behaviors of 3D printed lightweight engineered cementitious composites (ECC) slab with hollow sections. Eng. Struct. **299**, 117113 (2024). https://doi.org/10.1016/j.engstruct.2023.117113
31. ABAQUS Dassault Systemes Simulia Corp. P: ABAQUS Analysis User's Guide (6.17), 17th edn, (2017)
32. Zhou, J., Pan, J., Leung, C.K.Y.: Mechanical behavior of fiber-reinforced engineered cementitious composites in uniaxial compression. J. Mater. Civil. Eng. **27**(1), 04014111 (2015). https://doi.org/10.1061/(ASCE)MT.1943-5533.0001034
33. Tong-Seok Han, P.H.F., Sarah, L.B.: Simulation of highly ductile fiber-reinforced cement-based composite components under cyclic loading. ACI Struct. J. **100**(6), 749–757 (2003). https://doi.org/10.14359/12841
34. GB 50010–2010: Code for Design of Concrete Structures. China Architecture and Building Press, Beijing (2010)

Characterizing Shape Changes in 4D-Printed ABS Beams Under Thermal Stimuli

Grigorios Kostopoulos, Konstantinos Marneras, and Stelios K. Georgantzinos(✉)

Department of Aerospace Science and Technology, National and Kapodistrian University of Athens, 34400 Psachna, Greece
sgeor@uoa.gr

Abstract. The present study focuses on the shape morphing behaviour of 3D-printed beams manufactured using Acrylonitrile Butadiene Styrene (ABS) filament subjected to thermal stimulus. The aim of this study is to explore the impact of various printing parameters, that is, printing speed, layer height, layer width, and nozzle temperature, on the shape transformation of ABS beams. The beams were printed using a standard Fused Deposition Modeling (FDM) printer and then subjected to thermal stimulus (in a thermal chamber). The level of deformation of the beams was quantified by taking the following measurements: chord of deformed structure, height of arc, and internal arc length. Increased bending deformation was observed when the printing speed, layer height, and layer width were higher. On the other hand, bending responses decreased with an increase in nozzle temperature. The shape transformation was also temperature dependent. ANOVA (Analysis of Variance) was performed accordingly to confirm the statistically significant influence of these parameters on the transformation of shape.

Keywords: 3D Printing · Acrylonitrile Butadiene Styrene (ABS) · 4D Printing · Shape-Morphing

1 Introduction

Additive manufacturing (3D printing) has enabled the production of complex shaped structures with great accuracy and speed. 3D printing is used in many fields, such as aerospace, healthcare, automotive, and consumer products, due to its flexibility and ability to create custom parts [1]. For instance, in aerospace, 3D printing is used to manufacture lightweight, complex parts that improve fuel efficiency, while in healthcare, it has enabled the production of patient-specific implants and prosthetics [2, 3]. Similarly, in the automotive sector, manufacturers use 3D printing for rapid prototyping and the creation of highly specialized components, reducing lead times and production costs [4].

In 4D printing the dimension of time is introduced, as the printed structures can change shape with time when exposed to external stimuli such as heat, moisture, light, or magnetic fields. This transformative capability opens up exciting possibilities in the design and development of adaptive and functional structures [5, 6]. For example, self-assembling components can simplify complex assembly processes by transforming into

N. D. Lagaros et al. (Eds.): ADDOPTML 2024, LNCE 646, pp. 46–56, 2025.
https://doi.org/10.1007/978-3-031-92029-5_5

their final configuration post-production, reducing the need for additional labor or tools [7]. Deployable structures, such as foldable shelters or expandable aerospace components, can be transported in compact forms and later expanded into their operational state, enhancing portability and efficiency [7]. Similarly, smart textiles that adjust their properties based on environmental conditions, such as temperature or moisture, hold promise for advanced wearable technology and adaptive clothing [8]. In architecture, 4D-printed materials could create structures capable of responding to environmental changes, such as shifting sunlight or temperature, optimizing energy use and enhancing sustainability [9]. In robotics, soft actuators made through 4D printing could enable the creation of lightweight, flexible robots capable of performing tasks in confined or sensitive environments [10]. Even in the energy sector, 4D printing could contribute to advancements such as solar panels that reorient themselves for maximum efficiency or materials that alter thermal conductivity based on need [11].

However, the shape-morphing behavior is affected by the material properties, the stimuli applied, and the manufacturing parameters. In this study, ABS filament has been selected for this study. When used in 4D printing, ABS structures can exhibit programmable shape-morphing behaviors under thermal stimuli, making them ideal candidates for applications requiring controlled deformations [12].

Nevertheless, the properties of 4D-printed ABS components are highly sensitive to the parameters used during the printing process. Factors such as printing speed, layer height, layer width, and nozzle temperature significantly influence the stresses and deformations that accumulate within a material during the printing process [13, 14].

To investigate how such factors can influence the shape-morphing behavior, this study employs Taguchi Design of Experiments (DoE). Taguchi DoE is a powerful statistical tool designed to optimize processes by reducing the number of experimental trials while simultaneously evaluating multiple variables [13–15]. By strategically varying printing parameters, this approach helps identify the factors that have the most significant impact on the shape-morphing behavior of ABS beams.

2 Methodology

2.1 Materials and Equipment

The dimensions of the specimens utilized in the experiments are L = 60 mm, b = 3.5 mm, and h = 1.2 mm (Fig. 1a). The fabrication of test specimens was carried out using a Raise3D Pro2 3D printer (Raise3D Technologies, Shanghai, China) with a nozzle diameter of 0.4 mm. For this study, an ABS filament was selected, with a 1.75 mm diameter, presented in an ivory white color (3D Edge, Metamorfosi, Greece). The CAD model of the test specimen is depicted on Fig. 1b, while the corresponding 3D-printed test part is shown in Fig. 1c, providing a practical representation of the geometry used in this study.

The experimental process involved a thermal chamber capable of reaching and maintaining the activation temperature required for triggering shape transformation in ABS. The thermal chamber used for the experiments was a SNOL 6,7/1300 LSM01 (SNOL, Narkūnai, Lithuania). This chamber provided precise temperature control to ensure consistency across all tests.

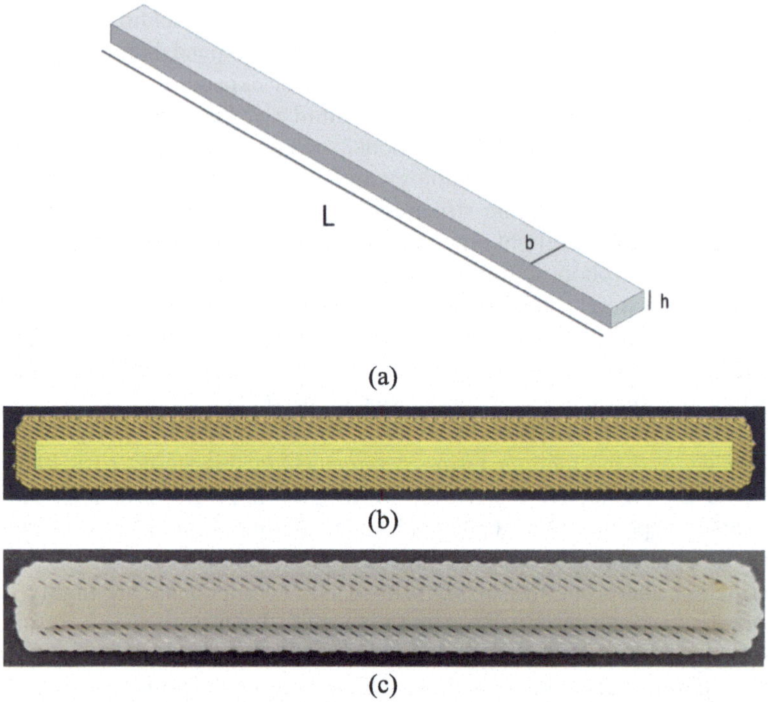

Fig. 1. Test specimen (**a**) dimensions, (**b**) CAD, and (**c**) 3D printed part.

2.2 Experimental Design Using the Taguchi Method

This study employed the Taguchi Design of Experiments methodology, a statistical approach designed to optimize process parameters while minimizing the number of trials needed. The Taguchi method uses orthogonal arrays to systematically evaluate the effects of multiple factors on response variables. An L18 orthogonal array was applied, allowing for the analysis of five key parameters, i.e. printing speed, layer height, layer width, nozzle temperature, and activation temperature, at various levels.

The selected levels for each factor included printing speeds of 75 mm/s, 90 mm/s, 105 mm/s, 120 mm/s, 135 mm/s, and 150 mm/s; layer heights of 0.15 mm, 0.2 mm, and 0.3 mm; layer widths of 0.4 mm, 0.48 mm, and 0.52 mm; nozzle temperatures of 220 °C, 230 °C, and 240 °C; and activation temperatures of 140 °C, 150 °C, and 160 °C. A total of 18 unique experimental setups were created using this design, as shown in Table 1. This design enabled a thorough assessment of how printing parameters influenced the deformation behavior of ABS beams under thermal conditions. The beam geometry was selected to effectively evaluate the impact of thermal stimulus on shape transformation, with measurements focusing on the chord of the deformed structure, height of the arc, and internal arc length (Fig. 2).

By systematically varying the parameter levels, the influence of each parameter on the shape transformation could be quantified. This systematic approach allowed for

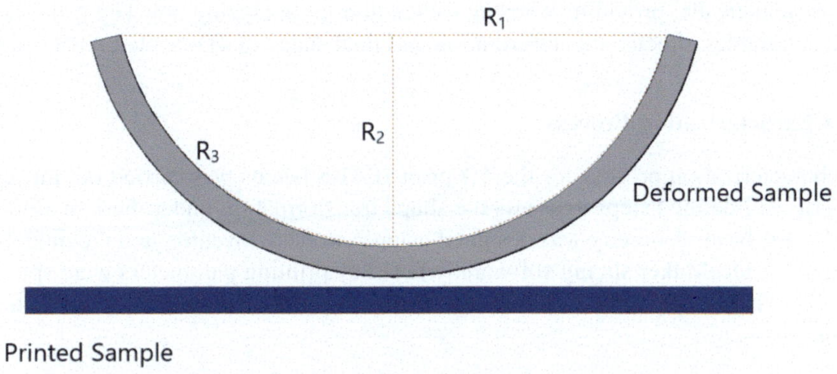

Fig. 2. Schematic representation of the deformed shape parameters, *R1* is the chord of the deformed structure, *R2* indicates the height of the arc, and *R3* is the internal arc length.

identifying the optimal printing and thermal conditions while reducing experimental time and resource consumption.

Table 1. L18 orthogonal array based on using the Taguchi DoE methodology.

Exp. run	Printing speed (mm/s)	Layer height (mm)	Layer width (mm)	Nozzle temperature (°C)	Activation temperature (°C)
1	75	0.15	0.40	220	140
2	75	0.20	0.48	230	150
3	75	0.30	0.52	240	160
4	90	0.15	0.40	230	150
5	90	0.20	0.48	240	160
6	90	0.30	0.52	220	140
7	105	0.15	0.48	220	160
8	105	0.20	0.52	230	140
9	105	0.30	0.40	240	150
10	120	0.15	0.52	240	150
11	120	0.20	0.40	220	160
12	120	0.30	0.48	230	140
13	135	0.15	0.48	240	140
14	135	0.20	0.52	220	150
15	135	0.30	0.40	230	160
16	150	0.15	0.52	230	160
17	150	0.20	0.40	240	140
18	150	0.30	0.48	220	150

The experimental setup was structured to evaluate all relevant parameter combinations systematically, providing a strong foundation for analyzing how key printing and thermal variables affected the deformation and final shape of 4D-printed ABS beams.

2.3 Characterization Process

The characterization process for the 4D-printed ABS beams was carried out through a series of well-defined steps to assess the shape transformation under thermal stimulus. Initially, the beam geometry was designed using Autodesk Inventor, and the model was sliced using IdeaMaker slicing software, where key printing parameters were specified (Fig. 3a). This slicing process generated the necessary G-code instructions for the 3D printer.

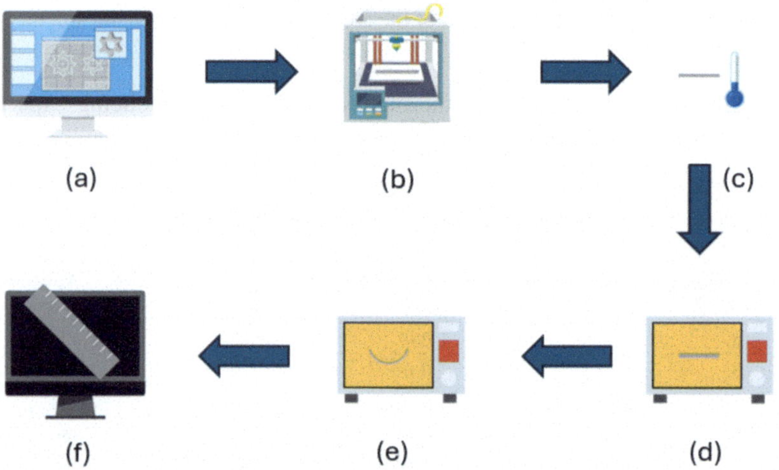

Fig. 3. Representation of characterization process, (**a**) design and slicing of the sample, (**b**) printing of the sample, (**c**) cooling of the sample to room temperature, (**d**) placement of the sample in the laboratory oven, (**e**) activation and deformation of the sample, and (**f**) Cooling of the deformed sample to room temperature, measurements and data acquisition.

The sample was then printed, with the ABS filament extruded layer by layer according to the predefined parameters, resulting in a solid beam structure (Fig. 3b). After printing, the sample was allowed to cool naturally at room temperature, stabilizing in its initial, undeformed state (Fig. 3c). Subsequently, the sample was placed inside a laboratory oven, where it was heated at activation temperature. The sample was maintained at this temperature for a set duration of fifteen minutes to ensure complete deformation (Fig. 3d, e).

Following the activation and deformation process, the sample was removed from the oven and allowed to cool at room temperature, during which it retained its deformed shape, achieving a 'locked-in' final form. To ensure the reliability and repeatability of the results, each experimental run was conducted five times, with specimens fabricated and tested under identical conditions. Finally, precise measurements were taken using Autodesk Inventor to quantify the degree of deformation, including the chord of the deformed structure, *R1*, the height of the arc, *R2*, and the internal arc length, *R3* (Fig. 3f). These measurements were used for data analysis, enabling the evaluation of the impact of printing parameters on the shape transformation behavior of the ABS beams.

3 Results and Discussion

3.1 Main Effect Analysis

The main effect analysis for *R1, R2, and R3* reveals the influence of key printing and activation parameters on the shape transformation of the printed ABS beams. As depicted in Fig. 4a, one can observe that increase in printing speed tends to lead to a decrease in *R1* mean response. Accordingly, increase in layer height and layer width leads to a decrease in the mean response value. Furthermore, an increase in nozzle temperature leads to an increase in mean response value. Finally, activation temperature tends to decrease the mean response value, with a maximum deformation at 150 °C, as reflected by a shorter chord length.

As for the height of arc, *R2* (Fig. 4b), higher printing speed tends to result in higher *R2* mean response, indicating higher deformation. Moreover, higher layer height and layer width leads to an increase in *R2* mean response value. Nozzle temperature also affects the shape-morphing behavior of the beam, with higher nozzle temperature values leading to a decrease in response value. Lastly, for activation temperature, maximum deformation (i.e. maximum *R2* mean) is observed at 150 °C, as evidenced by the highest *R2* values. At 140 °C and 160 °C deformation decreases, indicating reduced shape transformation.

Lastly for *R3* (Fig. 4c), one can observe that higher printing speed tends to result in a lower *R3* mean response, although with a spike to a higher *R3* mean value at 135 mm/s. Layer height and nozzle temperature demonstrate a significant effect, with an increase of the response values as the factor levels increase. For layer width, a similar effect can be observed, although with a maximum mean *R3* value at 0.48 mm. Lastly, activation temperature influences response value distinctly, with the maximum mean internal arc length observed at 140 °C and the minimum at 160 °C. This suggests that 160 °C provides optimal conditions for maximizing deformation, that is minimizing beam length.

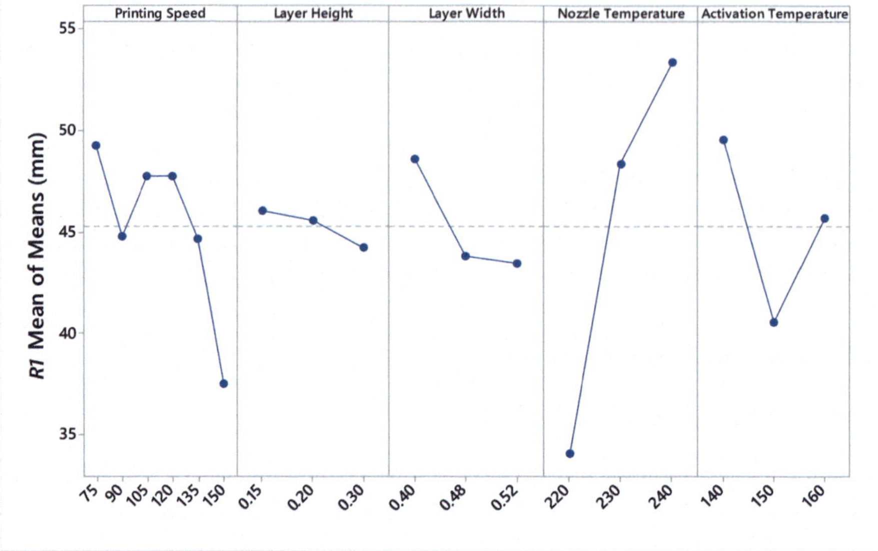

(a)

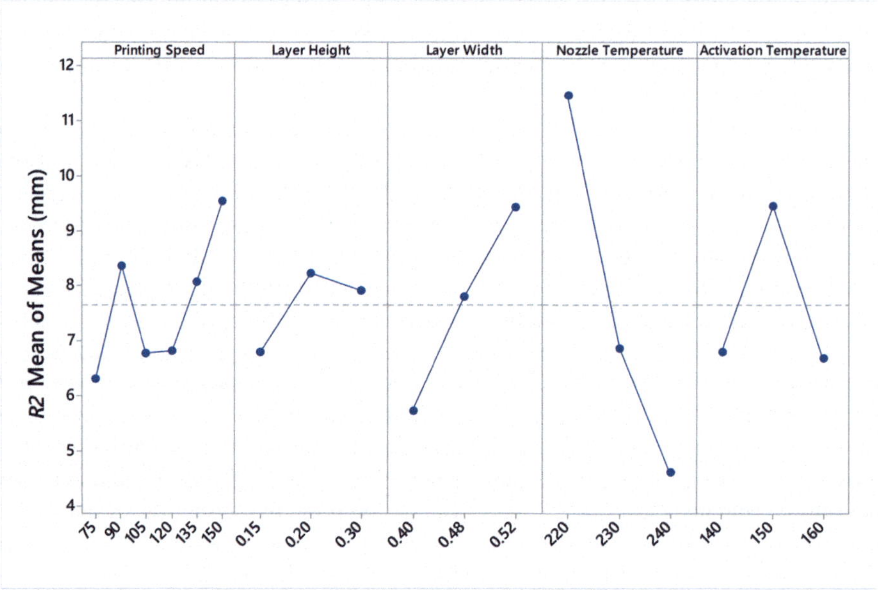

(b)

Fig. 4. (**a**) *R1*, (**b**) *R2*, and (**c**) *R3* mean of means graph.

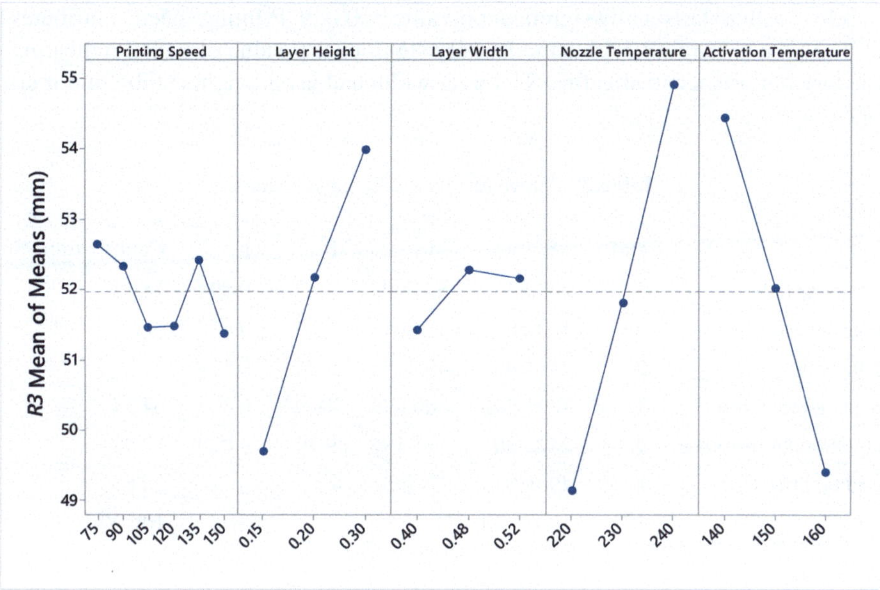

(c)

Fig. 4. (*continued*)

3.2 Analysis of Variance (ANOVA)

In this study, ANOVA is utilized to assess the influence of each control parameter on the response variables *R1*, *R2*, and *R3*. By partitioning the total variability into components attributable to the control factors, ANOVA provides a clear understanding of how much each factor influences the outcome. This analysis enables the identification of the most significant factors, quantifies their contribution to the overall variability, and helps distinguish genuine effects from random noise. The ANOVA was conducted with a 95% confidence level (significance level of 5%).

The following ANOVA tables consist of several key components. Degrees of freedom (DoF) represent the number of independent values available to estimate variability for each factor, calculated as the number of factor levels minus one. The Sum of Squares (SS) quantifies the total variability in the response variable, which is divided into the variability that is explained by the factors (factor SS) and the unexplained variability (residual error SS). The Mean Squares (MS) is the average variability for each factor, while the F-value is the ratio of the variability explained by the factor (MS of the factor) to the unexplained variability (MS of the error). The p-value evaluates the statistical significance of the factor's effect. Finally, the contribution percentage of a factor in ANOVA is calculated by determining the proportion of the Sum of Squares for that factor relative to the Total Sum of Squares.

For *R1*, the ANOVA results (Table 2) indicate that nozzle temperature is the most significant factor, contributing 64.34% to the total variability, with a statistically significant p-value of 0.002. The activation temperature contributes by 12.98% to the total variability, with a statistically significant p-value of 0.029. Printing speed contributes by 14.13% to the variability, with a moderately significant p-value of 0.094, indicating its secondary importance in affecting *R1*. Layer width and layer height exhibit minor effect on *R1*.

Table 2. Anova results for *R1* mean values.

	DoF	SS	MS	F	p	Contribution (%)
Printing speed	5	263.840	52.769	4.22	0.094	14.13
Layer height	2	10.650	5.326	0.43	0.680	0.57
Layer width	2	98.970	49.485	3.96	0.113	5.30
Nozzle temperature	2	1201.390	600.695	48.07	0.002	64.34
Activation temperature	2	242.390	121.196	9.70	0.029	12.98
Residual error	4	49.980	12.495	–	–	2.68
Total	17	1867.230	–	–	–	100.00

For *R2*, the ANOVA results (Table 3) reveal that nozzle temperature is the most significant factor, contributing 58.91% to the total variability with an extremely high F-value of 131.59 and a p-value of 0.000. Layer width and activation temperature also significantly influence *R2*, contributing 16.54% and 11.82% to the total variability, with a p-value of 0.003 and 0.005 respectively. Printing speed and layer height exhibit less influence on the response value.

Table 3. Anova results for *R2* mean values.

	DoF	SS	MS	F	p	Contribution (%)
Printing speed	5	22.553	4.511	8.12	0.032	9.08
Layer height	2	6.852	3.426	6.16	0.060	2.76
Layer width	2	41.060	20.530	36.94	0.003	16.54
Nozzle temperature	2	146.281	73.140	131.59	0.000	58.91
Activation temperature	2	29.340	14.670	26.39	0.005	11.82
Residual error	4	2.223	0.556	–	–	0.90
Total	17	248.309	–	–	–	100.00

For *R3*, the ANOVA results (Table 4) show that nozzle temperature is the most significant factor, contributing 41.62% to the total variability and with a statistically significant p-value of 0.000. Activation temperature and layer height also have significant influence on *R3*, with a contribution of 31.74% and 23.03% respectively and a p-value of 0.000. In contrast, printing speed and layer width show minimal impact on the response variable *R3*.

Table 4. Anova results for *R3* mean values.

	DoF	SS	MS	F	p	Contribution (%)
Printing speed	5	4.902	0.980	3.14	0.145	2.04
Layer height	2	55.315	27.657	88.57	0.000	23.03
Layer width	2	2.514	1.257	4.03	0.110	1.05
Nozzle temperature	2	99.949	49.975	160.03	0.000	41.62
Activation temperature	2	76.222	38.111	122.04	0.000	31.74
Residual error	4	1.249	0.312	–	–	0.52
Total	17	240.151	–	–	–	100.00

4 Conclusions

In conclusion, this study provides significant advancements in understanding the shape-morphing behavior of ABS beams by analyzing the influence of the control factors. The main effect analysis and ANOVA highlight the critical role of the printing and activation parameters in achieving desired deformation outcomes, offering valuable insights for optimizing stimuli-responsive structures. Key findings include the following:

- Nozzle temperature was the most influential factor across all response variables (*R1*, *R2*, and *R3*), with higher values generally increasing *R1* and *R3* response but decreasing *R2* response.
- Activation temperature had also a significant impact on deformation, with 150 °C being critical for maximizing *R2* response, while 150 °C and 160 °C minimized *R1* and *R3* response respectively.
- Layer height had significant effects on *R3*, where increasing its value led to an increase in the internal arc length.
- Layer width demonstrated significant effects for *R2*, where increasing its value led to an increase in the height of the arc, indicating a higher deformation.
- Printing speed had moderate effects across the responses, with higher speed generally reducing *R1* and *R3* but increasing *R2*.

These results emphasize the importance of precise control over printing and thermal parameters, as they can greatly influence the shape-morphing behavior of the additively manufactured structures.

References

1. Mahmood, A., Akram, T., Shenggui, C., Chen, H.: Revolutionizing manufacturing: A review of 4D printing materials, stimuli, and cutting-edge applications. Compos. Part B Eng. **266**, 110952 (2023)
2. Aldawood, F.K.: A comprehensive review of 4D printing: state of the arts, opportunities, and challenges. Actuators. **12**, 101 (2023)

3. Ramezani, M., Mohd Ripin, Z.: 4D printing in biomedical engineering: advancements, challenges, and future directions. J. Funct. Biomater. **14**, 347 (2023)
4. Nayeem, A.M., Hossain, M.M.N.: Usage of additive manufacturing in the automotive industry: a review. Bangladesh J. Multidiscip. Sci. Res. **8**, 9–20 (2023)
5. Georgantzinos, S.K., Giannopoulos, G.I., Bakalis, P.A.: Additive manufacturing for effective smart structures: the idea of 6D printing. J. Compos. Sci. **5**, 119 (2021)
6. Sajjad, R., Chauhdary, S.T., Anwar, M.T., Zahid, A., Khosa, A.A., Imran, M., Sajjad, M.H.: A review of 4D printing—technologies, shape shifting, smart polymer based materials, and biomedical applications. Adv. Ind. Eng. Polym. Res. **7**, 20–36 (2024)
7. Vatanparast, S., Boschetto, A., Bottini, L., Gaudenzi, P.: New trends in 4D printing: a critical review. Appl. Sci. **13**, 7744 (2023)
8. Li, S.: Review on development and application of 4D-printing technology in smart textiles. J. Eng. Fiber. Fabr. **18**(2), 155892502311774 (2023)
9. Niazy, D., Ashraf, M., Bodaghi, M., Zolfagharian, A.: Resilient city perspective: 4D printing in art, architecture and construction. Mater. Today. Sustain. **26**, 100708 (2024)
10. Soleimanzadeh, H., Rolfe, B., Bodaghi, M., Jamalabadi, M., Zhang, X., Zolfagharian, A.: Sustainable robots 4D printing. Adv. Sustain. Syst. **7**, 2300289 (2023)
11. Momeni, F., Ni, J.: Nature-inspired smart solar concentrators by 4D printing. Renew. Energy. **122**, 35–44 (2018)
12. Mendenhall, R., Eslami, B.: Experimental investigation on effect of temperature on FDM 3D printing polymers: ABS, PETG, and PLA. Appl. Sci. **13**, 11503 (2023)
13. Kostopoulos, G., Stamoulis, K., Lappas, V., Georgantzinos, S.K.: Shape morphing of 4D-printed polylactic acid structures under thermal stimuli: an experimental and finite element analysis. Aerospace. **11**, 134 (2024)
14. Georgantzinos, S.K., Kostopoulos, G., Stamoulis, K.P.: Characterizing the shape-morphing behaviour of 4D printed PLA structures. J. Phys. Conf. Ser. **2716**, 012034 (2024)
15. Rasheed, A., Hussain, M., Ullah, S., Ahmad, Z., Kakakhail, H., Riaz, A.A., Khan, I., Ahmad, S., Akram, W., Eldin, S.M., Khan, I.: Experimental investigation and Taguchi optimization of FDM process parameters for the enhancement of tensile properties of Bi-layered printed PLA-ABS. Mater. Res. Express. **10**, 095307 (2023)

Optimization of 3D Printing Parameters on Surface Roughness and Flatness of PLA Using Taguchi Design of Experiments

George Dimitrakopoulos[1(✉)], Grigorios Kostopoulos[1], Nikos D. Lagaros[2], and Stelios K. Georgantzinos[1]

[1] Department of Aerospace Science and Technology, National and Kapodistrian University of Athens, 34400 Psachna, Greece
gdimitrakop@aerospace.uoa.gr, sgeor@uoa.gr

[2] Department of Structural Engineering, School of Civil Engineering, Institute of Structural Analysis and Antiseismic Research, National Technical University Athens, Heroon Polytechniou Str. 9, Zografou Campus, 15780 Athens, Greece

Abstract. The objective of this research was to evaluate the influence of some parameters of 3D printing on the rough surface and flatness of the parts. For the study, all samples were produced from polylactic acid (PLA), following the Taguchi experimental design method. The printing parameter settings that optimised the roughness and flatness were firstly determined. Layer height, print speed and nozzle temperature were the key printing parameters used to devise the experiments. The experiments were defined by varying these according to the Taguchi L9 orthogonal array. Surface roughness in different orientations (X, Y, Z) and angles (15°, 30°, 45°, 60°, 75°) was then measured whereas flatness was evaluated based on the standard deviation from the value of a reference plane. This study potentially opens a new door to manufacturers to enhance surface finish and dimensional accuracy for various 3D printed components.

Keywords: 3D Printing · Polylactic Acid (PLA) · Surface Roughness · Flatness · Taguchi Design of Experiments · Additive Manufacturing

1 Introduction

Due to its versatility, economy and adaptation it is cutting edge in production technology. That makes it advantaged in prototyping and low-volume object production [1] PLA is a common polymer amongst other materials and it is due to its biodegradable nature, user-friendly nature and low cost [2, 3]. PLA is obtained from renewable sources, such as corn starch or sugarcane, which makes it [4] an eco-friendly substitute for petroleum-based plastics. PLA in many areas of 3D printing has provided opportunities in the fields of education, health, consumer goods and prototyping because of its low melting point, less distortion and user-optimizing properties [5, 6].

Detailing the printing parameters to optimize part quality is a challenge to establish good surface finish and dimensional accuracy. As an example, poor surface quality

N. D. Lagaros et al. (Eds.): ADDOPTML 2024, LNCE 646, pp. 57–67, 2025.
https://doi.org/10.1007/978-3-031-92029-5_6

will have a negative impact on visual appearance, frictional and surface properties, and post-processing requirements, whereas deviations in flatness will lead to fit issues and mechanical strength reductions in parts designed to be integrated into an assembly [7–11]. High flatness is essential for applications in sectors like aerospace, automotive and medical devices, where dependability and precision are vital [13].

The process parameters like layer height, printing speed, nozzle temperature and filling density have a great influence on the quality of 3D printed PLA parts and their mechanical properties [14]. Layer height determines the thickness of each layer that is extruded and applied to the print which directly affects both smoothness and visual quality of the part. If not calibrated well, the printing speed, which builds up how fast the material is compressed, can cause the layer not to adhere properly or generate types of defects like spots or voids. The nozzle temperature is a critical parameter for sufficient fusion and adhesive connection of PLA layers and the filler density determines the printed article's inner structure and mechanical property [15, 16].

The first research identifies optimal parameter settings in 3D printing based on Taguchi DoE method which is quite preferable experimental design method. By utilizing an orthogonal array, the Taguchi method reduces the amount of experiments needed, while obtaining important information about how several factors affect the output properties [17, 18]. These approaches aid in perfecting adjust parameters to achieve optimum surface roughness and flatness, resulting in reduced costs and enhanced efficiency [19]. The present study attempts to bridge the gap between parameter interactions during 3D printing and their effect on the surface quality of PLA components. Results provide practical recommendations for enhancing the overall quality and performance of 3D printed PLA components and demonstrate the viability of Taguchi method. Overall, these results are not just relevant for enhancing the surface and dimensional stability of PLA as additively manufactured parts, but also add to the project objective of sustainable and efficient manufacturing.

2 Methodology

2.1 Materials and Equipment

It was found that the printing parameters have a significant effect on surface roughness and flatness, which were determined using a specific prototype for controlled experiments. Figure 1a, CAD model of prototype is ablated using oblique planes angles 15°, 30°, 45°, 60° and 75° (similar surface orientations during printing). Figure 1b shows the 3D printed test part that corresponds to the one tested.

The test parts were printed using a desktop FDM (Fused Deposition Modeling) printer. The filament used for printing was PLA with a diameter of 1.75 mm. A 0.4 mm diameter nozzle was used for extrusion with appropriate layer heights to ensure even deposition of material. To avoid impact of environmental factors like temperature or humidity, all the printing experiments were performed in open air conditions.

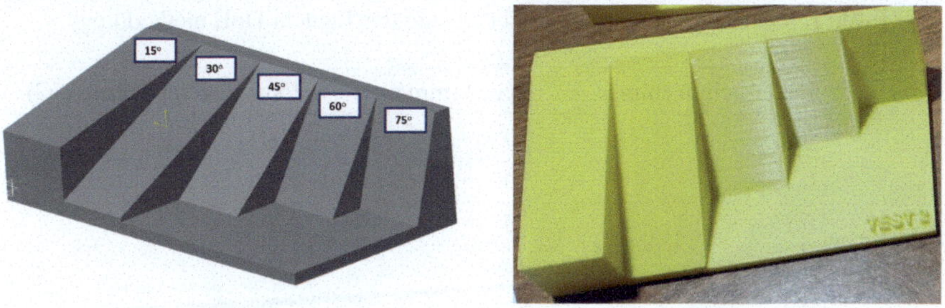

(a) (b)

Fig. 1. A custom-designed part (**a**) CAD, and (**b**) 3D printed part.

Surface roughness measurements were performed with a profilometer. It gave quantitative readings of roughness like Ra and Rz at different orientations and angles. Flatness was assessed with a flatness meter determining deviations from a reference plane to high accuracy. Prior to use, both instruments were calibrated for accuracy and to ensure reliable measurements.

This experimental setting establishes a strong foundation for the study of relevant printing variables that affect the surface roughness and flatness of 3D printed PLA components. Using advanced tools along with a systematic design, the accuracy and reproducibility of the findings were ensured.

2.2 Experimental Design Using the Taguchi Method

The Taguchi experimental design procedure was utilised to determine the experimental design for this study, as the statistic is minimal in the experimental tests needed to optimise the process parameters. In this study, an L9 rectangular array was adopted to evaluate 3 printing parameters (layer height, printing speed and extrusion temperature) in three levels. The chosen parameter levels were determined from preliminary studies and general settings used for PLA 3D printing. The parameters included layer heights [mm] (0.1 mm, 0.2 mm and 0.3 mm), print speeds [mm/s] (30 mm/s, 50 mm/s and 70 mm/s) and extrusion temperatures [°C] (200 °C, 210 °C, 220 °C); therefore, the nine combinations result in nine unique experimental conditions as shown in Table 1.

The specially designed test fixture was constructed using each combination of parameters. The geometry of test piece was designed into inclined planes at 15°, 30°, 45°, 60° and 75° angles. Such a design enabled a holistic assessment of the influence of process parameters on the quality of the printed part in terms of surface quality and dimensional stability. This experimental design allowed systematic evaluation of all relevant combinations of parameters, yielding reliable data from which the effect of key process variables on 3D printed PLA quality could be examined.

Table 1. L9 orthogonal array based on using the Taguchi DoE methodology.

Test part	Layer height (mm)	Speed (mm/s)	Extrude temperature (°C)
1	0.1	30	200
2	0.1	50	210
3	0.1	70	220
4	0.2	30	210
5	0.2	50	220
6	0.2	70	200
7	0.3	30	220
8	0.3	50	200
9	0.3	70	210

2.3 Measurement Procedure

All instruments were calibrated before measuring process to ensure proper operation. Then the surface roughness was characterized by means a profilometer (Fig. 1). Figure 2a, b, an apparatus capable of accurate surface texture measurement in micrometer scale. It should be mentioned that measurements were done in different modes (X, Y, Z) and in inclined planes at angles of 15°, 30°, 45°, 60° and 75° to account for different geometries of the surface.

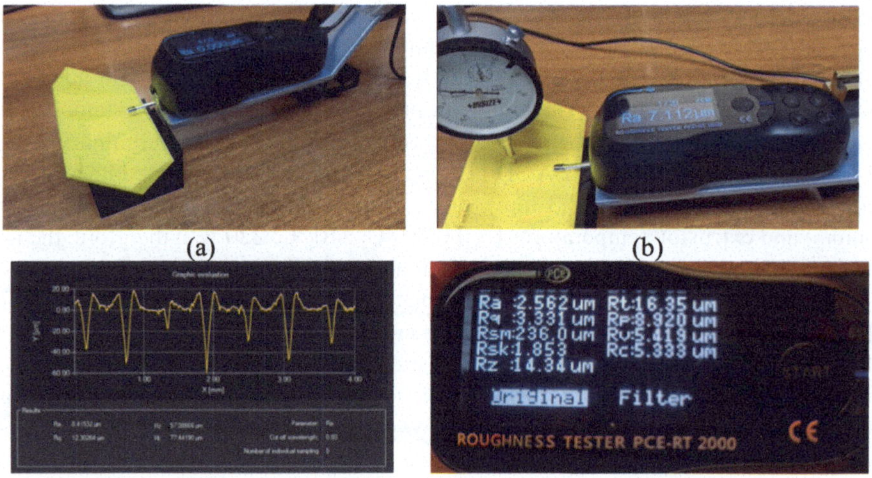

(a) (b) (c) (d)

Fig. 2. Measurement of roughness with roughness tester PCE RT 2000.

To account for possible variations in surface quality, each part of the test was measured at different locations. 3 Record and measure the various roughness parameters e.g. Ra (average roughness), Rz (maximum peak-to-valley height), etc. Localized deviations were reduced by using average values from multiple measurements for analysis. Representative outputs of the roughness tester are shown in Fig. 2c, d.

Flatness, defined as the extent to which a surface deviates from a perfectly planar reference, was assessed with a flatness gauge (Fig. 3). Using a reference plane, this device measures the deviation of the surface of the test part. Measurements were conducted thrice for each test part for reproducibility and accuracy. To reduce experimental errors, the mean values of the measurements were used for further analysis.

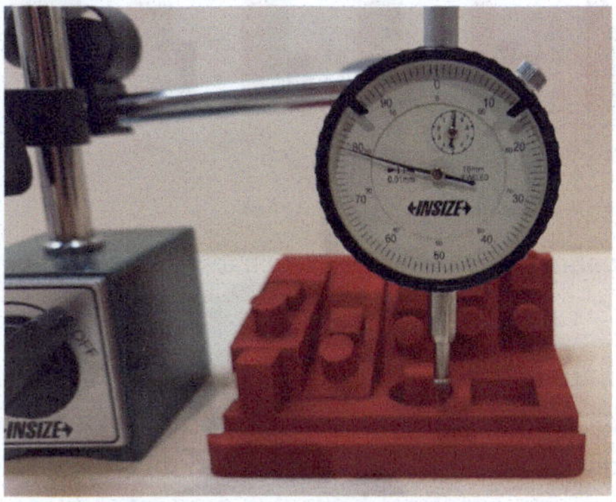

Fig. 3. Measurement of flatness.

To guarantee consistency and accuracy, all measurement instruments in these experiments were calibrated as per standard procedures. This requires the calibration of the profilometer and the flatness gauge to standard reference values. This was a crucial step in removing any systematic errors when collecting data.

3 Results and Discussion

3.1 Experimental Results

Using the bar chart (Fig. 4) demonstrates the values of the surface roughness (Ra) parameters on all walls of the Y, X, and Z axis for every test piece, which indicates the impact of the parameters of 3D printing on surface properties. Across the three axes, the X-axis maintains the lowest surface roughness, suggesting regular deposition of material and lower sensitivity to parameter alterations. All tests tested in strips, with the Y-Axis showing the worst surface roughness across all samples, indicating that this axis is highly affected by layer stacking and issues related to support structures during the print. This is Z axis with intermediate roughness values to help characterize the physical depiction of the output data, usually ranging from high means or low means as they are designed around a layer of 3D printer, where the transition produces a surface with non-plane topography.

These results underscore the importance of parameter optimization in obtaining desired surface finishes. Although the X-axis is generally strong, the Y-axis is a parameter

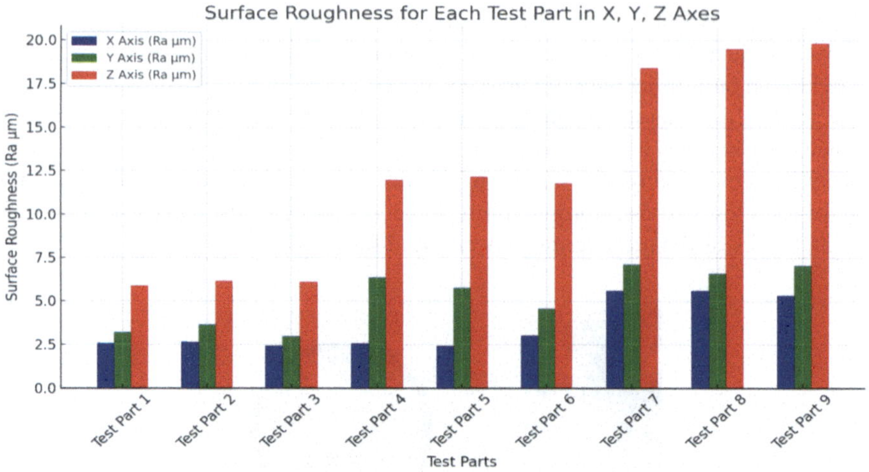

Fig. 4. Surface roughness (Ra) values for each test part along the X, Y, and Z axes.

that may need selection to alleviate bumps. Overall, we found that lower layer heights, lower printing speeds and moderate extrusion temperatures mainly support the reduction of surface roughness and were most effective at the angles where the process requires the most change. This means practical hints are presented to enable improved surface quality in 3D printed parts.

However, the experimental results shown in the line graph (Fig. 5) give a detailed look at the evolution of the roughness (Ra) of the surface of each test piece in a function of the angles (15°, 30°, 45°, 60° and 75°). The obtained results show that the increase of thickness affects the profile production can become more significant for a higher angle (60° and 75°) where the deposition process has more irregularities. The most significant factor for angular surface quality, is layer height, as can be seen by the test parts 7–9 printed with the highest layer height (0.3 mm) exhibiting the most roughness at all angles. On the contrary, test piece 1 which has the least layer height (0.1 mm) always shows a lower surface roughness at all angles, showing the benefit of having thinner layers to accomplish a smoother surface. Although low layer heights and moderate temperatures are positively correlated with improved surface finish in general, the data points out that optimizing printing speed and thermal control is needed in order to reduce roughness at steep angles. With careful consideration of a parameter's purpose relative to geometric orientation, manufacturers can achieve the same surface quality across complex geometries.

Table 2 shows the flatness deviation (in mm) that was measured on each of the test pieces. The data show a clear distribution along the Y axis, with deviations in flatness growing as layer height increases; hence, layer height is the most impactful parameter. It is always the case that lower layer heights have better flatness than higher layer heights, while higher layer heights have greater deviations. The experimental data strongly support selection of parameter settings to minimize deviations from flatness in 3D printed components.

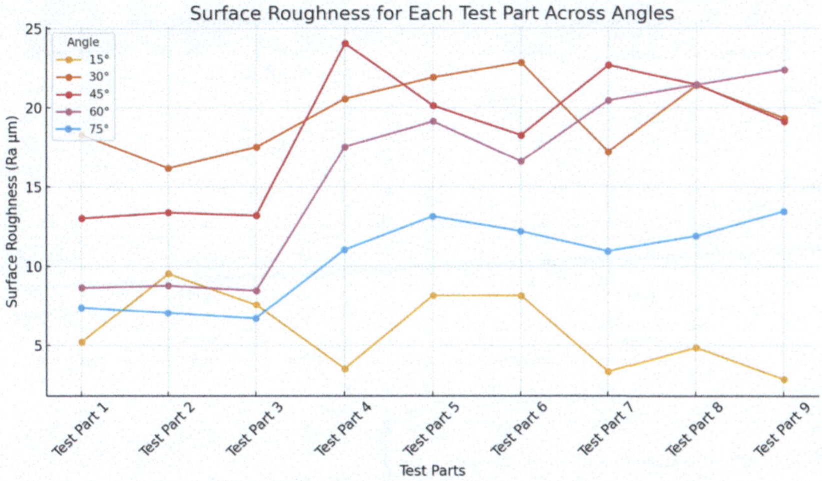

Fig. 5. Surface roughness (Ra) values for each test part measured across angles (15°, 30°, 45°, 60°, and 75°).

Table 2. Flatness results in mm for every test part.

1	2	3	4	5	6	7	8	9
0.060	0.090	0.100	0.200	0.150	0.140	0.350	0.400	0.500

3.2 Main Effects Plots for Roughness

The main effects plots in Fig. 4 illustrate the influence of the three primary parameters—layer height, print speed, and extrusion temperature—on surface roughness (Ra) across multiple orientations (X, Y, Z) and angles (15°, 30°, 45°, 60°, and 75°). Layer height has a significant influence on surface roughness through all orientations and angles. Smoother surfaces are usually obtained with a smaller layer height as evidenced in the steady decrease of Ra values with decreasing layer height levels (0.1 mm). This influence is especially significant for Ra-X, Ra-Y, and Ra-Z, where smaller layers always yield a better surface finish. Although not a significant improvement, at higher angles from the Ra-75 plot it can be observed that the effect of layer height is minimised. The third factor is print speed, where slower print speeds can reduce roughness. This phenomenon can be observed for most of the plots, with a particularly notable example being in Ra-15, Ra-30, and Ra-X where a decrease in the feeding speed (70 mm/s down to 30 mm/s) results in having a lower value for Ra. The reduced surface formations at lower speeds lead to a better surface quality because the material is less deposited. On the other hand, the impact of print speed starts to fade out in some orientations (e.g., Ra-45, where the aforementioned trends are less noticeable) (Fig. 6).

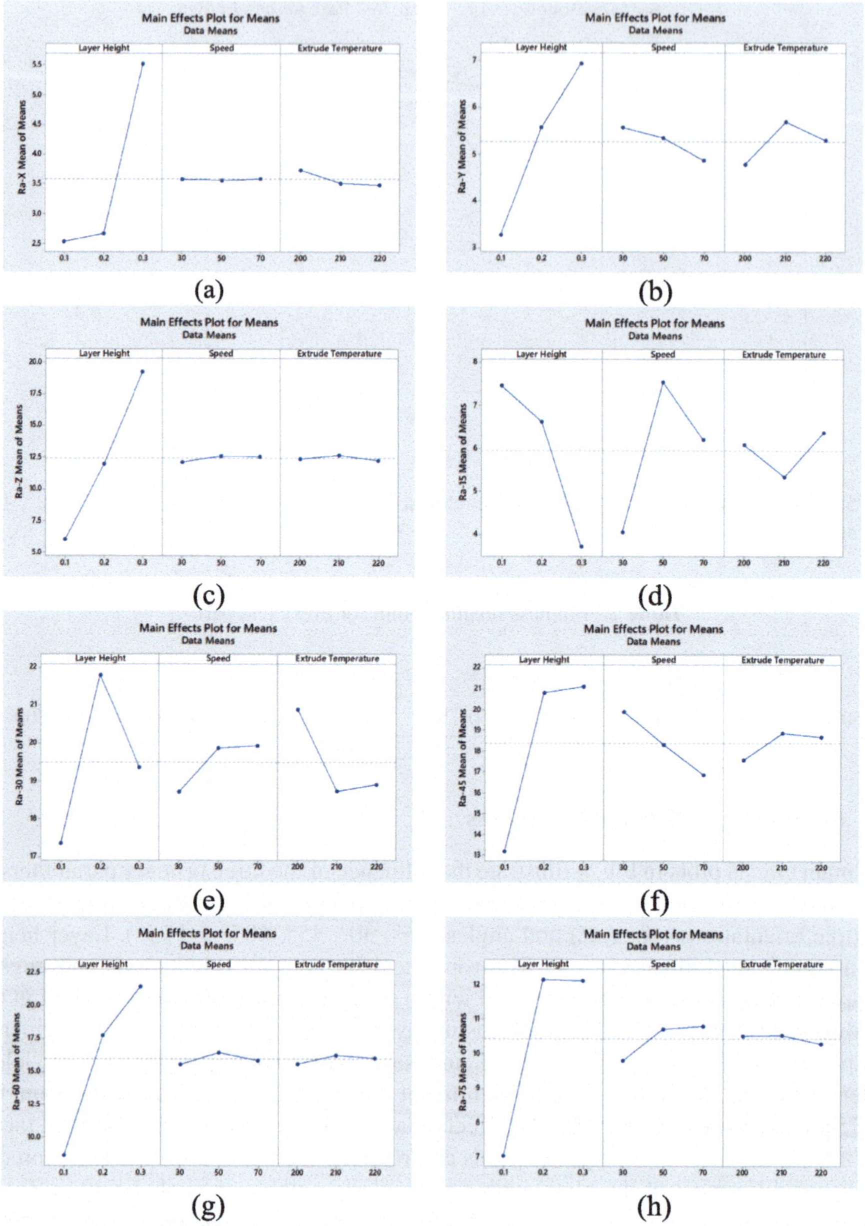

Fig. 6. Surface roughness (Ra) main effect plots for (**a**) orientation X, (**b**) orientation Y, (**c**) orientation Z, (**d**) angle 15°, (**e**) angle 30°, (**f**) angle 45°, (**g**) angle 60°, (**h**) angle 75°.

Extrusion temperature shows an inconsistent effect on the surface roughness. In several plots, eg.: Ra-Y and Ra-45 moderate extrusion temperatures (210 °C) provided better surface finish, conversely higher temperatures (220 °C) cause higher Ra values when the material is overheated or is flowed out of control. On the other hand, insufficient temperatures (200 °C) may result in poor adhesion of the layers, consequently providing rough surfaces. At steep angles (e.g., Ra-60 and Ra-75), this effect is pronounced too, since the right temperature settings are decisive for the surface quality.

The parameters also have different impact depending on the angle of inclination. At lower angles (e.g. Ra-15 and Ra-30) there is a clear trend in all three parameters, where the layer height and the velocity have the most significant effect. However, for higher angles (such as Ra-60 and Ra-75) the contribution of layer height becomes less significant as the extrusion temperature plays a more decisive role in the structural quality of the surface. Such variation implies that the parameter optimization should consider workpiece geometry to ensure uniform quality across all orientations.

Conclusion: Layer height and printing speed are the parameters with the highest impact on surface roughness for horizontal or slightly inclined surfaces, the study finds. While secondary to the feed angle, the extrusion temperature is also crucial to maintaining consistency at steeper angles. Simultaneously, the results emphasize the need to balance these variables in order to achieve the most effective surface roughness.

3.3 Main Effects Plots for Flatness

The main effects plot for flatness shows the impact of layer height, printing speed and extrusion temperature on the means of flatness. The results indicate that layer height plays the most prominent role and its rise (layer height) results in the increased flatness deviation. Layer height influences how strongly the object sticks to the working platform, as higher height is capable of generating larger flatness deviations, such as those test pieces printed with a layer height of 0,3 mm while better flatness is achieved by parts when a layer height of 0,1 mm is used, which is a benefit of using low layer heights in terms of dimensional stability. Printing speed has a small effect on flatness, with a slight tendency for higher flatness deviations at higher speeds. Notably however, this would also cause higher level shrinks at high print speed, where layer consistent deposition may be compromised, thus yielding higher flatness deviations.

Printing speed has a similar effect, but the overall effect is much less from speed than layer height. Extrusion temperature has a more non-linear effect on the other hand. As can be seen from the results, in the case of a moderate temperature, namely, 210 °C, the deviations are lower than both lower (200 °C) and higher (220 °C) temperatures. The reason can be weak bonding in layers due to low temperature and thermal deformation at high temperature.

These results indicate that a very low height and a moderate extrusion temperature need to be used to obtain the best flatness for 3D printed parts. Printing speed has a secondary effect, but even so we should consider its influence for dimensional stability for geometry-sensitive parts (Fig. 7).

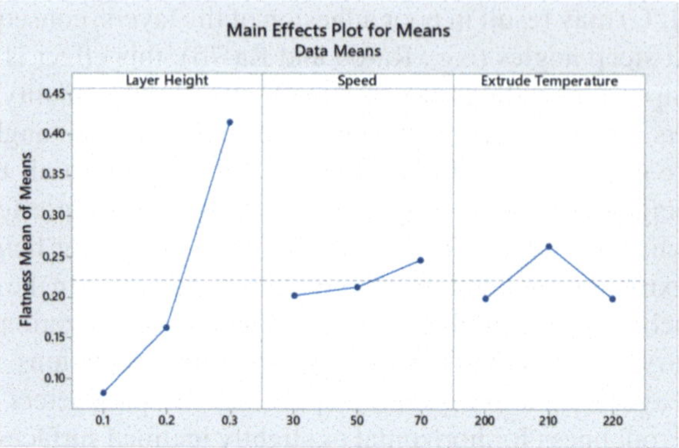

Fig. 7. Main effects plot showing the influence of layer height, print speed, and extrusion temperature on the flatness mean values.

4 Conclusions

This study examined the influence of 3D printing parameters including layer height, printing speed and extrusion temperature on surface roughness and flatness of PLA parts by means of analyzing the experiments according to Taguchi methodology. These findings differentiated the value of these parameters in relation to the dimensional stability and quality of the printed parts. Meanwhile, they were able to demonstrate that the most relevant parameter related to surface roughness and flatness is the layer height. Especially with lower values, such as 0.1 mm, better surface smoothness and flatness was achieved while an increase in layer height showed clear undulations and deviations from flatness. Surface roughness was also influenced by the printing speed, with lower speeds enhancing the surface finish due to better control over the deposition of material. But, on flatness, its impact was less severe. The influence of extrusion temperature was also non-linear, the use of a moderate temperature of 210 °C provided the best result with regards to both, roughness and flatness. Inadequate bonding would occur at lower temperatures while thermal deformation would occur at higher temperatures, affecting overall part quality.

Using orthogonal arrays to find the optimal parameter settings was confirmed to be effective with Taguchi analysis. The parameters of low layer height (0.1 mm), normal printing speed (30 mm/s), and moderate extrusion temperature (210 °C) produced the best results in terms of reducing the surface roughness and maximizing the flatness.

Acknowledgments. This research was supported by the OrThOP3Dics project: "Topology optimization of 3D printed patient-specific spinal braces" (No: TAEDK-06191) belonging to the National Recovery and Resilience Plan, Greece 2.0 Project.

References

1. Sachs, E., Cima, M., Cornie, J., Brancazio, D., Bredt, J., Curodeau, A., Fan, T., Khanuja, S., Lauder, A., Lee, J., et al.: Three-dimensional printing: the physics and implications of additive manufacturing. J. Manuf. Sci. Eng. **42**(2), 257–260 (1992)
2. Kruth, J.P., Leu, M.C., Nakagawa, T.: Progress in additive manufacturing and rapid prototyping. CIRP Ann. **47**(2), 525–540 (1998)
3. Hopkinson, N., Dickens, P.: Rapid prototyping for direct manufacture. Rapid Prototyp. J. **7**(4), 197–202 (2001)
4. Schniederjans, D.G.: Adoption of 3D-printing technologies in manufacturing: A survey analysis. Int. J. Prod. Econ. **183**, 287–298 (2016)
5. Ngo, T.D., Kashani, A., Imbalzano, G., Nguyen, K.T.Q., Hui, D.: Additive manufacturing (3D printing): a review of materials, methods applications and challenges. Compos. Part B Eng. **143**, 172–196 (2018)
6. Shahrubudin, N., Lee, T.C., Ramlan, R.: An overview on 3D printing technology: technological, materials, and applications. Procedia. Manuf. **35**, 1286–1296 (2019)
7. Murr, L.E.: Frontiers of 3D printing/additive manufacturing: from human organs to aircraft fabrication. J. Mater. Sci. Tech. **32**(11), 987–995 (2016)
8. Zhou, J.G., Herscovici, D., Chen, C.C.: Parametric process optimization to improve the accuracy of rapid prototyped stereolithography parts. Comput. Aided Des. **40**(2), 363–379 (2000)
9. Anitha, R., Arunachalam, S., Radhakrishnan, P.: Critical parameters influencing the quality of prototypes in fused deposition modelling. J. Mater. Process. Technol. **118**(1), 385–388 (2001)
10. Armillotta, A.: Assessment of surface quality on textured FDM prototypes. Rapid Prototyp. J. **12**(1), 35–41 (2006)
11. Sood, A.K., Ohdar, R.K., Mahapatra, S.S.: Improving dimensional accuracy of Fused Deposition Modelling processed part using grey Taguchi method. Mater. Des. **30**(10), 4243–4252 (2009)
12. Byun, H.S., Lee, K.H.: Design of a new test part for benchmarking the accuracy and surface finish of rapid prototyping processes. In: Kumar, V., Gavrilova, M.L., Tan, C.J.K., L'Ecuyer, P. (eds.) Computational Science and Its Applications—ICCSA 2003, Lecture Notes in Computer Science, vol. 2668, pp. 731–740. Springer, Berlin (2003)
13. Kumar, K., Singh, H.: Parametric optimization of the 3D printing process for dimensional accuracy of biopolymer parts using the Grey–Taguchi method. Iran. J. Sci. Technol. Trans. Mech. Eng. **48**, 1–13 (2023)
14. Alafaghani, A., Qattawi, A.: Investigating the effect of fused deposition modeling processing parameters using Taguchi design of experiment method. J. Manuf. Process. **32**, 164–174 (2018)
15. Mahmood, S., Qureshi, A.J., Talamona, D.: Taguchi based process optimization for dimension and tolerance control for fused deposition modelling. Addit. Manuf. **21**, 183–190 (2018)
16. Zisopol, D.G., Portoaca, A.I., Tanase, M.: Dimensional accuracy of 3D printed dog-bone tensile samples: a case study. Eng. Technol. Appl. Sci. Res. **13**(4), 11400–11405 (2023)
17. Kostopoulos, G., Stamoulis, K., Lappas, V., Georgantzinos, S.K.: Shape morphing of 4D-printed polylactic acid structures under thermal stimuli: an experimental and finite element analysis. Aerospace. **11**(2), 134 (2024)
18. Georgantzinos, S.K., Kostopoulos, G., Stamoulis, K.P.: Characterizing the shape-morphing behaviour of 4D printed PLA structures. J. Phys. Conf. Ser. **716**(1), 012034 (2024)
19. Azhikannickal, E., Uhrin, A.: Dimensional stability of 3D printed parts: effects of process parameters. Ohio J. Sci. **119**(2), 9 (2019)

Prediction of Layered Soil Permeability Through Artificial Intelligence Optimization Procedure

Abdulla A. Sharo[1,2](✉), Saja Al-Khateeb[2], Areen M. Al-Ababneh[2], and Yazan Alahmed[3]

[1] Civil Engineering Program, Al Ain University, Abu Dhabi campus, P.O. Box 112612, Abu Dhabi, UAE
aasharo@just.edu.jo

[2] Department of Civil Engineering, Jordan University of Science and Technology, P.O. Box 3030, Irbid 2210, Jordan

[3] Software Engineering, Al Ain University, Abu Dhabi campus, P.O. Box 112612, Abu Dhabi, UAE

Abstract. In this study, different AI approaches have been used to predict layered soil permeability, Linear Regression, Gaussian Process Regression (GPR), Artificial Neural Network (ANN), and Support Vector Machine (SVM). To find the model that best predicted the permeability of layered soil, the results underwent additional research and comparison because measuring soil permeability is a tedious and time-consuming process. This research builds upon prior investigations. The data sets literature (originated from an earlier in-lab investigation) containing 102 observations (70% and 30% of the remaining data were used for training and testing respectively) used to build up the models. It has been learned that AI is promising in presenting models that can measure layered soil permeability with a high degree of precision without requiring laboratory tests. The outcomes demonstrate that ANN is the best model for describing outcomes and for predicting the permeability coefficient of layered soil k_{eq}.

Keywords: Permeability of layered soil · Support Vector Machine (SVM) · Artificial Neural Network (ANN) · Gaussian Process Regression (GPR) · Linear Regression

1 Introduction

The various sized and shaped particles that make up the soil provide continuous or linked gaps. The permeability of soil refers to the properties of the soil that permit water to pass through.

Due to the lengthy procedures involved in laboratory methods—such as determining soil permeability—different researchers have developed unique strategies and methodologies.

The findings from literature review indicate that artificial intelligence (AI) approaches can predict soil permeability (k). The majority of researchers of the studies and papers reviewed used artificial intelligence artificial intelligence (AI) [1].

N. D. Lagaros et al. (Eds.): ADDOPTML 2024, LNCE 646, pp. 68–75, 2025.
https://doi.org/10.1007/978-3-031-92029-5_7

This particular study considered soil condition in the natural habitat unlike the published studies which had been conducted on difference datasets. Naturally, soil deposits exist in layers in the field and not as masses that are homogeneous. The soil system has various layers each with unique characteristics and permeability coefficient for layered deposits. Therefore, the direction of flow and soil stratification affects the value of coefficient of permeability. If Darcy's law can be extended, it is possible to compute an equivalent coefficient of permeability (k_{eq}). However, this depends on the direction flow of the bedding planes, either perpendicular or parallel to the bedding plans [2].

A lot of studies have been carried out to estimate layered soils' permeability coefficient in cases of normal flow direction to the bedding planes, considering that frequent observed and straightforward situation occurs anytime the direction of the flow and stratified deposit are perpendicular to each other [3].

This study extensively investigates layered soil system coefficient of permeability (k_{eq}). In addition, the study investigates permeability of three-layered soil systems and two-layered soil systems that is either different or repeated layers of soil with different perpendicular orientations flow which include downward flow, horizontal flow, and upward flow.

2 Theoretical analysis

Regression Learner is a tool for training regression models, like support vector machines and neural networks. By use of the tool, a researcher can do data exploration, methods definition and validation, choose features, and outcomes assessment in addition to models training. To utilize a model with fresh data, export it to the workspace. Alternatively, MATLAB® code can be generated to gain knowledge about programmatic regression.

There are Four AI approaches that have been used in the study to carry out soil permeability prediction. The approaches are Gaussian Process Regression (GPR), Artificial Neural Network, Support Vector Machine, and Linear Regression which are associated with machine learning.

SVM; Continuous responses are predicted by modifying SVM regression algorithms. To minimize sensitivity to mistakes, they identify a model that departs from the measured data.

LR: Linear regression is a statistical modeling technique used to express continuous response variable expressed as a linear function of single or a variety of predictor variable. When dealing with a new data set, linear regression models are given the first priority in trials when handling new data set because they are straight forward, easy to understand and train.

GPR: The value of a continuous response variable can be predicted using nonparametric machine learning models called GPR models. Using covariance with each input variable, the response variable is modeled as a random Gaussian process. The uncertainty of the response is also modeled by the machine learning model. When interpolating in the face of uncertainty, these models are frequently utilized in spatial analysis. Another name for GPR is Kriging.

ANN: A neural network, which gets its inspiration from the human brain, is made up of linked nodes or neurons arranged in layers that connect inputs to desired outputs. To train the machine learning model to map provided inputs to the correct response, the connections' strengths are changed iteratively. Hidden layers in a neural network are the layers of neurons that lie between the input and output layers. Generally speaking, shallow neural networks have few hidden layers. Compared to shallow and deep neural networks include more hidden layers. They may have hundreds of layers that are buried. By adding an output layer for classification or regression at the network's end, neural networks can be set up to handle classification or regression challenges.

The validated model's findings are shown in the app. The validated outcomes of the model are shown a diagnostics metric like model accuracy and residuals or response plots graph. One or more regression models can be automatically trained.

3 Data

A set of soil permeability data from literature studies [2, 6–8], to measure layered soil system coefficient of permeability, that illustrates three scenarios of water movement in the soil—whether in two or three layers—are included. The following Table 1 gives a sample of the studied data.

Table 1. Summary of data taken from the literature review; k_{cal} (calculated permeability coefficient) and r_i[a] were the input values, and k_{exp} (experimental coefficient of permeability) was the outcome.

Study by	Parameter
[2] Sharo, A. A., Al-Zoubi, M. S., Al-Ababneh, A. M. (2021)	$k_{cal} = 106.23$ $r_i = 0.082$ $k_{exp} = 128.21$
[6] Srivastava, P., Verma, R., Singh, D. (2023)	$k_{cal} = 0.0406$ $r_i = 0.00708$ $k_{exp} = 0.715$
[7] Sridharan, A., Prakash, K. (2002)	$k_{cal} = 4.333$ $r_i = 250$ $k_{exp} = 5.12$
[8] Prakash, K., Sridharan, A. (2013)	$k_{cal} = 177.9$ $r_i = 2.826$ $k_{exp} = 139.3$

[a] r_i: the ratio between the coefficient of permeability of the inlet layer and the exit layer.

The measurement of multilayer soil systems' permeability is the goal of the research presented in this publication. Three types of flow have been considered: upward, horizontal, and downward. Additionally, a formula for predicting the experimental coefficient of permeability value was offered in this article using empirical data [6]. Permeability behavior of multilayer soil systems is being studied. It looks at the factors that affect permeability, the methods for characterizing soils, and the approaches taken to forecast and monitor fluid flow in stratified soils.

The study found that the permeability of soil varies with depth and layer. Soil layering can have a significant impact on soil permeability, with the top layer often having a higher permeability than the lower layers [7]. It has been noted that the theoretically estimated equivalent coefficient of permeability of a stratified soil system differs from the directly measured, in order to explain this deviation, a hypothesis has been put out in this inquiry, which states that the exit layer permeability determines whether the measured permeability is higher or lower than the value that was calculated theoretically.

In addition, the void ratio, soil type in layered systems, thickness, and the interaction between the soil and surrounding soil it comes into touch with are all thought to have an impact on soil permeability [8]. A comparison has been done on three-layer soil sediments between the measured equivalent coefficient of permeability and the theoretically calculated values. The study found that for the most part, the exit layer's coefficient of permeability is critical at determining whether the measured value of the equivalent coefficient of permeability's is higher or lower than its theoretically calculated value. This observation informs that any layered soil deposit equivalent coefficient of permeability is dependent on the relative positioning of the layers within the system as well as the values of K of the individual layers comprising the deposit.

The data offers results for coefficients of permeability made by theoretically estimated values k_{cal} and experimental measurements k_{exp} with the inlet layer and exit layer ratio r_i for the cases from literature studies.

The collected data, which included 102 observations, was split into the model's input and output. k_{cal} and r_i were the input values, and k_{exp} was the outcome. More values can be predicted by the anticipated model.

4 Results and Discussion

4.1 Theoretical Results

In the study, the training, validation and testing of the AI models was done using 102 datasets. The datasets were organized in form of 70% to 30% subset to train and test respectively. According to the literature survey carried out on artificial intelligence approaches, it is possible to forecast layered soil permeability. The Linear regression, Gaussian process, Support vector machine, regression, and the artificial neural network. The models have been developed with different hyper parameters in MATLAB R2024a to predict layered soil permeability. Figure 1 Shows results of the different models. The proposed AI models' performance and results have been analyzed and discussed in this section. The proposed AI models' performance has been determined using MAE and RMSE, R. The following are the formulas for the model of RMSE, R, and MAE are shown in Eqs. (1), (2), and (3):

$$\mathbf{RMSE} = \sqrt{\frac{1}{N}\sum_{i=1}^{N}(T_i - P_i)^2} \tag{1}$$

$$\mathbf{R} = \frac{\sum_{i=1}^{N}(T_i - \overline{T})(P_i - \overline{P})}{\sqrt{\sum_{i=1}^{N}(T_i - \overline{T})^2 \sum_{i=1}^{N}(P_i - \overline{P})^2}} \tag{2}$$

$$\mathbf{MAE} = \frac{1}{N}\left(\sum_{i=1}^{N} abs(T_i - P)_i\right. \tag{3}$$

The optimal architecture AI model has been determined for each AI approach based on the models' performance AI models are shown in Figs. 2, 3, 4 and 5. The current research effort has identified the top four architectures to determine which architecture AI model performs best, and the best models' performances have been compared. The outcomes demonstrate that ANN is the best model for describing outcomes and for predicting k_{exp}.

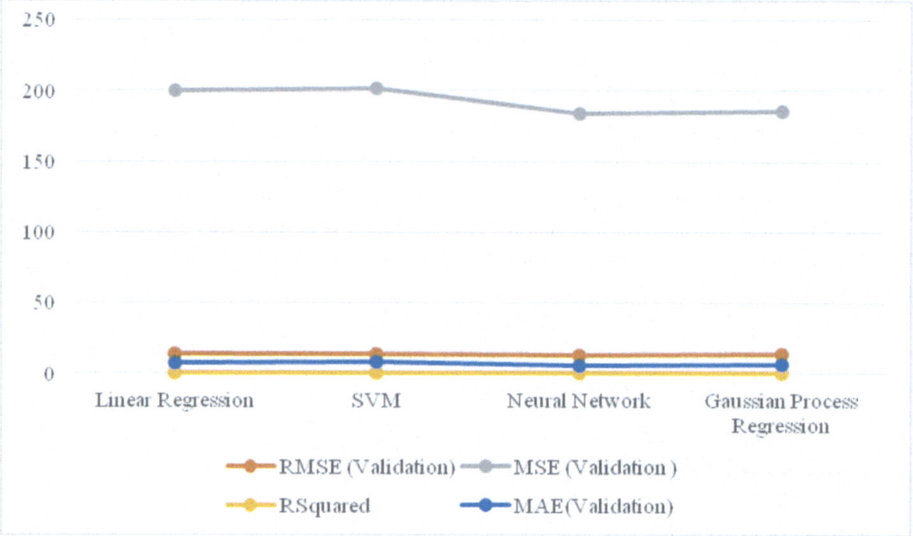

Fig. 1. Models results

To end with, the literature survey has shown that AI can forecast layered soil permeability. There are many types of models that have been developed to predict the permeability of the layered soil. The models have been developed with various MATLAB R2024a hyperparameters to predict the permeability of layered soil. In addition, this section of the study shows the proposed AI models performance and results. R, MAE, and RMSE have been used to determine the AI models. The ANN model is the best-fit AI model based on performance.

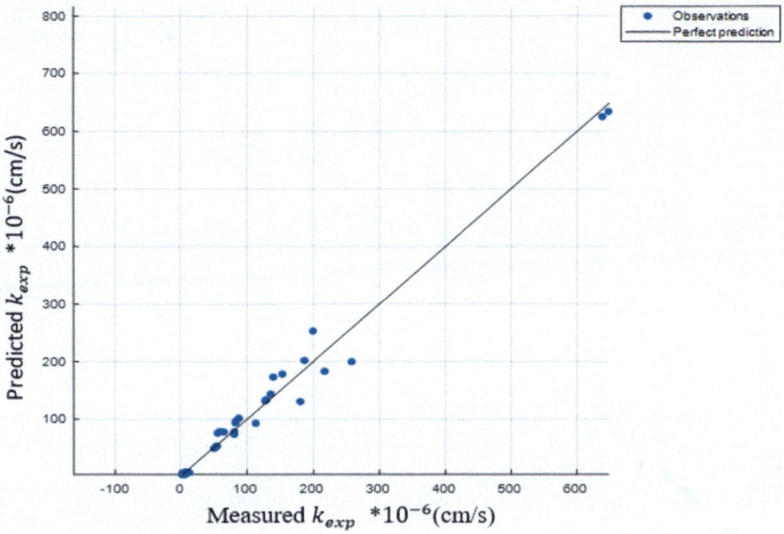

Fig. 2. Predicted k_{exp} vs measured using LR Model

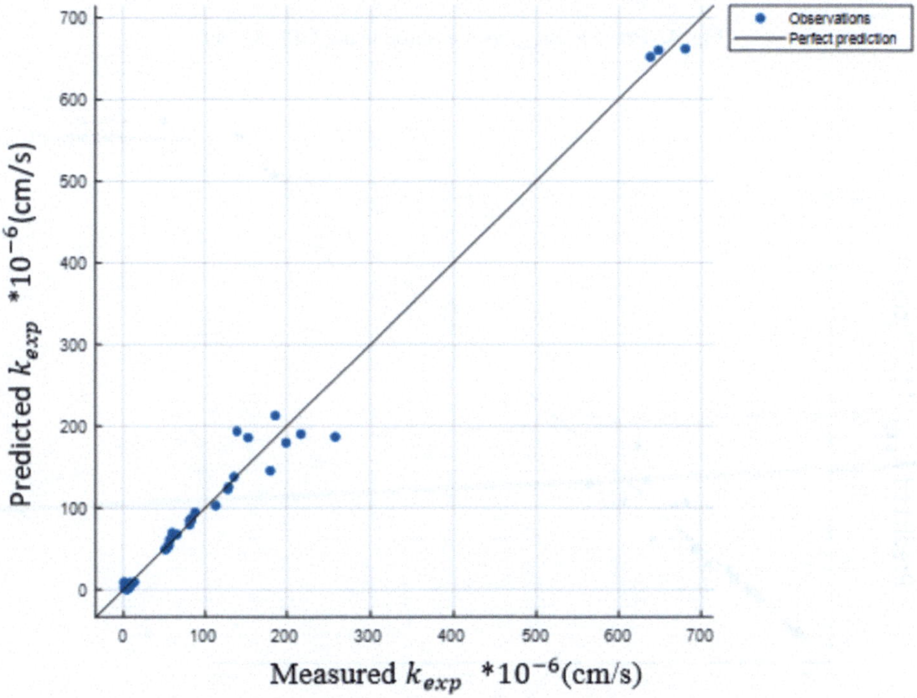

Fig. 3. Predicted k_{exp} vs measured using SVM Model

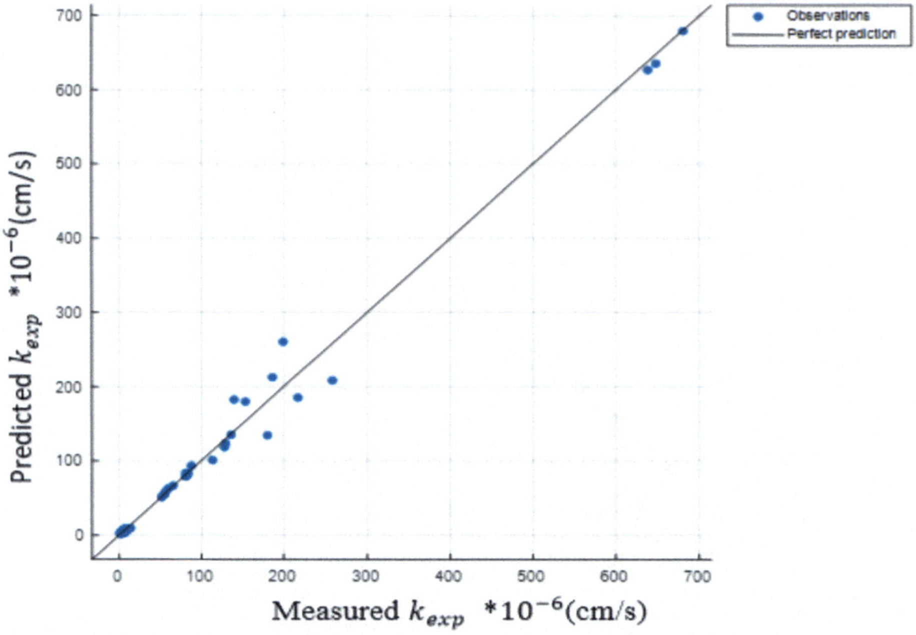

Fig. 4. Predicted k_{exp} vs measured using GPR Model

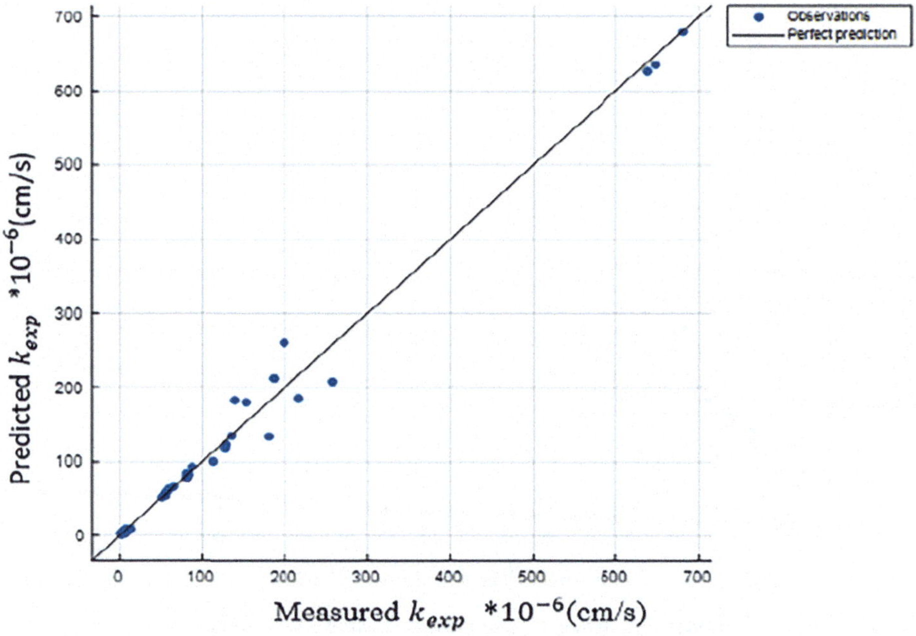

Fig. 5. Predicted k_{exp} vs measured using ANN Model

5 Conclusions

Following the work carried out in research and adopting the AI MATLAB-based theoretical research for the determination of the permeability of layer soils, the following conclusions can be drawn:

1. Rather than using a laboratory, models can be utilized to predict the permeability coefficient.
2. The artificial neural network, which can predict more values more accurately, was the best model.
3. The arrangement of the layers is one of the numerous variables that affect the permeability coefficient of stratified soil; this is the primary element that accounts for the discrepancy between the calculated and measured values. Nevertheless, further research can identify a coefficient that accounts for this discrepancy.

References

1. Khatti, J., Grover, K.: Computation of permeability of soil using artificial intelligence approaches. Int. J. Eng. Adv. Technol. **11**, 257–266 (2021). https://doi.org/10.35940/ijeat.A3220.1011121
2. Sharo, A.A., Al-Zoubi, M.S., Al-Ababneh, A.M.: On the permeability of layered soil system: experimental study. Int. J. Geotech. Eng. **16**(2), 176–186 (2021). https://doi.org/10.1080/19386362.2021.1972607
3. Dange, A.: Laboratory Study on Permeability of Layered Soil. (2015). http://ethesis.nitrkl.ac.in/7974/
4. Park, H.I.: Development of neural network model to estimate the permeability coefficient of soils. Mar. Georesour. Geotechnol. **29**(4), 267–278 (2011). https://doi.org/10.1080/1064119X.2011.554963
5. Uncuoglu, E., Citakoglu, H., Latifoglu, L., Bayram, S., Laman, M., Ilkentapar, M., Oner, A.A.: Comparison of neural network, Gaussian regression, support vector machine, long short-term memory, multi-gene genetic programming, and M5 Trees methods for solving civil engineering problems. Appl. Soft Comput. **129**, 109623 (2022). https://doi.org/10.1016/j.asoc.2022.109623
6. Srivastava, P., Verma, R., Singh, D.: Permeability of layered soil. Int. Res. J. Modern. Eng. Technol. Sci. (2023). https://doi.org/10.56726/irjmets38869
7. Sridharan, A., Prakash, K.: Permeability of two-layer soils. Geotech. Test. J. **25**(4), 443–448 (2002). https://doi.org/10.1520/gtj11293j
8. Prakash, K., Sridharan, A.: Permeability of layered soils: An extended study. Geotech. Geol. Eng. **31**(5), 1639–1644 (2013). https://doi.org/10.1007/s10706-013-9684-9

Evaluating the Structural Performance of 3D Printed FRCC Beams with Anchoring Reinforcement: Material, Geometry, and Loading Perspectives

Bara'a R. Alnemrawi[1], Rajai Z. Al-Rousan[1](✉), Khairedin M. Abdalla[1], and Nikos Lagaros[2]

[1] Department of Civil Engineering, Engineering, Jordan University of Science and Technology, P.O. Box 3030, Irbid 22110, Jordan
rzalrousan@just.edu.jo

[2] School of Structural Engineering, National Technical University of Athens (NTUA), 15780 Zografou, Greece

Abstract. Increasing attention was paid to three-dimensional (3D) printing technology in the construction, architecture, and medical care sectors. However, the utilization of cementitious composites in 3D concrete printing was increased to upgrade the structural behavior of the printed members. In contrast, difficulties and uncertainties still exist regarding the reinforcement method of the printed members with other induced challenges regarding the cost efficiency and sustainability concerns. This study evaluates the structural performance of 3D printed beams with fiber-reinforced cementitious composite (FRCC) and anchoring reinforcement using the method of finite element analysis (FEA). A total of eighteen beams were simulated with different shear spans (400, 500, and 575 mm), overall depth (120, 200, and 276 mm), and FRCC compressive strengths (23.2 and 43.2 MPa). The behavior was reported for all models in terms of the ultimate load and its corresponding deflection, initial stiffness, toughness, and failure modes. Results were presented as general guidelines to highlight the effect of the examined parameters to provide the scientists and engineers with a wide knowledge of the printing field, mitigating the required time and cost. It was found that increasing the shear span increases the beam's ultimate deflection and reduces its capacity. Moreover, increasing the depth of the beams has various effects on the failure mode and the deflection values where beams with lower depths sustain larger deflections compared to the deeper ones.

Keywords: Finite element modeling · Fiber reinforced cementitious composite · 3D printing · Flexural beams · Anchoring reinforcement

1 Introduction

The inclusion of automation and digitalization within the construction industry has been pace increased in the last decades, such as the method of three-dimensional (3D) concrete printing which is also known as the additive manufacturing (AM) technique [1]. The

N. D. Lagaros et al. (Eds.): ADDOPTML 2024, LNCE 646, pp. 76–86, 2025.
https://doi.org/10.1007/978-3-031-92029-5_8

AM technology is mainly based on the layer-by-layer method formed using extrusion or spraying and controlled using dual control and digital models. Generally, the method of 3D printing has been used in many sectors within the last decades, especially within the construction industry and building structures, especially for complex geometries [2, 3]. 3D concrete printing has many advantages compared to traditional cast concrete where the need for formworks was compensated with the cost and the time required for construction being significantly reduced [4]. Additionally, the inclusion of computer-aided design (CAD) efficiently increased productivity within a finite time frame. On the other hand, 3D printing technology plays a major role in reducing the amount of global carbon emissions where less energy is consumed [5, 6].

The behavior of different printing materials was examined in the literature such as the cementitious where their mechanical properties [7], printability, and feasibility were investigated. It was found that the yield stress was reduced for the fresh cementitious material upon increasing the water amount in the mix and this was applied for both static and dynamic testing [8]. However, this could be significantly eliminated by the inclusion of micro-crystalline cellulose where the viscosity and the physical properties of the printable mix were also improved [9]. In the same context, the cementitious mix printability was also investigated to figure out the parameters that affect the mixture flow and concluded that the aggregate content, fiber content, and the yield stress of the cement paste have major contributions among all parameters [10]. On the other hand, the intrusion of micro- and macro-fibers appears to shed light on the durability and the ductility characteristics of the printed concrete along with elevating its tensile strength [11]. Different fiber types, diameters, and lengths were investigated such as carbon fiber, polyethylene fibers (PE), and chopped fibers where the strength of the mix was increased but on the contrary, the ductility was decreased [12–15].

The increased utilization of 3D printing technology within the construction industry makes it a revolution and inflection point toward automation using the AM methodology where infrastructure development becomes a choice. Besides, addressing the consistency and rheological properties of the printed materials is only one challenge that limits the usage of the 3D printing methodology while others exist which necessitates more studies and tests to fill the gap [16]. Another main challenge is the reinforcement method of the printed structural components which is done using conventional methods such as manual rebar placement [17]. Consequently, many difficulties were induced, including the incompatibility between the automated layering print and the manual reinforcement placement. Therefore, the expected time and cost-saving could be relatively hard reached [18]. Cai et al. [19] investigated the behavior of 3D printed beams and compared it to the conventional beam behavior with similar longitudinal and transverse reinforcement ratios. It was concluded that the anchoring method resulted in printed beams with higher load capacity compared to the traditional beam but with lower ductility due to the interfacial shear failure between the printed layers.

Generally, few research exists considering the structural behavior of 3D printed structural members under various circumstances to shed light on the differences between the printed and traditional beams. This study adopted the finite element modeling (FEM) method to fulfill the research objectives using ABAQUS software [20]. The effect of the shear span (400, 500, and 575 mm), beam overall depth (120, 200, and 276 mm), and the

FRCC compressive strength (23.2 and 43.2 MPa) using eighteen FEM models. Results were reported in terms of failure modes, energy absorption, ultimate load, initial stiffness, and ultimate deflection.

2 Experimental Work Review

The flexural behavior of 3D printed beams made of FRCC concrete was experimentally carried out by Cai et al. [19] using a total of five specimens. The behavior between the 3D printed and traditional beams was compared along with examining the efficiencies of different reinforcement configurations. Specimens were of 1600 mm length, 200 mm height, and 120 mm width with all beams having constant longitudinal and transverse reinforcement ratios of 0.011% and 0.018%, respectively.

Three types of reinforcement were examined (joint reinforcement, anchoring reinforcement, and inserted steel reinforcement) with the anchoring reinforcement type showing higher efficiency. The FRCC cementitious matrix was made of PE fibers with 6 mm length and 24 mm diameter resulting in a 23.2 MPa compressive strength mixture. The printed beams were cast using the layer-by-layer technique with various methodologies depending on the installed reinforcement type. The beam geometry and reinforcement details are presented in Fig. 1.

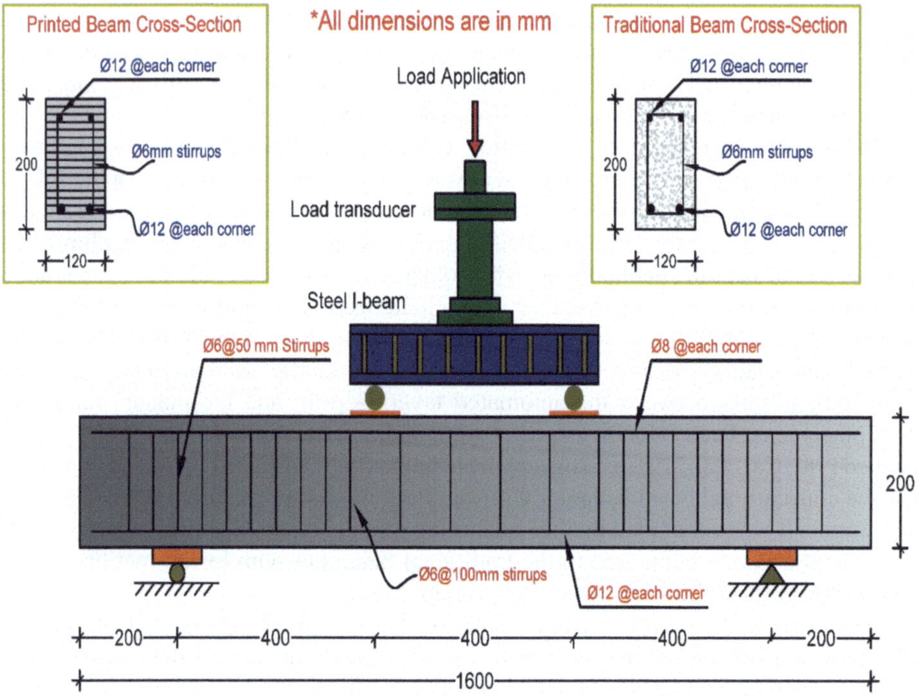

Fig. 1. Beam detailing

3 Numerical Modeling

3.1 Material Modeling and Element Types

The finite element modeling was utilized to fulfill the objectives of this study using ABAQUS software [20]. The FRCC was modeled using the (C3D8R) element while steel reinforcement was simulated using the (T3D2) element. Beams were simulated using the four-point loading with a 400 mm distance between each two constitutive steel plates. The geometry and the reinforcement details of the conventional and the printed beams are illustrated in Fig. 2.

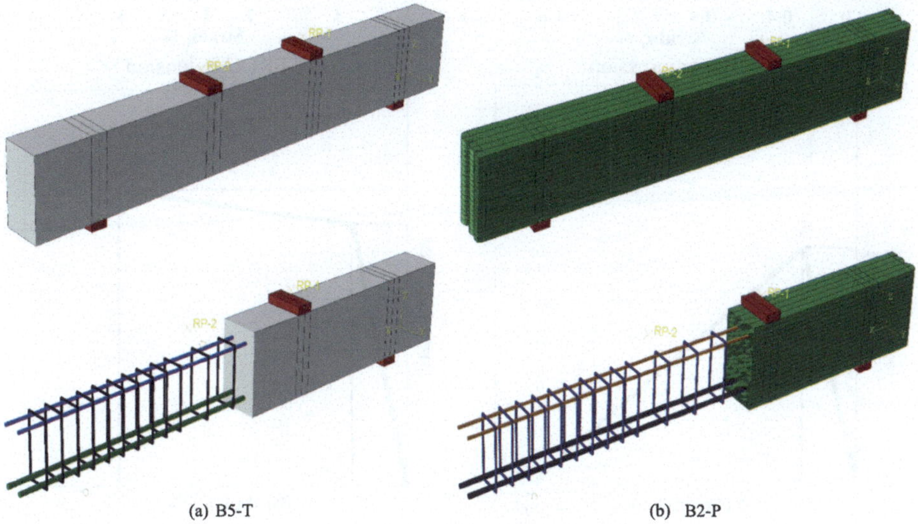

Fig. 2. FEA beams modeling

The FRCC material behavior was modeled in compression and tension [21] and plotted in Fig. 3a, b, respectively. In addition, steel was modeled using the bilinear strain hardening with yielding and ultimate strength values of 395 MPa and 524 MPa. Finally, the printed beam was constructed using the layer-by-layer method with the interaction between the layers being modeled using the bilinear traction-separation definition [22–26]. Loading was applied using two reference points coupled to the loading steel plates with the reinforcement cage having an embedded constraint.

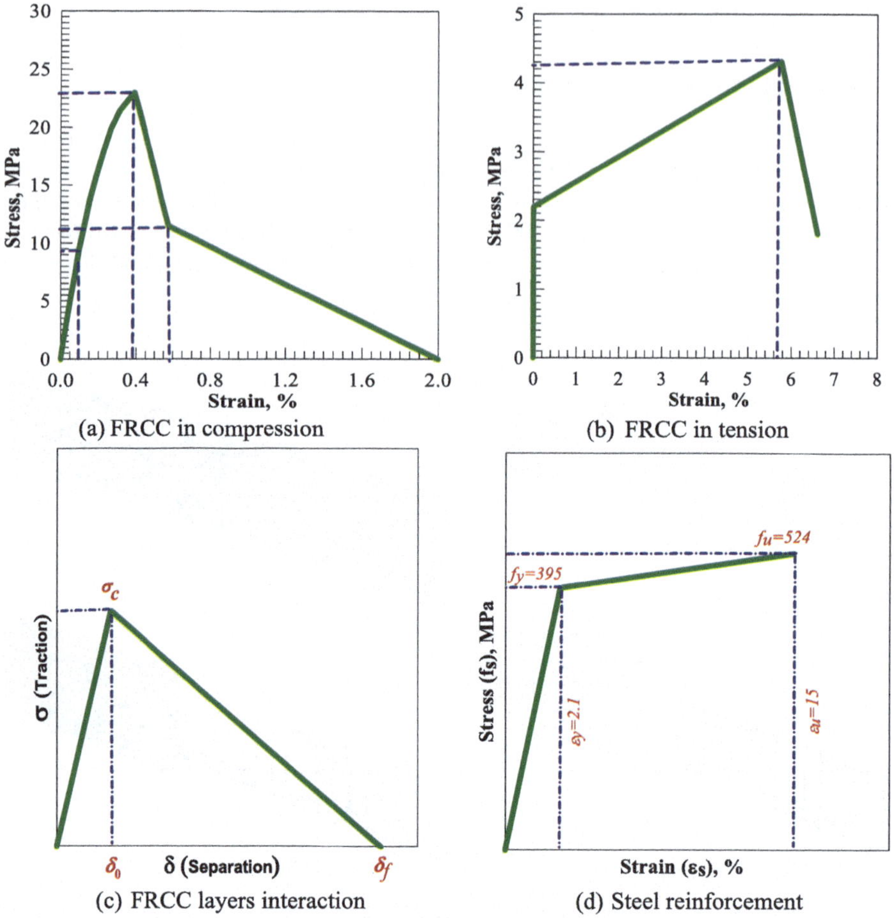

Fig. 3. Modeling characteristics

3.2 Validation

The validation of the FEA models was done using two specimens as printed and traditional beams (B2-P and B5-T), including their load-deflection curves and failure modes (Fig. 4). However, it was observed that the experimental and the simulation results matched well with a minimum error range ($<2\%$) including the ultimate strength, ultimate deflection, toughness, and initial stiffness. In addition, the traditional beam failed due to excessive bending action while the printed beam failed due to the interfacial shear stresses induced between the layers and extends from the supporting plates toward the loading plates location, as illustrated in Fig. 4.

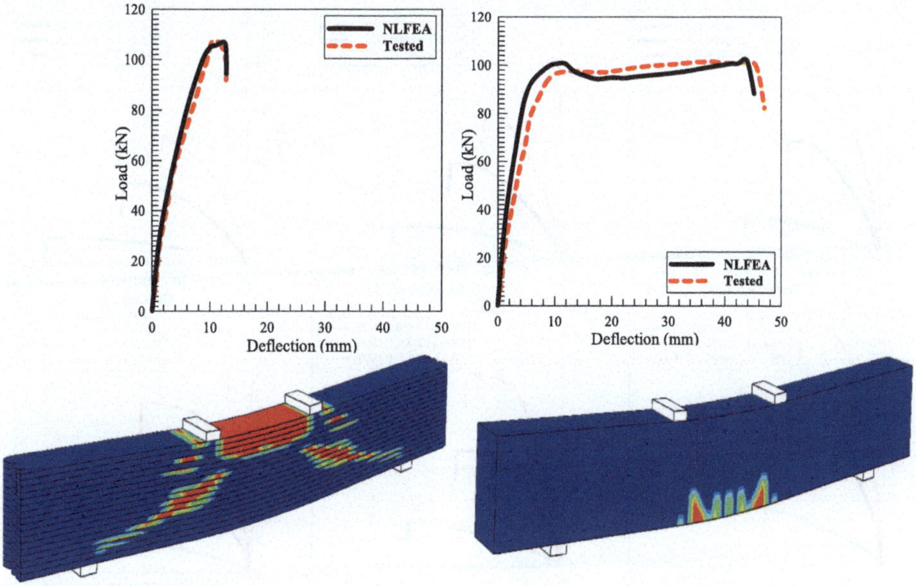

(a) B5-T (b) B2-P

Fig. 4. Load-deflection curves and failure modes validation

4 Results and Discussion

4.1 Load-Deflection Curves

The load-deflection curves of the eighteen printed beams are presented in Fig. 5 within six main groups. The model's designation appears as follows: the (PB) refers to a printed beam, the second letter (C) refers to the FRCC compressive strength in MPa, the third letter (S) refers to the shear span value in mm, and the final letter (H) refers to the height of the beam in mm. Observing Fig. 5 reveals that a significant effect was recorded by the FRCC compressive strength where an increased load capacity and reduced ultimate deflection were associated with increasing the FRCC strength or increasing the beam depth. Moreover, increasing the shear span reduces the failure brittleness and increases the beam ductility.

4.2 Failure Modes

The failure modes of the printed beams with various conditions are presented in Fig. 6 and compared to each other using different nine cases. However, beams with different FRCC compressive strengths experienced similar cracking patterns while the shear span and the beam depth have nonuniform behavior. Generally, beams with smaller depths (120 mm) show more ductile behavior with a concrete compression failure mode between the loading points. Besides, increasing the depth of the beam increases the induced shear stresses and results in interlayer shear failure modes. Finally, increasing the shear span widens the failure area and, in some cases, the beam bottom sides were also crushed.

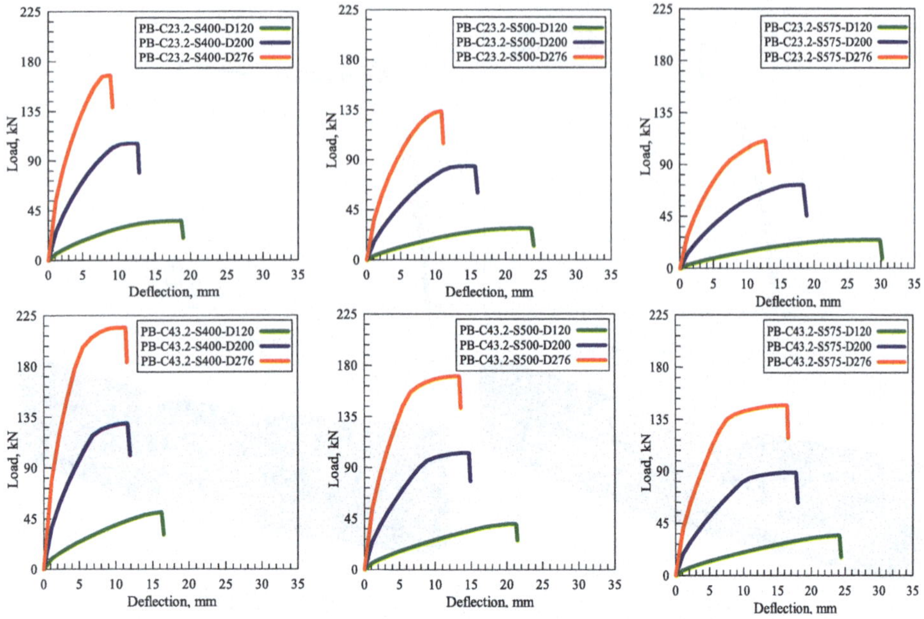

Fig. 5. Load-deflection curves

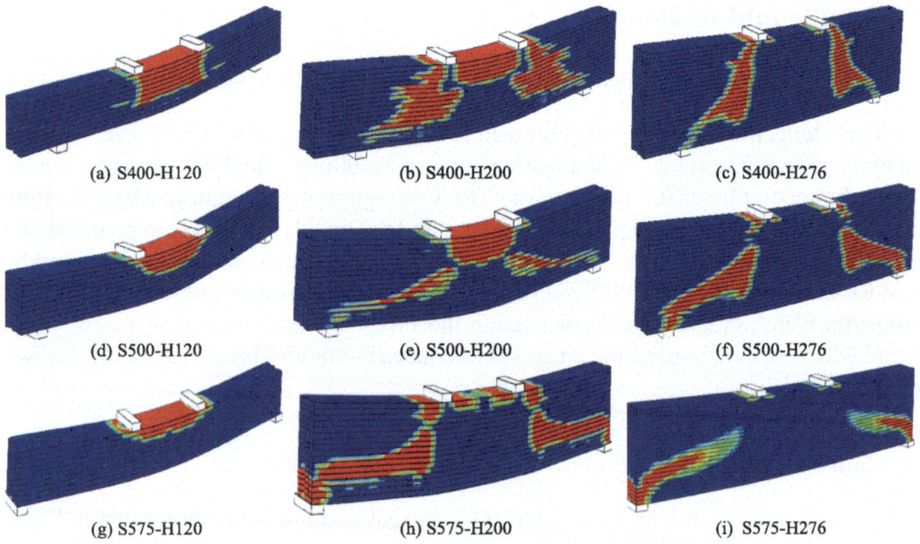

Fig. 6. Failure modes of the printed beams

4.3 Shear Span Effect

The structural characteristics of the printed beams were evaluated with respect to the effect of their shear span, including the ultimate load and their corresponding deflection, initial stiffness, and toughness, as shown in Fig. 7a. Toughness was calculated as the area under the load-deflection curves (Fig. 5) while initial stiffness reflects the slope of the linear part.

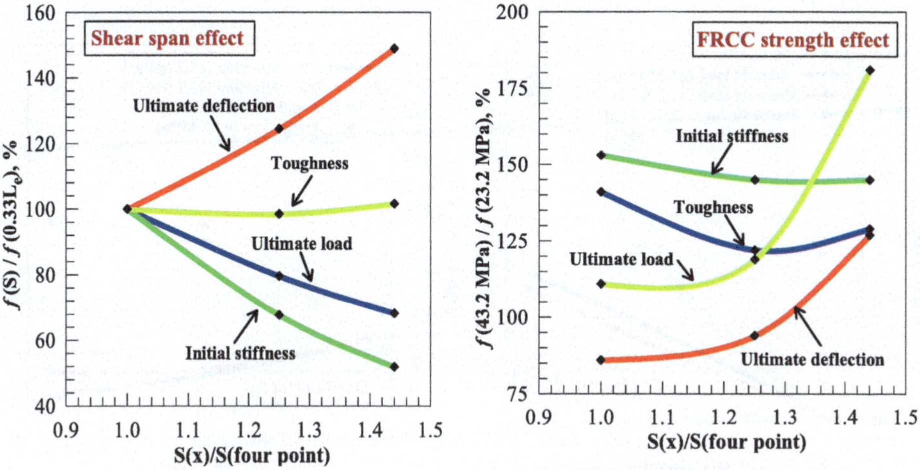

(a) Shear span effect (b) FRCC compressive strength effect

Fig. 7. Shear span and FRCC strength effect

Generally, the shear span of the four-point loading is located at a distance equal to one-third of the loading span. Therefore, the value of each structural measure was normalized with respect to the value corresponding to the four-point lading, as appears in Fig. 7a. It could be stated that increasing the shear span increases the ultimate deflection values up to 60% compared to an approximate toughness capacity regardless of the shear span value. In contrast, the ultimate load values and the initial stiffness were significantly reduced up to 30% and 50%, respectively.

4.4 FRCC Compressive Strength Effect

The compressive strength effect of the FRCC layers utilized in the printed beams was evaluated in Fig. 7b for all the structural performance characteristics at different shear span values. Results were normalized between the beam capacities corresponding to a 20 MPa FRCC compressive strength increase (from 23.2 MPa to 43.2 MPa). It could be stated that increasing FRCC strength has a general behavior depending mainly on the shear span value where the ultimate load and the ultimate deflection are affected similarly. Generally, larger percentages were recorded for the ultimate load up to a 90% increase compared to only a 25% increase for the ultimate deflection. On the other hand, the initial stiffness and toughness were minorly affected.

4.5 Beam Depth Effect

The behavior of the 3D printed beams was evaluated depending on the depth effect where an effect was observed for the FRCC compressive strength. The values were normalized with respect to a printed beam with 120 mm depth and plotted in Fig. 8. However, large normalized percentages were recorded for the initial stiffness capacity of up to ten times corresponding to a depth increase of 2.3 times. In addition, the ultimate load was increased by four times compared to a 75% reduction for the ultimate deflection.

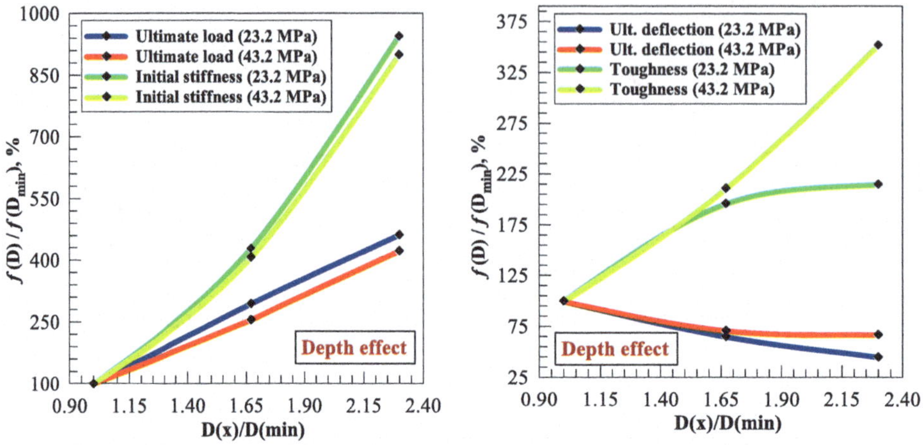

Fig. 8. Beam depth effect

5 Conclusions

The behavior of 3D printed beams was evaluated under different shear spans, overall depths, and FRCC compressive strength using the FEA method. Based on the obtained results, the following could be concluded:

1. The failure mode of the 3D printed beams depends on the overall depth of the beam where an interfacial shear failure was observed for beams with high depths and turned into concrete crushing for beams with lower depths.
2. The FRCC compressive strength has a negligible effect on the structural behavior of 3D printed beams with various shear spans where the normalized percentages ($f(S)/f(0.33L_e)$) were similar.
3. The amount of absorbed energy was constant for beams with different shear span ratios where the ultimate load and ultimate deflections were increased and decreased proportionally to keep a similar area under the load-deflection curves.
4. The overall depth of the beam has a major role in controlling the beam capacity that could be increased up to four times upon increasing the printed beam depth by 2.3 times compared to ten times for the initial stiffness.

Acknowledgments. Part of this research was supported by the ADDOPTML (ADDitively Manufactured OPTimized Structures by Means of Machine Learning) project (No: 101007595) belonging to the Marie Skłodowska-Curie Actions (MSCA) Research and Innovation Staff Exchange

(RISE) H2020-MSCA-RISE-2020. Their support is highly acknowledged. The authors would also like to thank the Jordan University of Science and Technology (JUST) for its support as a partner organization in the ADDOPTML project.

References

1. Liu, D., Zhang, Z., Zhang, X., Chen, Z.: 3D printing concrete structures: State of the art, challenges, and opportunities. Constr. Build. Mater. **405**, 133364 (2023). https://doi.org/10.1016/j.conbuildmat.2023.133364
2. García de Soto, B., Agustí-Juan, I., Hunhevicz, J., Joss, S., Graser, K., Habert, G., Adey, B.T.: Productivity of digital fabrication in construction: cost and time analysis of a robotically built wall. Autom. Constr. **92**, 297–311 (2018). https://doi.org/10.1016/j.autcon.2018.04.004
3. Al-Rousan, R.Z., Alnemrawi, B.R.: The influence of prestress level on the behavior of prefabricated precast concrete bridge deck panel systems with different overlay thicknesses. Structure. **58**, 105478 (2023). https://doi.org/10.1016/j.istruc.2023.105478
4. Al-Rousan, R.Z., Abdalla, K.M., Alnemrawi, B.R.: The behavior of heat-damaged RC beams reinforced internally with CFRP strips. In: Ungureanu, V., Bragança, L., Baniotopoulos, C., Abdalla, K.M. (eds.) 4th International Conference "Coordinating Engineering for Sustainability and Resilience" & Midterm Conference of CircularB "Implementation of Circular Economy in the Built Environment", pp. 165–174. Springer Nature Switzerland, Cham (2024). https://doi.org/10.1007/978-3-031-57800-7_15
5. Agustí-Juan, I., Müller, F., Hack, N., Wangler, T., Habert, G.: Potential benefits of digital fabrication for complex structures: environmental assessment of a robotically fabricated concrete wall. J. Clean. Prod. **154**, 330–340 (2017). https://doi.org/10.1016/j.jclepro.2017.04.002
6. Al-Noaimat, Y.A., Ghaffar, S.H., Chougan, M., Al-Kheetan, M.J.: A review of 3D printing low-carbon concrete with one-part geopolymer: engineering, environmental and economic feasibility. Case. Stud. Constr. Mater. **18**, e01818 (2023). https://doi.org/10.1016/j.cscm.2022.e01818
7. Al-Rousan, R.Z., Alnemrawi, B.R., Sawalha, H.M.: The ultimate capacity of geopolymer recycled aggregate concrete filled steel tubular columns: numerical and theoretical study. J. Build. Eng. **96**, 110365 (2024). https://doi.org/10.1016/j.jobe.2024.110365
8. Liu, Z., Li, M., Weng, Y., Wong, T.N., Tan, M.J.: Mixture design approach to optimize the rheological properties of the material used in 3D cementitious material printing. Constr. Build. Mater. **198**, 245–255 (2019). https://doi.org/10.1016/j.conbuildmat.2018.11.252
9. Long, W.-J., Tao, J.-L., Lin, C., Gu, Y.-c., Mei, L., Duan, H.-B., Xing, F.: Rheology and buildability of sustainable cement-based composites containing micro-crystalline cellulose for 3D-printing. J. Clean. Prod. **239**, 118054 (2019). https://doi.org/10.1016/j.jclepro.2019.118054
10. Carcassi, O.B., Maierdan, Y., Akemah, T., Kawashima, S., Ben-Alon, L.: Maximizing fiber content in 3D-printed earth materials: printability, mechanical, thermal and environmental assessments. Constr. Build. Mater. **425**, 135891 (2024). https://doi.org/10.1016/j.conbuildmat.2024.135891
11. Cao, D.: Increasing strength and ductility of extruded polylactic acid matrix composites using short polyester and continuous carbon fibers. Int. J. Adv. Manuf. Technol. **130**(7), 3631–3647 (2024). https://doi.org/10.1007/s00170-023-12887-9
12. Zhou, Y., Jiang, D., Sharma, R., Xie, Y.M., Singh, A.: Enhancement of 3D printed cementitious composite by short fibers: a review. Constr. Build. Mater. **362**, 129763 (2023). https://doi.org/10.1016/j.conbuildmat.2022.129763

13. Al-Rousan, R., Alnemrawi, B.R.: NLFEA of the behavior of polypropylene-fiber-reinforced concrete slabs with square opening. Buildings. **14**(2), 480 (2024). https://doi.org/10.3390/buildings14020480
14. Al-Rousan, R.Z.: Influence of macro synthetic fibers on the flexural behavior of reinforced concrete slabs with opening. Civ. Eng. J. **8**(9) (2022). https://doi.org/10.28991/CEJ-2022-08-09-016
15. Al-Rousan, R.Z.: Failure analysis of polypropylene fiber reinforced concrete two-way slabs subjected to static and impact load induced by free falling mass. Lat. Am. J. Solids Struct. **15**(1), e05 (2018). https://doi.org/10.1590/1679-78254895
16. Giwa, I., Game, D., Ahmed, H., Noorvand, H., Arce, G., Hassan, M., Kazemian, A.: Performance and macrostructural characterization of 3D printed steel fiber reinforced cementitious materials. Constr. Build. Mater. **369**, 130593 (2023). https://doi.org/10.1016/j.conbuildmat.2023.130593
17. Robayo-Salazar, R., Mejía de Gutiérrez, R., Villaquirán-Caicedo, M.A., Delvasto, A.S.: 3D printing with cementitious materials: challenges and opportunities for the construction sector. Autom. Constr. **146**, 104693 (2023). https://doi.org/10.1016/j.autcon.2022.104693
18. Asprone, D., Auricchio, F., Menna, C., Mercuri, V.: 3D printing of reinforced concrete elements: technology and design approach. Constr. Build. Mater. **165**, 218–231 (2018). https://doi.org/10.1016/j.conbuildmat.2018.01.018
19. Cai, J., Sheng, Z., Wang, X., Fang, Y., Pan, J.: Effect of reinforcement configurations on the flexural behaviors of 3D printed fiber reinforced cementitious composite (FRCC) beams. J. Build. Eng. **46**, 103773 (2022). https://doi.org/10.1016/j.jobe.2021.103773
20. ABAQUS Dassault Systemes Simulia Corp. P: ABAQUS Analysis User's Guide (6.17), 17th edn, (2017)
21. Li, V.C.: Constitutive modeling of engineered cementitious composites (ECC). In: Li, V.C. (ed.) Engineered Cementitious Composites (ECC): Bendable Concrete for Sustainable and Resilient Infrastructure, pp. 139–175. Springer, Berlin, Heidelberg (2019). https://doi.org/10.1007/978-3-662-58438-5_5
22. Misseri, G., Rovero, L., Galassi, S.: Analytical modelling bond behaviour of polybenzoxazole (PBO) and glass Fibre Reinforced Cementitious Matrix (FRCM) systems coupled with cement and gypsum matrixes: effect of the Cohesive Material Law (CML) shape. Compos. Part B Eng. **223**, 109090 (2021). https://doi.org/10.1016/j.compositesb.2021.109090
23. Hosseini, E., Zakertabrizi, M., Korayem, A.H., Xu, G.: A novel method to enhance the interlayer bonding of 3D printing concrete: an experimental and computational investigation. Cem. Concr. Compos. **99**, 112–119 (2019). https://doi.org/10.1016/j.cemconcomp.2019.03.008
24. Al-Rousan, R.Z., Alnemrawi, B.R.: Influence of overlay strength degradation on bond stresses of bridge deck system. Heliyon. **10**(10) (2024). https://doi.org/10.1016/j.heliyon.2024.e31037
25. Alhassan, M.A., Al-Rousan, R.Z., Taha, H.M.: Precise finite element modelling of the bond-slip contact behavior between CFRP composites and concrete. Constr. Build. Mater. **240**, 117943 (2020). https://doi.org/10.1016/j.conbuildmat.2019.117943
26. Al-Rousan, R.Z., Abu-Elhija, A.M.: Predicting the bond-slip relationship between concrete and CFRP using anchoring holes technique. Case. Stud. Constr. Mater. **13**, e00462 (2020). https://doi.org/10.1016/j.cscm.2020.e00462

Conceptual Design of a Transformable Transitional Shelter in the Philippines

Weifeng He[1], Eman Pagala[1], Keiya Nakasato[2], Stavros Chatzieleftheriou[3], Tiago Nuncio[4], Fátima Zorro[4], Rita G. Bola[4], and Maria Matheou[1](✉)

[1] Institute for Lightweight Structures and Conceptual Design, University of Stuttgart, Stuttgart, Germany
maria.matheou@ilek.uni-stuttgart.de
[2] Department of Architecture and Architectural Engineering, Kyoto University, Kyoto, Japan
[3] Structures and Sensors, Athens, Greece
[4] European Federation for Welding, Joining and Cutting, Lisbon, Portugal

Abstract. This paper presents a conceptual design of a lightweight shelter inspired by the traditional Filipino house, Bahay Kubo, aiming to meet sheltering needs after a natural disaster in the Philippines. To tackle environmental conditions, the proposed shelter alters its shape in two different spatial configurations to accommodate the Philippines' rainy and dry seasons. For instance, during the rainy season, the shelter features a steep roof and a raised floor (vertically elliptical in shape) to protect against rain and flooding. During the dry season, the shelter can be configured to have a wider layout (horizontally elliptical in shape). Using locally available materials for the shelters' structural and envelope system, such as bamboo beams, amakan wall panels (woven panels of bamboo) and nipa palm leaves using the pawid thatching technique for the roof, this approach not only reduces the environmental impact of the shelter but also supports the local economy. Additionally, the design incorporates a conceptual design of ball-and-socket joints, telescopic locking mechanism and an adjustable steel clamp to connect the bamboo beams with the joints in avoiding using bolted connections. Auger anchors are used to support and ensure the shelter's stability. 3D printing technology is employed to maximize the potential of ball-and-socket joints while also considering manufacturing complexity when joining multiple links. Finally, the paper presents the structural and joint stress analysis, and a prototype in scale 1:6 which demonstrates the shelter's transformation from the rainy season to the dry season configuration.

Keywords: Transitional Shelter · Bahay Kubo · Transformable Structure · Bamboo structure · 3D-printed joints

1 Introduction

The Philippines is located in the heart of the Ring of Fire, a path in the Pacific Ocean known for intense geological activity [1]. This region is home to most of the world's earthquakes and about 75% of its active volcanoes. The Philippines is particularly susceptible to natural disasters such as earthquakes and volcanic eruptions. Additionally,

N. D. Lagaros et al. (Eds.): ADDOPTML 2024, LNCE 646, pp. 87–99, 2025.
https://doi.org/10.1007/978-3-031-92029-5_9

the country is hit by an average of 20 typhoons each year, with about five causing significant damage [2]. In fact, 60% of the country's land area is exposed to multiple hazards including tsunamis, sea-level rise, landslides, and various forms of flooding, putting 74% of its population at risk [3].

In 2020, the Philippines faced a series of destructive natural disasters, beginning with the sudden eruption of Taal Volcano [4]. In the same year, the country was hit by several major typhoons. Among the most destructive were Typhoon Molave (Quinta), Super Typhoon Goni (Rolly), and Typhoon Vamco (Ulysses), displacing almost three million people and forced many into temporary shelters. The ongoing COVID-19 pandemic compounded the challenges for those affected by these disasters. However, even before these events, the Philippines had experienced one of the most devasting natural disasters in its history, Typhoon Haiyan (Yolanda) in 2013. This catastrophic storm affected 14 million people and destroyed or severely damaged over a million homes. In response, the UN Resident and Humanitarian Coordinator for the Philippines along with the Humanitarian Country Team (HCT), prioritized efforts in providing shelter and restoring livelihoods. Yet, after 100 days, many individuals still found themselves living in temporary shelters such as tents located on the ground [5, 6]. Affected communities urgently needed shelters that provide immediate relief and support the transition to permanent housing.

This paper presents a conceptual design of a transformable transitional shelter in the Philippines designed in the frame of the European Project "ADDOPTML". The key features of the shelter include lightweight construction, modular design, adaptability, recyclability, reusability and transportability that can meet transitional sheltering needs after a natural disaster in the Philippines. The International Federation of Red Cross and Red Crescent Societies, as well as the Sphere Association, have been considered when designing the shelter.

2 Design of the Transformable Transitional Shelter

2.1 Filipino Bahay Kubo

Our design project offers an architectural and engineering solution tailored to the specific needs of the Philippines, inspired by the Bahay Kubo, the traditional Filipino house [7]. Elevated on stilts, this bamboo house protects its inhabitants from flooding, a common issue in the Philippines. The steep roof, often made of overlapping thatched leaves, efficiently channels rainwater away from the structure. Additionally, the use of breathable materials and large openings enables passive ventilation, ensuring the interior remains cool despite the tropical climate. Usually locally available materials such as bamboo beams, amakan wall panels (woven bamboo), and pawid thatching using nipa palm leaves are used for the roof, walls and structure of the Bahay Kubo. The key features of the Filipino traditional house used in the proposed conceptual design are shown in Fig. 1. Different projects inspired by the Filipino traditional house presented in [8, 9].

2.2 Dry and Rainy Season Configurations, and Adaptive Reuse

This paper presents a conceptual design of a lightweight shelter inspired by the traditional Filipino house, Bahay Kubo, aiming to meet sheltering needs after a natural

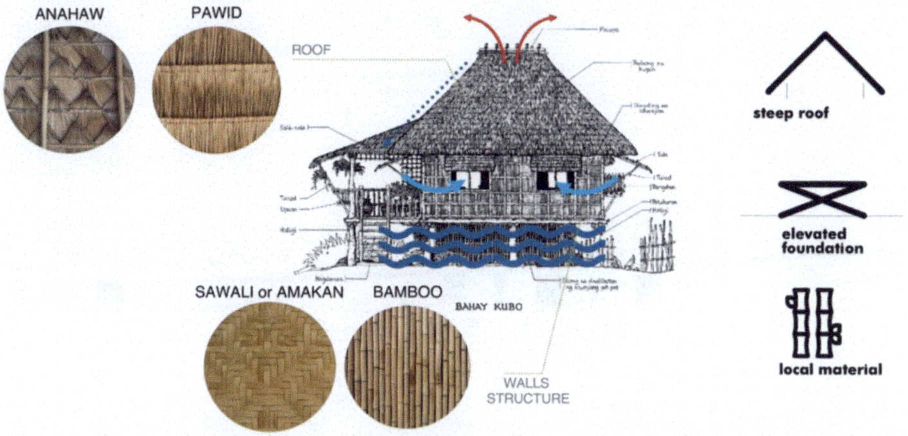

Fig. 1. Key features of the Filipino Bahay Kubo used in the proposed conceptual design. (Adapted from Francisco Mañosa [7]).

disaster in the Philippines. The transitional shelter aims to provide an architectural and engineering solution for post-disaster shelters in the Philippines, facilitating the transition to more permanent structures. The key idea of the proposed transitional shelter is a scissor-like deployable shelter that adapts to various weather conditions following a flood catastrophe.

The scenario of adaptivity is illustrated in Fig. 2 showing what happens after 30 days of dry and rainy season, after 100 days and after the shelter's end life. Under the umbrella of adaptive reuse, the shelter can be transformed with the existing structural members and joints into a water collector for the community. The Philippines has two major seasons, the rainy season, from June to November; and the dry season, from December to May (Fig. 3). For instance, during the rainy season, the shelter features a steep roof and a raised floor (vertically elliptical in shape) to protect against rain and flooding. During the dry season, the shelter can be configured to have a wider layout (horizontally elliptical in shape). Our shelter's components can be quickly assembled on site according to the different seasons as well as transformed into a water collector (Fig. 4).

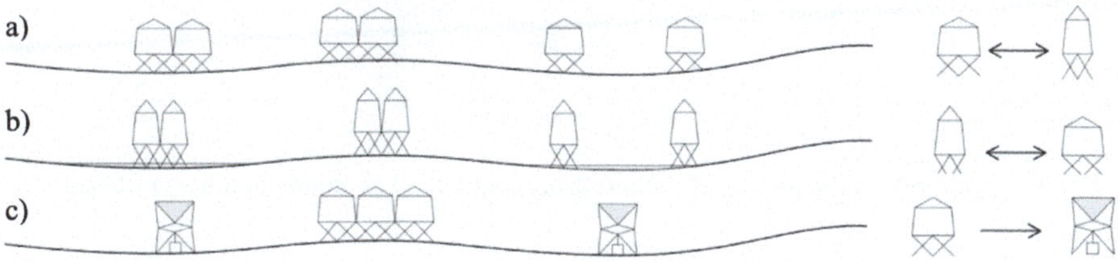

Fig. 2. Scenario of adaptivity. (**a**) after 30 days of dry season. (**b**) after 30 days of rainy season. (**c**) after 100 days.

Fig. 3. Dry and rainy season configurations of the proposed shelter.

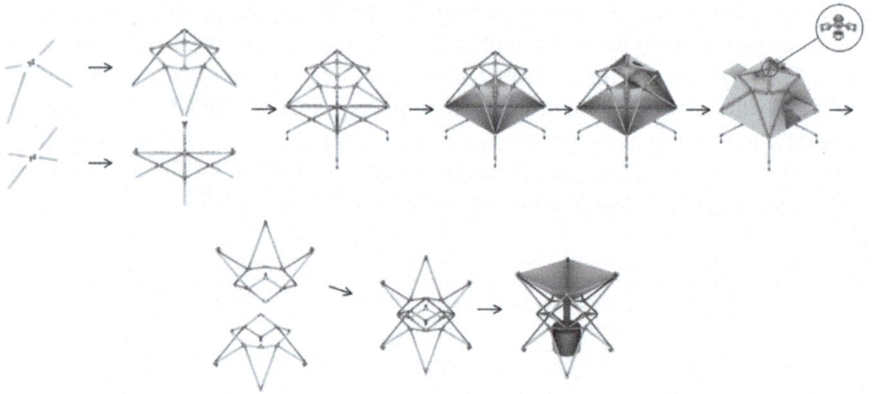

Fig. 4. Assembly of the proposed shelter, disassemble and assemble to a water collector.

3 Lightweight Construction

3.1 Bamboo Structural System and Structural Analysis

The structural system consists of four angular scissor-like bamboo beams for the shelter and four sets of bamboo scissors to raise and adjust the floor height (Fig. 5). All bamboo beams considered as a circular cross-section of 80 mm and a wall thickness of 20 mm with a bamboo density of 600 kg/m^3. In total, the weight of all the bamboo structural profiles is 142.05 kg. Peripheral and central telescopic bars with an integrated locking mechanism have been added to secure stability. The dry season configuration has 3.60 m height and the rainy season configuration has 5.0 m height, covering an area of 9 m^2 and 4.7 m^2 respectively. The shelter's walls are made of traditional Filipino woven split-bamboo mats known as amakan or sawali, which are lightweight and porous, allowing for air circulation to keep the interior cool and balance pressure during strong winds (Fig. 5). The roof is made of nipa palm leaves using the pawid thatching technique. To accommodate the varying surface areas between the dry and rainy season configurations, different envelope panels will be designed for each setup.

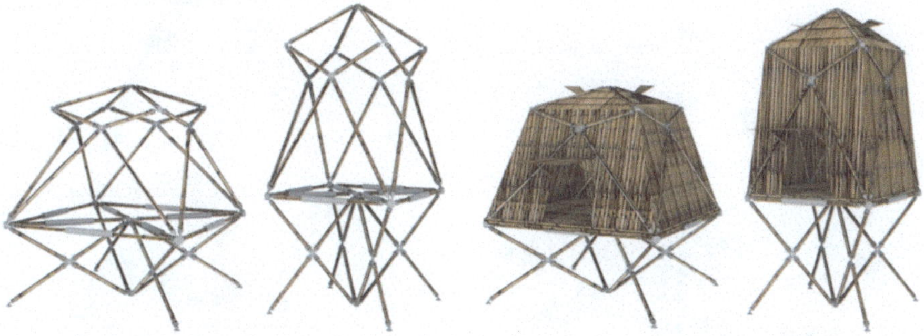

Fig. 5. Bamboo structural system of the dry and rainy season configurations and envelope.

A preliminary structural analysis was conducted using Karamba3D, a plugin for Grasshopper [11]. All supports were assumed to be pinned, while the joints of the shelter were considered rigid when the shelter takes its configuration either dry or rainy season configuration. To secure the foundation to the ground we proposed using auger anchors. The supporting capacity of each anchor is 1270 kg [10]. Bamboo material properties were assigned to all members to account for evenly distributed live loads of 1.92 kN/m^2 on the floor and 0.96 kN/m^2 on the roof (Fig. 6). Considering the climate in the Philippines and using a wind speed of 200 kph [12], the wind loads assigned to the structural system were 1.98 kN/m^2. The density of the bamboo used was based on previous studies of Bambusa blumeana, commonly found throughout the Philippines, with a value of 5.7 kN/m^3 [13].

The rainy season configuration, which has the greater height (h = 5.00 m) than the dry season configuration (h = 3.60 m), and a larger surface area exposed to wind, resulting in higher maximum stress on the members when considering the deformation of the shelter. The axial forces of the structural analysis for the dry and rainy season configurations are shown in Fig. 6. Compression forces are indicated in red and tensile forces in blue. The maximum bending moment for the sunny season configuration, 0.57 kNm occurs at the intersection of the floor members. In the sunny season configuration, vertical loads dominate and this results in bending moments around the y-axis. In contrast, the rainy season configuration is influenced by wind loads, leading to bending moments around the z-axis meaning the structure bends in the xy-plane. The maximum value of 0.35 kNm occurs at the intersection of the angular scissor units. Considering non-bolted connections, we assumed an intermediate fixity with k = 0.7 and the structure achieves a safety factor of 2.2 indicating that the bamboo structural system can withstand winds of up to approximately 180 kph while maintaining its structural integrity with fixed connections. The axial forces, bending moments and deformations for both configurations are shown in Table 1.

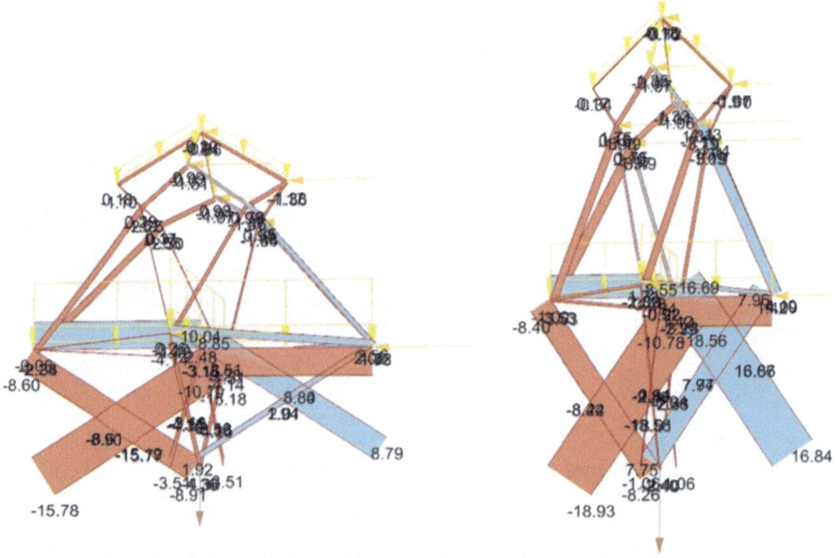

Fig. 6. Axial forces of the dry and rainy season configurations.

Table 1. Axial forces, bending moments and displacement of both season configurations.

Dry season configuration			Rainy season configuration		
N_{max} [kN]	M_{max} [kNm]	Δ_{max} [cm]	N_{max} [kN]	M_{max} [kNm]	Δ_{max} [cm]
15.78	0.35	0.93	18.93	0.57	1.38

Following the initial structural analysis using Karamba3D, we conducted a subsequent analysis using SOFiPLUS [14], an add-in for AutoCAD and SOFiSTiK [15] only for the rainy season configuration since it is the worst-case scenario. Two different models (Fig. 7), a base structure and with additional eight cables, were analysed to improve the structural integrity of the shelter and the connections. Since rigid connections will add more complexity to our design, we proposed to analyse 3D beam elements with pinned connections. The circular cross-sections with external diameter of 80 mm and wall thickness of 8 mm used for this study instead of 80 mm diameter and 20 mm thickness as previously studied. As this is a preliminary design and the Eurocodes do not cover bamboo as a structural material, the axial, shear forces and bending moments used as design values were adopted from a relevant article – guideline [16]. The members were not checked for combined values or buckling. Therefore, an adequate super resistance was sought. In this design, two most crucial combinations are investigated namely the ultimate limit state for live and permanent loads and a combination with wind loads (wind loads were calculated approximately from the general values given) for only one direction. The second model has eight cable elements with a cross-section of 10 mm.

While the connections were accurately modelled regarding degrees of freedom and relevant rotations, they were implemented as points where the members are pinned. In reality due to their size and geometry they may behave as rigid bodies allowing for movements that will make the structure unfit for service. It is therefore suggested that all the pinned connections (scissors) will have a way to get fixed after installation (possibly by tightening appropriately manufactured plates with bolts). Based on the simulation results, the use of cables is recommended as it reduces displacements by 74.01% (Table 2), as well as bamboo members' thickness from 20 mm to 8 mm.

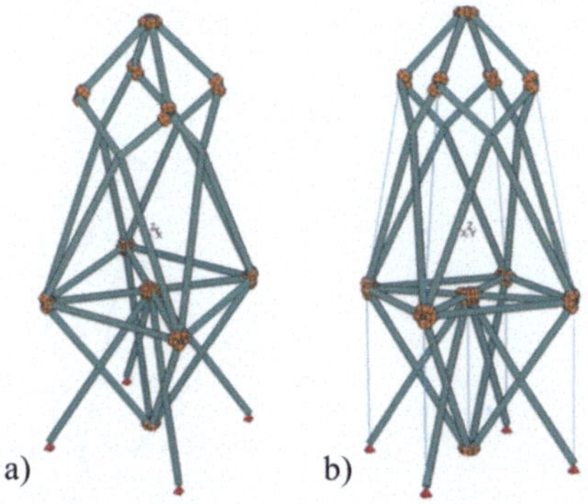

Fig. 7. (**a**) base structure. (**b**) additional eight cables.

3.2 Bamboo Joints with Adjustable Steel Clamps and 3D-Printed Joints

The shelter design incorporates four ball-and-socket joints and a telescopic locking mechanism, enabling two shape configurations to suit environmental conditions (Fig. 8). A key feature of this shelter is the adjustable steel clamps, designed to accommodate bamboo profiles of varying diameters. These clamps allow for easy insertion and fastening without relying on bolted connections. By avoiding bolts, the bamboo members retain

Table 2. Axial forces, bending moments and displacement for the two models.

No cables			Additional eight cables			
N_{max} [kN]	M_{max} [kNm]	Δ_{max} [cm]	N_{max} beams [kN]	N_{max} cables [kN]	M_{max} [kNm]	Δ_{max} [cm]
16.80	0.94	3.77	4.10	11.10	0.27	0.98

their natural integrity and behave as more continuous elements, potentially enhancing load distribution and minimizing stress concentrations at connection points.

Although mortar-infilled bolted bamboo connections effectively distribute loads not only within bearing area of the bolts but also across the entire bamboo node [17], a lightweight connection system was prioritized due to the ephemeral nature of this shelter. Alternatives to bolted joints, such as T-shaped bamboo connections requiring up to four screws proposed in [18] found less suitable as they increase assembly and disassembly time. A more feasible solution, explored in [19], employs steel clamps to improve the structural performance and versatility of bamboo structures. Building on this approach, our project further developed the use of steel clamps, complemented by 3D printing (3DP) technology. 3DP allows for the precise fabrication of ball-and-socket joints while addressing the manufacturing complexities associated with joining multiple links [20]. The integration of bamboo structures with 3D-printed joints, as initially proposed in [21], has been developed further in our design using ball-and-socket joints to create a novel, adaptable and lightweight construction system.

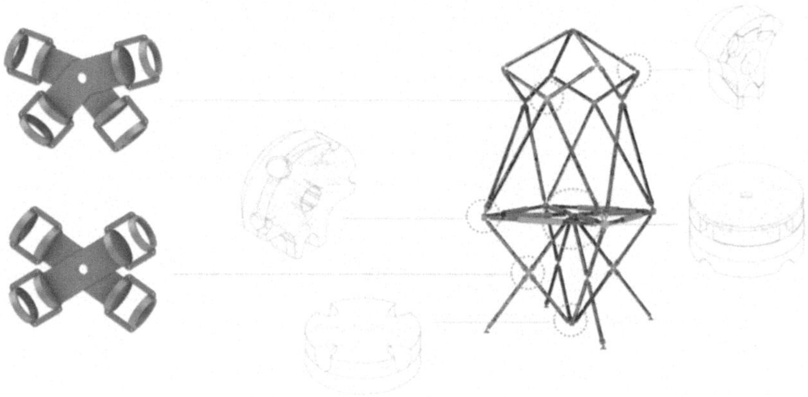

Fig. 8. Adjustable steel clamps (left) and 3D-printed joints (right).

The prototypes in scale 1:6 demonstrate the shelter's feasibility of reconfiguring its shape from dry to rainy season configuration. All the joints have been further improved except for the scissor joints. In particular, the lower side joint, as shown in Fig. 9, changed since a peripheral member was added to the structural system. Given that the

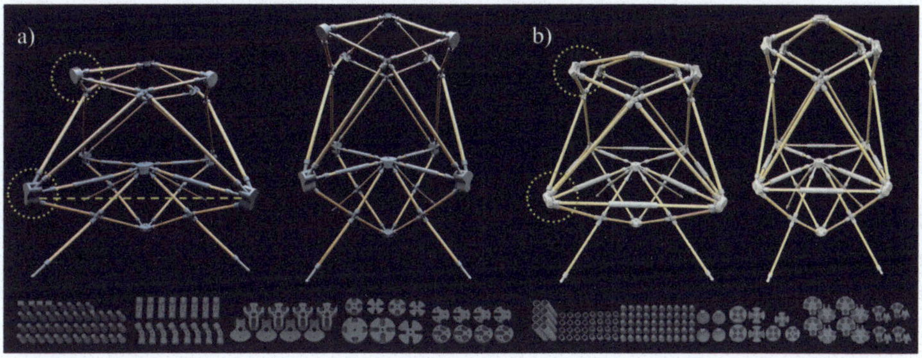

Fig. 9. Prototype in scale 1:6. (**a**) first generation joint design. (**b**) improved second generation.

joint components are comprised of two 3D-printed parts, tongue and groove features were designed for the assembly.

3.3 Stress Analysis on the Central Multiple-Link Joint

In the stress analysis using Autodesk Fusion [22] only the central joint was considered which is connecting eight bamboo beams. A vertical force of 5440 N was applied at the junction of the bamboo beams considering the combined effect of dead loads, live loads, wind loads, and rain loads. For instance, dead loads account for the self-weight of the shelter's structural materials, such as the bamboo profiles. Live loads reflect the weight of people, furniture, and equipment applied to the ground floor. Wind loads and rain loads are also considered due to the shelter's location in the Philippines. Due to the unavailability of bamboo in the database of materials, MDF (Medium Density Fibreboards) was used as a substitute material for the analysis. The MDF used in the analysis has a tensile strength of 7.72 MPa, which is inferior to bamboo's tensile strength, estimated at 82.62 MPa [23]. This may affect the overall strength of the structure. For the non-cylindrical pieces, Polyethylene terephthalate (PET) was selected as the material.

The safety factor is a numerical value that represents the ratio between a structure's ultimate strength (its capacity to resist failure) and the maximum expected load or stress it will experience during its lifetime. Lower safety factors are associated with higher applied stresses increasing the likelihood of failure or fracture. In Fig. 10a, the lower part of the shelter is shown, while Fig. 10b provides a zoomed-in view of the central section. The regions experiencing the highest stresses (in red) were identified within the joints, where the material is subjected to the greatest forces. These areas also correspond to the lowest safety factor, which is 0.652, as shown in Fig. 10c.

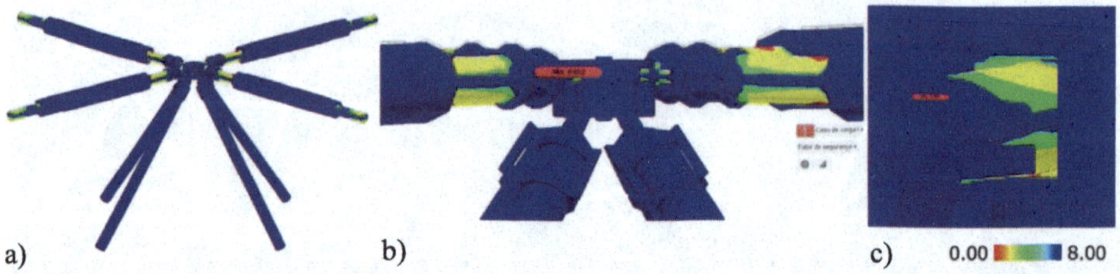

Fig. 10. Stress distribution and safety factor analysis of the shelter structure. (**a**) lower part of the shelter, (**b**) view of the central joint, (**c**) regions of highest stress and minimum safety factor within the joints.

The displacement analysis revealed that the areas furthest from the origin, marked in red in Fig. 11, experience the greatest deformation. These regions should be carefully considered in the design to prevent potential failure. The maximum deformation observed was 4.452 mm, which represents the maximum displacement within the system under the applied load.

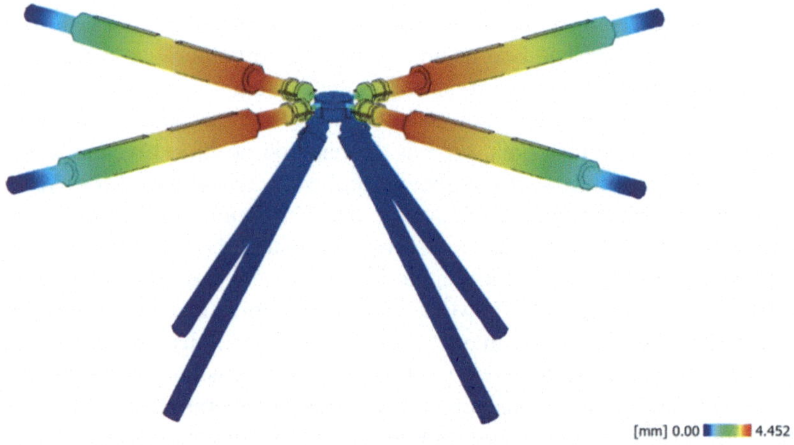

Fig. 11. Displacement analysis showing areas of maximum deformation.

The safety factor analysis revealed that the central joint experience the highest stresses indicating a higher risk of failure in these areas. The displacement analysis identified regions furthest from the origin, marked in red, as experiencing the greatest deformation, with a maximum displacement of 4.452 mm. This suggests potential vulnerabilities in the design, particularly in areas of higher deformation. To enhance the structural integrity of the shelter, it is recommended to round off the joints, especially at sharp corners, to reduce stress concentration and mitigate the risk of material failure (Fig. 12). These design improvements could contribute to a more robust and reliable shelter under the expected loads.

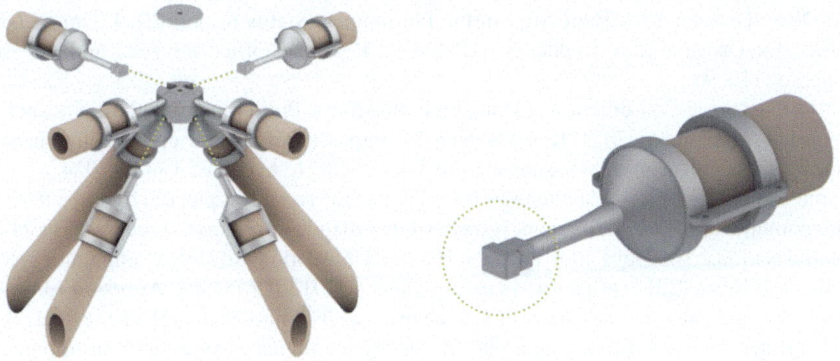

Fig. 12. Joint between the bamboo multi-links and the central connection component.

4 Conclusion

An interdisciplinary team of architects, civil engineers and mechanical engineers collaborated to design a transformable transitional shelter using bamboo cross-section profiles. Following a holistic approach with lightweight construction principles, the shelter's modular and adaptable design allows for the transformation between configurations suited for both the rainy and dry seasons. This transformation is facilitated by 3D-printed ball-and-socket joints. Sustainable strategies have been implemented in this design project utilizing locally sourced materials for the shelter's envelope and structural system, while implementing adaptive reuse concepts. Structural design and analysis were conducted in order to reduce the bamboo shelter's cross-section profiles while improving its structural integrity. Additionally, stress analysis was conducted on the central joint of the shelter, indicating a need for further refinement of the joint design and materiality. The prototype demonstrates the feasibility of reconfiguring its shape to adapt to environmental conditions or natural disasters such as flooding. Future research will focus on improving the shelter's structural performance and manufacturing process for bamboo connection systems.

Acknowledgements. The authors acknowledge the students' work of Wataru Kimura and Yagmur Bulut for their contributions to further improve joints. The project ADDOPTML was funded by the European Union's Horizon 2020 research and innovation program under the Marie Skłodowska-Curie Actions (MSCA) RISE initiative.

References

1. Britannica, The Editors of Encyclopaedia: Ring of fire. In: Encyclopedia Britannica (2024) https://www.britannica.com/place/Ring-of-Fire. Accessed 13 Oct 2024
2. Asian Disaster Reduction Center (ADRC): Information on Disaster Risk Reduction of the Philippines. Asian Disaster Reduction Center. https://www.adrc.asia/nationinformation.php?NationCode=608&Lang=en. Accessed 13 Oct 2024

3. UNDRR: Disaster Risk Reduction in the Philippines: Status Report 2019. United Nations Office for Disaster Risk Reduction (UNDRR), Regional Office for Asia and the Pacific, Bangkok (2019)
4. UN Office for the Coordination of Humanitarian Affairs: Philippines: 2020 Significant Events Snapshot—14 January 2021. ReliefWeb (2021) https://reliefweb.int/report/philippines/philippines-2020-significant-events-snapshot-14-january-2021. Accessed 13 Oct 2024
5. Camus, M.R.: Emergency Shelter Kits for Yolanda Survivors. Rappler (2013) https://www.rappler.com/moveph/44364-emergency-shelter-kits-yolanda-survivors/. Accessed 13 Oct 2024
6. United Nations: 100 Days after Typhoon Haiyan, UN Working to Help Communities Rebuild Lives. UN News (2014) https://news.un.org/en/story/2014/02/461962. Accessed 13 Oct 2024
7. Real Living Team: 5 Ways to Achieve a Filipino-Style Home. Real Living (2017) https://www.realliving.com.ph/lifestyle/arts-culture/5-ways-to-achieve-filipino-style-a00043-20170224. Accessed 13 Oct 2024
8. Ham, P.: Vernacular Architecture Inspires Flood-Proof Housing. TU Delft DeltaLinks (2018)
9. Cangas, J.L., Guevarra, K.G.F., Pascual, M.J.G., Tacata, Y.A., Bricia, M.V.D., Villano, I.J., Abana, E.C.: Disaster-resistant modern bahay kubo. Eng. Res. Bull. **9**, 13–20 (2013)
10. Earth Anchors and Tree Kits. https://bbponline.com/landscaping-and-nursery/earth-anchors-and-tree-kits/. Accessed 13 Oct 2024
11. Preisinger, C.: Linking structure and parametric geometry. Archit. Des. **83**, 110–113 (2013). https://doi.org/10.1002/ad.1564
12. Tanzo, W.T., Pacheco, B.M.: New Wind Load Provisions in Philippine Structural Code. National Structural Code of the Philippines. https://www.wind.arch.t-kougei.ac.jp/APECWW/Report/2004/PHL.pdf. Accessed 13 Oct 2024
13. Salzer, C., Wallbaum, H., Alipon, M., Lopez, L.: Determining material suitability for low-rise housing in the Philippines: physical and mechanical properties of the bamboo species Bambusa blumeana. Bioresources. **13**(1), 346–369 (2018)
14. SOFiSTiK AG: SOFiPLUS User's Guide. SOFiSTiK documentation (2022) https://docs.sofistik.com/2022/en/sofiplus/index.html
15. SOFiSTiK AG. (2023). SOFiSTiK FEA Manual. https://docs.sofistik.com/2022/en/fea/index.html
16. Kaminski, S., Lawrence, A., Trujillo, D.: Structural use of bamboo. Part 1: Introduction to bamboo. Struct. Eng. **94**(8), 40–43 (2016). https://doi.org/10.56330/PNSC8891
17. Correal, J.F., Prada, E., Suárez, A., Moreno, D.: Bearing capacity of bolted-mortar infill connections in bamboo and yield model formulation. Constr. Build. Mater. **305**, 124597 (2021). https://doi.org/10.1016/j.conbuildmat.2021.124597
18. Malkowska, D., Trujillo, D., Toumpanaki, E., Norman, J.: Study of screwed bamboo connection loaded parallel to fibre. Constr. Build. Mater. **398**, 132532 (2023). https://doi.org/10.1016/j.conbuildmat.2023.132532
19. Moran, R., García, J.J.: Bamboo joints with steel clamps capable of transmitting moment. Constr. Build. Mater. **216**, 249–260 (2019). https://doi.org/10.1016/j.conbuildmat.2019.05.025
20. Liao, Y., Krishnan, S.: Deployable scissor structures: classification of modifications and applications. Autom. Constr. **165**, 105547 (2024). https://doi.org/10.1016/j.autcon.2024.105547
21. Kladeftira, M., Leschok, M., Skevaki, E., Tanadini, D., Ohlbrock, P.O., D'Acunto, P., Dillenburger, B.: A study on bamboo, 3D printed joints, and digitally fabricated building components for ultralight architectures. In: Proceedings of the 42nd Annual Conference of the Association for Computer-Aided Design in Architecture, pp. 406–417. ACADIA (2022)
22. Autodesk, Inc. Autodesk Fusion [2.0.20981] (2022). https://www.autodesk.com/products/fusion-360/overview

23. Ghavami, K., Marinho, A.B.: Propriedades físicas Propriedades físicas e mecânicas do colmo inteiro e mecânicas do colmo inteiro do bambu da espécie do bambu da espécie Guadua angustifólia. Rev. Bras. Eng. Agríc. Ambient. **9**(1), 107–114 (2005)

Enhancing Onshore Wind Tower Foundations: A Comprehensive Automated Design Approach

Emmanouil Vougioukas, Athanasios Stamos, Charikleia Pappa, and Nikos D. Lagaros(✉)

School of Civil Engineering, Institute of Structural Analysis and Antiseismic Research, National Technical University of Athens, 9, Heroon Polytechniou Str., Zografou Campus, 15780 Athens, Greece

{manolis,stamthan,nlagaros}@central.ntua.gr

Abstract. The need of green energy production is very intense the last years and it rises the attention of all people who are interested in this sector. A lot of studies were conducted so as to reduce cost, time and at the same time keep the relevant functions operational. This study focuses on the way to study deeper and enhance the foundation of onshore wind tower foundation while being aligned with the regulation standards. The procedure that leads to this result includes the analysis of the forces that are applied on the system wind turbine – foundation, the identification of the soil conditions, the receival of the wind turbine characteristics and the alignment with the regulations. More specifically, the foundation is initially considered of cubic shape which acts as a counterweight to the overturn force that is predominant at the whole system. At a second phase, the shape of the foundation is considered rectangular and finally it is designed as a beam slab whose voids are filled in with soil which contributes to the counterweight needed to overcome the overturn forces. After implementing the proposed methodology and comparing it with a compatible wind turbine foundation, it is derived that there is high reduction is cost, materials and time while the precision is increased, and the regulations are followed Through this way the installation of the wind turbine foundation will be quicker and more precise leading to increased efficiency in time and cost.

Keywords: Onshore wind turbine · Foundation · Optimization · Structure · Engineering

1 Introduction

The demand for renewable source of energy is increasing and all parties involved in this process are targeting the most efficient and cost-effective way to this direction. Wind energy and wind turbines are a major part of this attempt and a lot of studies were conducted for the optimization of their functionality or constructability. It is noted that a lot of effort was invested for the optimization and enhancement of the wind turbine itself so as their energy production is higher and their parts, like the blades, the tower or the hub are working at their maximum capacity. As far as the wind turbine foundations is concerned, while a wide study was done for the offshore wind turbines, a restricted

N. D. Lagaros et al. (Eds.): ADDOPTML 2024, LNCE 646, pp. 100–109, 2025.
https://doi.org/10.1007/978-3-031-92029-5_10

study was conducted for the onshore wind turbines foundation. This triggers the interest of the current study to deepen more into this and explore the path to its enhancement and further development both on the design phase and on the construction phase.

This study targets the advancement of the design in the first place and the site implementation at a next level of the onshore wind turbine foundation. The current study aims to minimize the cost and time of the foundation by following steps to this direction while taking into consideration and satisfying all relevant regulations. In the first place the study identifies which are the parameters that need to be identified for the wind turbine. It identifies the shape and configuration while keeping the concrete and cost quantity low. In order to reach the targets, an automated design approach was developed by entering two variables, the width of the wind turbine foundation and the length of it. The results verify the cost reduction while time is highly eliminated.

2 Presentation of the Proposed Design

This study focuses on the presentation and development of the wind turbine foundation. More specifically, the steps that are followed so as to reach the final design are as described below: In the first place the wind turbine that needs to be installed is selected. It should be mentioned here that the wind turbine that this study is focused on is the onshore wind turbine with stable axis. This plays a significant role in the shape of the wind turbine foundation, as wind turbines of stable axis require square or rectangular foundation, while wind turbine of non-stable axis require circular wind turbine foundation. Along with the wind turbine selection, the allocation of the wind turbine in the field is explored. Usually both the wind turbine characteristics that are derived from the previous step and the soil characteristics, which are derived from the current step based on the area allocation, are given by the parties who cope with such subject. Based on the area, the soil characteristics can be identified. The first two steps are, actually, the data that need to be inserted at the proposed algorithm. Following the afore mentioned steps, the forces that act on the wind turbine are identified. It should be highlighted that the main force that is applied at the system wind turbine – foundation is the wind force. Furthermore, the self-weight of the wind turbine and the foundation are calculated. After identifying the forces, it is deeply analyzed how these forces affect the system. More specifically, the phenomena developed around the foundation and at the wind turbine are described and explained so as to understand under which phenomena and situation the under-examination foundation is. Finally, the excess of concrete whose contribution at the system could be replaced by a cheaper material is identified and the foundation takes the shape of a beam slab which is derived after the removal of the concrete excess and the formation of the steel reinforcement so that it can undertake all forces applied and phenomena developed [1–10].

Having identified the foundation shape and steel, the concrete and steel quantity are calculated along with the soil which fills the voids of the foundation. It is of high importance to mention that the soil contributes to the counterweight that is needed so as to overcome the overturn phenomenon which is the predominant of all phenomena developed and against which the attention is focused on (Fig. 1).

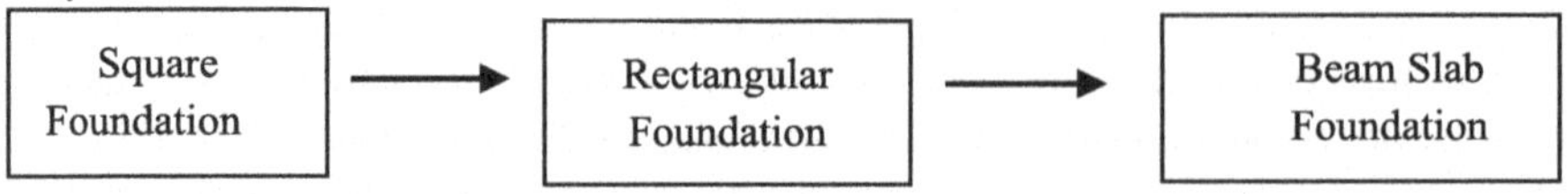

Fig. 1. Progress of foundation design according to the proposed analysis.

3 Initial Information Required for the Proposed Analysis

The distinction of the wind turbines is based on two categories. The first is the location and more specifically whether they are onshore or offshore and secondly based on the blade's axis and specifically if it is stable or rotated. In case the wind turbine is on shore the foundation types are different from the ones for the offshore. At the onshore the foundation may be shallow or with piles. In both cases the connection between the foundation and the wind turbine should be stable to certify the connectivity and stability of the system. For an offshore wind turbine, the foundation types may be three, including the monopile, jacket and gravity foundation. There is further effort in order to develop more the offshore wind turbine foundations and introduce new systems like tripods, tripiles and floating platforms. All offshore foundations face high forces, while they are keeping the stability on the top.

The second distinction which is based on the axis of the blasé identifies whether the foundation will be rectangular or circular. In case of stable axis, the foundation is rectangular and in case of rotated axis the foundation is circular. The current study focuses on the onshore wind turbines with stable axis. As a result, the foundation is either gravity or pile and it is rectangular (Fig. 2).

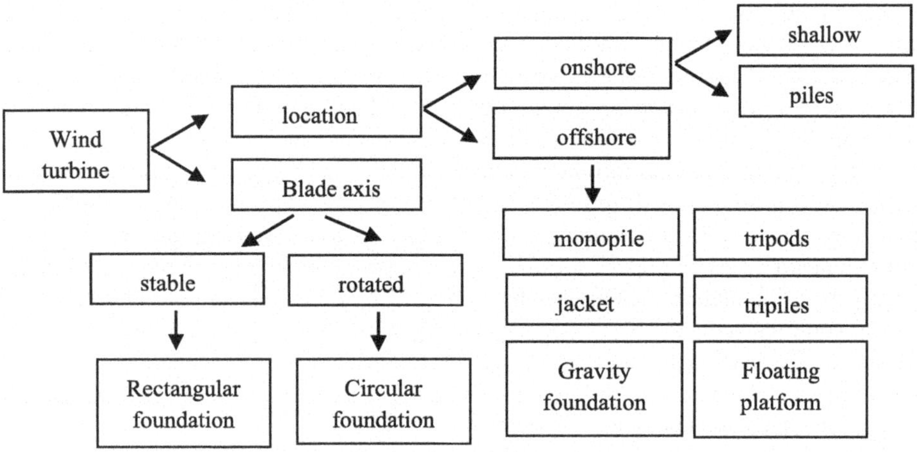

Fig. 2. Wind turbines categorization.

The current study focuses on the onshore wind turbines of stable axis. As a result, the foundation is rectangular and the type of foundation that is examined is shallow gravity foundation. In order for the analysis to commence, there is a series of data that are needed to be inserted. The data that should be inserted have to do with the wind turbine itself and with the area where the wind turbine is placed. The data that refer to the wind turbine include the total height of the wind turbine, the tower height, the wind turbine diameter at the base, the reduction of the wind turbine diameter with the height and the self-weight of the wind turbine itself. These data are mandatory and are specified each time according to the wind turbine that is examined. These data are provided by the manufacturer and the access is usually simple. Concerning the data needed for the area that the wind turbine is allocated, a geotechnical study is needed in the first place. Through this process the soil characteristics will be derived which are mandatory for the soil identification through which the forces and stresses applied on the foundation can be calculated. Furthermore, the allocation area defines the wind force that is applied on the wind turbine. This data is derived through relevant authorities link and according to Eurocodes. After receiving the data required for the analysis, the whole process can commence, and the relevant results can be derived.

4 Forces and Phenomena Identification

After completing the receival of all data needed for the analysis to commence, the next step is to analyze the forces that are applied on the system wind turbine – foundation. The forces are the wind turbine self-weight, W_W, which is according to the Eurocodes and is a vertical load. Additionally, the self-weight of the foundation itself, W_G, is calculated. The self-weight is derived based on the shape of the foundation and the force is vertical. Apart from this, the horizontal force of the wind, F_W, is applied on the wind turbine tower. Due to the wind force, there is the tendency for overturn of the foundation and as a result passive stresses, P_p, are developed. Last but not least, soil stress, σ_{max} and σ_{min}, under the foundation are developed according to the foundation soil.

The scope of the current study is to maintain the stability of the structure and even enhance it while optimizing the use of the materials used in construction, like the soil or concrete used on site. It is effectively managed to distribute the forces transmitted from the wind turbine under the foundation and solve the foundation as a beam slab which consists of a slab that uniformly distributes the loads of the upper structure at the soil, of beams that undertake the bending and shear forces and of soil that contributes with its self-weight to the resistance against overturn [11–15] (Fig. 3).

Forces applied on the system wind turbine – Foundation:

1) Wind turbine self-weight, W_w
2) Foundation self-weight, W_G
3) Wind force, F_w
4) Passive soil stresses at the lateral sides of the foundation, P_p
5) Soil stresses developed under the foundation, σ_{max} and σ_{min}

Fig. 3. Forces applied on the system wind turbine – foundation.

The forces applied on the system wind turbine – foundation result in the development of some phenomena and checks that should be conducted. As a result, the foundation is checked against:

- overturn in the x and y direction,
- bending in the x and y direction,
- shear,
- perforation,
- verification that the soil stress developed is within the approved soil limits.

It should be highlighted that the crucial check to be conducted is the check against overturn. If all other checks are met and the check against overturn is not, the system wind turbine – foundation is not operation setting the check against overturn the most crucial of all. In order to comply with this check, high mass should be gathered at the foundation area rather than strength. This happens because what is needed is a counterweight to the overturn tendency the wind force causes. As a result, the current study focuses on this part and more specifically removes all concrete quantity that does not contribute with its strength to overturn phenomenon and replace it with soil which is of lower cost and at the same time contributes to the check against overturn.

5 Objective Function

The current study focuses on the optimization of the wind turbine foundation aiming to the highest decrease of the cost. The cost of the foundation, C_{total} is a sum of the below costs:

The excavation cost, $C_{excavation}$, which refers to the cost to excavate the area where the wind turbine foundation will be constructed. This cost depends on the soil quality, so if the soil is rocky the cost will be higher, while for soft soil the cost decreases.

The concrete pouring cost, $C_{concrete}$, which refers to the cost of the foundation framing and concrete pouring in it including procurement and installation. The fact that the excess of the concrete that does not contribute with its strength to the system stability is removed, contributes to the cost minimization as less concrete quantity is used and it is replaced with soil whose cost is much lower, especially if the procurement is achieved through the recycle of the excavation material derived through the excavation phase. Even if, the recycle is not possible, soil material is being transferred to the site by soil deposit areas, for which case the cost is still lower than the concrete cost.

The steel cost, C_{steel}, which includes the steel procurement and installation. The steel quantity is reduced compared with the conventional wind turbine foundation, as at this case only at the base slab and the beams the steel is installed. The type of steel is derived according to the loads applied on the system which led to the need of reinforcement.

The backfilling cost, $C_{back\text{-}filling}$, which refers to the cost related to the soil that needs to be put in the voids created between the concrete beams of the beam slab foundation. As already mentioned, this soil reaches the site either by the already excavated material for the needs of the foundation or by soil deposit areas. In case the soil is rocky, the cost is reduced as the compression of it can be omitted while in case of soft soil the compression is mandatory and is included in the cost.

The recycling cost, $C_{recycling}$, which refers to the recycling of the excavated materials at the area where the wind turbine will be installed (Fig. 4).

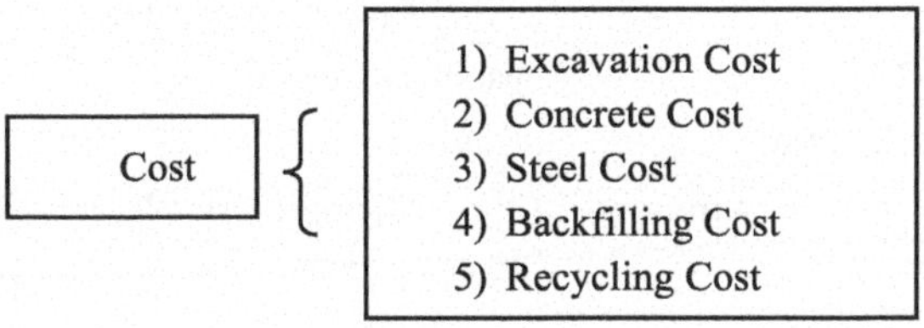

Fig. 4. Cost related to the wind turbine foundation construction.

In order to meet the minimum cost needed for the foundation construction, a repetitive procedure is followed taking various sizes for the wind turbine foundation, till the cost is minimized. The repetitive procedure incudes the satisfaction in the first place of the overturn check and at a second phase of the rest checks described at the previous paragraph.

It should be mentioned that in case of rocky aggregates, the formwork of the foundation can be omitted as the rock aggregates work as a framework. As a result, the procedure of construction includes in the first place the excavation of the foundation area, the lean concrete pouring, the steel installation, the framework installation in case we have soft soil, the rocky aggregates filling (and omitting the framework) and finally the concrete procurement and pouring. It is important to mention that the little quantity of concrete that will enter the adjacent layer of the steel is acceptable as it is too low. Taking all the above into consideration we conclude that a different way to use the same materials on site can lead to a more effective and stable structure. Since, as mentioned above, the wind turbines that are examined at the current study are of stable axis and not rotated, all the checks at the foundation are done at the horizontal and vertical axis [16–19].

6 Application of the Proposed Design

The above analyzed design was implemented on an actual wind turbine so as to check its usability and how effective the method is on an actual case. The Table 1 shows its characteristics.

The wind turbine chosen is from Vestas Company and is V52–850 kW. It is an onshore wind turbine, and its characteristics will be used for the implementation of the proposed analysis and method. Through this process, it will be derived how the proposed method can comply with the needs of the real wind turbine at the construction phase and what is the advantage of using it.

Since the wind turbine characteristics are received, the area allocation characteristics are needed and given the fact that the case study is being implemented in Greece, the relevant features for the soil are received from the geotechnical study while the wind force is as per the Eurocodes.

Table 1. Wind turbine V52-850 (Vestas) characteristics [20].

Rotor	Gearbox
Diameter: 52 m	Type: 1 planet step/2-step
Area swept: 2124 m2	parallel axle gears
Nominal revolutions: 26 rpm	**Control**
Operational interval: 14.0–31.4 rpm	Type: Microprocessor-based monitoring of all
Number of blades: 3	**Weight**
Air brake: Full blade pitch	Nacelle: 22 t
Tower	Rotor: 10 t
Hub height: 40 m, 44 m, 49 m, 55 m,	Towers: IEC IA IEC IIA DIBt II DIBt III
60 m, 65 m, 74 m, 86 m	**Hub height**
Operational data	40 m 40 t – – –
Cut-in wind speed: 4 m/s	44 m 45 t – – –
Nominal wind speed: 16 m/s	49 m 50 t – – –
Cut-out wind speed: 25 m/s	55 m 55 t 50 t – –
Generator	60 m 70 t 70 t – 70 t
Nominal output: 850 kW	65 m 75 t 75 t – 75 t
Operational data: 50 Hz/60 Hz	74 m – – 95 t –
690 V	86 m – – 110 t –

According to the equations and constraints specified in Eurocode 1 the wind force is calculated as follows: $F_{wx} = 254.80$ kN and $F_{wy} = 228.84$ kN. It should be mentioned that the wind force is greater in one direction because it acts on both the tower and the blades, while in the perpendicular direction, it only acts on the tower.

After the analysis implementation, the foundation dimensions are specified as follows: the foundation height (H_F) is equal to 2.5 m, the width (B) is equal to 11 m, and the length (L) is 13 m. The proposed methodology results in a rectangular foundation, with the longer dimension parallel with the predominant wind force. The construction phase includes also excavating the foundation area to the specific dimensions as per the foundation needs. It is important to note that additional 10 cm is excavated in depth to allow for the pouring of lean concrete. Due to the fact that extra concrete will be needed to fill some recesses in case the rocky aggregates are used for framework, an extra 2% of concrete is calculated. On the other hand, in case of soft soil, where actual framework is used, extra 10 cm are excavated at the perimeter of the foundation so as the workability of the framework installation is achieved.

After installing the steel, pouring the concrete and filling with rocky aggregates or soft soil the voids between the beams, light concrete is poured over the voids filling in order to protect for any potential case of reaching or removing it and at the same time protect it from whether conditions.

In order to check how the objective function, which is related to the cost, is satisfied, actual market prices were received and more specifically: (Table 2)

Table 2. Cost calculation of beam foundation

Task	MU
Excavation	13–18
Concrete	100
Steel price	1.5
Stone filling	0.003–0.010
Light concrete	67

It should be mentioned that for a single foundation the overall cost was reduced by 3% in case of rock soil. More specifically, the excavation cost is increased in rock soil up to 27.7% due to the difficulty in excavating the material, the concrete cost is reduced by 20% while the steel remains the same for both cases. Last, the recycle cost is increased by 37.5%.

After implementing the proposed analysis, as far as the excavation is concreted, the cost remains the same as with single foundation. The concrete cost is reduced by 14% and 31% respectively for rock and soft soil. The steel is reduced by 90% in both rock and soft soil and the recycling is 3.9% and 3.7% of the total cost for rock and soft soil respectively. It should be mentioned that the voids filling is 4.3% and 13.7% of the total cost for rock and soft soil respectively in case of beam foundation.

It should be highlighted that despite the material reduction by 31% for concrete and 90% for steel, the stability and resistance of the structure increased.

Having assessed the results of the implementation of the proposed study on an actual wind turbine, it can be derived that a lot of advantages can be counted through the proposed analysis. In the first place, the proposed method can be implemented at any kind of soil. This gives the advantage that after getting the required information concerning the soil and the wind of the area, the proposed analysis acquires the data needed to be inserted so as to optimize the structure. Furthermore, the concrete quantity is reduced and as a result the cost is also reduced. The parallel advantage that this has is the reduction of the environmental footprint. Following the concrete, the steel quantity is decreased and as a result the cost decreases also. This leads also to the simplification of the construction process. Additionally, the construction time is decreased and the whole process is accelerated. It is proved that the proposed study enhances the feasibility, the stability and the performance of the system wind turbine – foundation.

7 Conclusions

This study analysis a method that aims to enhance the structural approach of the wind turbine foundation of onshore wind turbines. It uses the same material as a conventional construction site for foundation construction but enhances the role of each material by maximizing its contribution to the phenomena developed at the system wind turbine – foundation. The proposed method was implemented on an actual wind turbine in the area of Greece and the results were encouraging. After analysis and deconstruction of each phenomenon developed and each force applied on the system, a series of conclusions were derived.

In the first place, it is proved that the optimum shape for the foundation of a wind turbine of stable axis is the rectangular one, consisting of beams at both directions and achieving a solid resistance against overturn. Furthermore, the excess of concrete is identified through the resistance against shear. The final concrete quantity is significantly reduced while the structure aligns with the requirements against bending, shear, perforation and overturn check.

The steel reinforcement is also reduced while the checks as per the regulations are met and the foundation is structurally solid and stable.

All the above result in an optimized foundation whose configuration is a rectangular beam slab that is filled in with soil material leaving the minimum environmental footprint and resulting in a much more cost-efficient structure while the construction process is highly accelerated.

Acknowledgement. The authors acknowledge the support of the Hellenic Foundation for Research and Innovation (H.F.R.I.) under the "2nd Call for H.F.R.I. Research Projects to support Faculty Members and Researchers", AMOSS project: "Additively Manufactured Optimized 3D Printed Steel Structures", (Project Number: 02779).

References

1. Abdelmoteleb, S.E., Mendoza, A.S.E., dos Santos, C.R., Bachynski-Poli'c, E.E., Griffith, D.T., Oggiano, L.: Preliminary sizing and optimization of semisubmersible substructures for future generation offshore wind turbines. J. Physics: Conf. Ser. **2362**, 012001 (2022). https://doi.org/10.1088/1742-6596/2362/1/012001
2. Al-Sanad, S., Parol, J., Wang, L., Kolios, A.: Design optimisation of wind turbine towers with reliability-based calibration of partial safety factors. Energy Rep. **9**, 2548–2556 (2023). https://doi.org/10.1016/j.egyr.2023.01.090
3. Asim, T., Islam, S.Z., Hemmati, A., Khalid, M.S.U.: A Review of recent advancements in offshore wind turbine technology. Energies. **15**, 579 (2022). https://doi.org/10.3390/en15020579
4. Bošnjakovi'c, M., Katini'c, M., Santa, R., Mari'c, D.: Wind turbine technology trends. Appl. Sci. **12** (2022). https://doi.org/10.3390/app12178653
5. Byron, M., Denis, K., Charles, W.C.: Loading and dynamic response considerations for the design of wind turbine foundations on South African soils. Geotech. Eng. **48**, 110–117 (2017) Cited by: 0
6. Chen, X., Xu, J.Z.: Structural failure analysis of wind turbines impacted by super typhoon Usagi. Eng. Fail. Anal. **60**, 391–404 (2016). https://doi.org/10.1016/j.engfailanal.2015.11.028
7. De Anda, J., Ruiz, S.E., Bojórquez, E., Inzunza-Aragon, I.: Towards optimal reliability-based design of wind turbines towers using artificial intelligence. Eng. Struct. **294**, 116778 (2023). https://doi.org/10.1016/j.engstruct.2023.116778

8. IEA: Executive Summary – Renewables 2023 – Analysis and Forecast to 2028. International Energy Agency (IEA) (2023) https://www.iea.org/reports/renewables-2023/executive-summary. Accessed 3 Aug 2024
9. Kallioras, N.A., Lagaros, N.D., Karlaftis, M.G., Pachy, P.: Optimum layout design of onshore wind farms considering stochastic loading. Adv. Eng. Softw. **88**, 8–20 (2015). https://doi.org/10.1016/j.advengsoft.2015.05.002
10. Lagaros, N.D., Karlaftis, M.G.: Life-cycle cost structural design optimization of steel wind towers. Comput. Struct. **174**, 122–132 (2016). https://doi.org/10.1016/j.compstruc.2015.09.013
11. Lago, B.D., Flessati, L., Marveggio, P., Martinelli, P., Fraraccio, G., di Prisco, C., di Prisco, M.: Experimental tests on shallow foundations of onshore wind turbine towers. Struct. Concr. **23**, 2986–3006 (2022)
12. Lavassas, I., Nikolaidis, G., Zervas, P., Efthimiou, E., Doudoumis, I., Baniotopoulos, C.: Analysis and design of the prototype of a steel 1-MW wind turbine tower. Eng. Struct. **25**, 1097–1106 (2003). https://doi.org/10.1016/S0141-0296(03)00059-2
13. Lehmann, P., Reutter, F., Tafarte, P.: Optimal siting of onshore wind turbines: Local disamenities matter. Resour. Energy Econ. **74**, 101386 (2023). https://doi.org/10.1016/j.reseneeco.2023.101386
14. Matos, R.P., Pinto, P.L., Rebelo, C.S., Gervásio, H.S., Veljkovic, M.: Improved design of tubular wind tower foundations using steel micropiles. Struct. Infrastruct. Eng. **12**, 1038–1050 (2016). https://doi.org/10.1080/15732479.2015.1076853
15. Satymov, R., Bogdanov, D., Breyer, C.: Global-local analysis of cost-optimal onshore wind turbine
16. Srikakulapu, R., Vinatha, U.: Optimal design of collector topology for offshore wind farm based on ant colony optimization approach. In: Proceedings of the 2016 IEEE International Conference on Power Electronics, Drives and Energy Systems (PEDES), 14–17 December 2016, pp. 1–6. IEEE, Trivandrum (2016). https://doi.org/10.1109/PEDES.2016.7914458
17. Sun, Y., Fan, Y., Lu, J., Hou, J., Wang, Y., Bai, X.: Optimal design method for topology of onshore wind power DC collection system. Dianli Xitong Zidonghua/Autom. Electr. Power Syst. **47**, 142–152 (2023). https://doi.org/10.7500/AEPS20220417007
18. Tafarte, P., Lehmann, P.: Quantifying trade-offs for the spatial allocation of onshore wind generation capacity – a case study for Germany. Ecol. Econ. **209**, 107812 (2023). https://doi.org/10.1016/j.ecolecon.2023.107812
19. Global Wind Energy Council: Wind Industry Statement on Economic Recovery from COVID-19. Gwec.Net. https://gwec.net/wind-industry-statement-on-economic-recovery-from-covid-19/. Accessed 27 June 2024
20. www.Vestas.com

A Continuum Finite Element-Based Micromechanical Approach for the Optimal Material Design of Graphene-Reinforced Composites

Stelios K. Georgantzinos(✉) and Georgios P. Papa

Aerospace Science and Technology, National and Kapodistrian University of Athens, 34400 Psachna, Greece
sgeor@uoa.gr

Abstract. The finite element method (FEM) is an essential tool for exploring the complex behaviors of advanced materials and nanocomposites, particularly when experimental analysis is challenging. This study refines the modeling of graphene-reinforced nanocomposites by developing and validating 3D representative volume elements (RVEs) using FEM. The RVE geometry is designed with an embedded-like structure, which optimizes load distribution and enhances mechanical performance. In these models, graphene is simulated as surface geometry with shell elements, while the matrix and interphase regions are represented as 3D continua, with the interphase modeled at the theoretical thickness of graphene. This approach significantly improves the accuracy of mechanical property predictions by enhancing load transfer mechanisms and minimizing assumptions. The 3D RVE models were validated against other numerical data and employed to predict key elastic properties, including elastic and shear moduli, across various graphene volume fractions and sizes. Additionally, linear dynamic simulations were conducted to investigate the material's vibrational characteristics, offering deeper insights into the mechanical performance of the nanocomposites. The results suggest that this modeling approach could be particularly beneficial for applications in industries requiring high-performance materials, such as aerospace and automotive sectors.

Keywords: Finite Element Method (FEM) · Graphene-reinforced nanocomposites · Representative Volume Elements (RVEs) · Mechanical performance · Load distribution · Elastic properties

1 Introduction

In recent years, there has been a rapid growth of composite structures in various sectors such as aerospace, construction and automotive industries. In this context, it is necessary to develop reliable models for the calculation of mechanical characteristics and prediction of their behavior in order to optimally design advanced structures and systems. The simultaneous rapid development of nanotechnology in recent years has led to the

N. D. Lagaros et al. (Eds.): ADDOPTML 2024, LNCE 646, pp. 110–121, 2025.
https://doi.org/10.1007/978-3-031-92029-5_11

discovery of innovative materials and nanostructures. Nanostructures exhibit excellent properties and multi-functionality. Their efficient utilization in the layout of structures and systems is one of the most important issues in science and technology nowadays. The distinct microstructure, which is due to the nanoscale, leads to a different behaviour compared to macromaterials, so that classical mechanical and strength of materials does not describe it with the required accuracy. In order to describe the behaviour accurately, the geometry of the nanostructure must be taken into account. This creates considerable complexity and long calculation times. Therefore, it is necessary to develop new computational analysis methods that take into account the different scales involved in nanostructure reinforced composite structures while being computationally efficient. Since the finite element method is one of the most basic techniques for structure analysis, this study focuses on the development of efficient finite element method that can yield cost-free optimal designs.

Polymer matrix composites, known for their exceptional strength and stiffness-to-weight ratio, are extensively utilized in aerospace, automotive, construction, and medical industries [1, 2]. To further enhance their mechanical properties, carbon nanotubes (CNTs) and graphene, among the strongest and most rigid nanomaterials, have been integrated into polymer matrices. Over the last decade, graphene and its associated nanostructures (GANS) have been extensively studied using both experimental and numerical approaches. Numerous studies highlight the transformative potential of these nanofillers. Shahbaz [3] reported significant enhancements in flexural strength by incorporating CNTs into carbon fiber-glass hybrid composites, while Tüzemen et al. [4] observed increases of 54% in flexural strength and 27.4% in tensile strength with 0.3 wt% CNTs in composite plates. Similarly, Eskizeybek [5] demonstrated improved flexural strength in glass fiber-epoxy nanocomposites with chemically bonded multi-walled CNTs, and Gouda et al. [6] showed that graphene nanoparticles enhance elastic modulus and hardness, while CNTs improve flexural modulus and impact resistance. Even minimal CNT incorporation, as examined by Ciecierska et al. [7], produced measurable gains in strength and elastic modulus in foam composites.

Finite element modeling (FEM) has become the benchmark for analyzing the mechanical and thermal properties of graphene-reinforced composites, with significant progress made in modeling these materials using representative volume elements (RVEs) [8]. While atomistic models based on interatomic potentials are well-established, there remains a critical need for cost-effective and accurate continuum-based models. These models have the potential to bridge the gap between nanoscale behaviors and macroscopic material properties, advancing the understanding of graphene's influence on composite performance.

2 Numerical Modeling Framework

This chapter presents the numerical modeling framework employed in this study. The computational approach is based on micromechanical theory and utilizes representative volume elements (RVEs) for analysis. The RVE model developed in this work comprises a single graphene sheet embedded in a polymer resin, with an interphase region characterized by the theoretical thickness of the graphene. The graphene sheet is modeled as a surface body, while the matrix and interphase region are represented at the macroscale using three-dimensional solid elements.

2.1 Micromechanical Analysis

The micromechanical analysis is based on several key assumptions. The polymer matrix is assumed to be uniformly reinforced with short-length graphene sheets, and the interaction between graphene and the matrix is modeled through an imperfect interphase. Also, small graphene volume fractions are considered, allowing the assumption of negligible interaction between adjacent graphene sheets. Additionally, for validation and comparative purposes, a simplified RVE model, based on a sandwich configuration previously proposed in the literature [9], is employed. This alternative model simplifies geometry and interactions while adhering to the same micromechanical principles as the proposed model.

The length and width of the RVE correspond to the matrix length, $\mathbf{L_m}$, and width, $\mathbf{W_m}$, respectively, while the graphene dimensions are $\mathbf{L_g}$, $\mathbf{W_g}$, and $\mathbf{t_g}$. The matrix thickness is $\mathbf{t_m}$. The volume fraction of the graphene sheet in the composite is calculated as follows:

$$V_{fr} = \frac{V_g}{V_m + V_g} \tag{1}$$

Considering the dimensions of the constituents in the embedded RVE (Fig. 1a), the volume fraction can be expressed as:

$$V_{fr} = \frac{W_g \cdot L_g \cdot t_g}{W_m \cdot L_m \cdot t_m} \tag{2}$$

For the sandwich RVE model:

$$V_{fr} = \frac{W_g \cdot L_g \cdot t_g}{(W_m \cdot L_m \cdot t_m) + \left(W_g \cdot L_g \cdot t_g\right)} \tag{3}$$

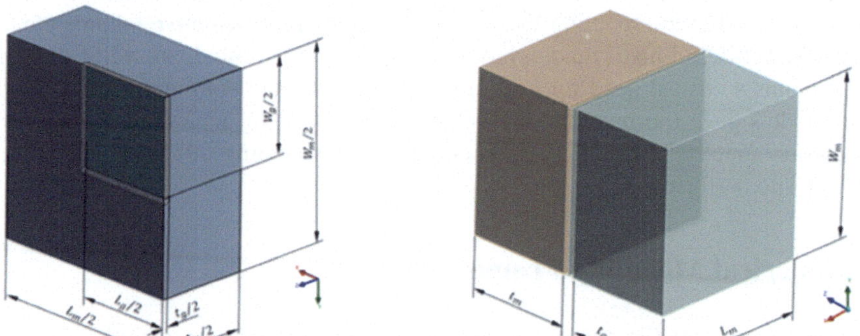

(a) (b)

Fig. 1. (**a**) the Embedded RVE model of the Graphene-reinforced Composite and, (**b**) the sandwich RVE model of the Graphene-reinforced Composite.

Numerical results derived from this model will be presented and compared with similar models. Since simplified RVE is based on similar principles, it is expected to provide insights into the behavior of related models.

2.2 Finite Element Modeling of Graphene

The graphene sheet is represented as a three-dimensional surface using quadratic shell elements, each consisting of eight nodes with six degrees of freedom per node (three translations and three rotations). Unlike traditional approaches that use beam or spring elements to simulate carbon atom interactions, this study adopts a homogeneous shell representation [8]. This significantly reduces design and preprocessing time while allowing orthotropic material properties to be directly assigned through the shell element's material definition.

The elemental triads are aligned with the global coordinate system to effectively model the orthotropic moduli in each material direction. The Young's modulus of the constructed graphene was calculated and validated against published data to ensure model accuracy.

2.3 Macroscale Modeling of the Matrix

The polymer matrix is modeled as a continuum isotropic elastic medium characterized by Young's modulus ($\mathbf{E_m}$) and Poisson's ratio ($\boldsymbol{\nu}$). While a discrete molecular-level representation could have been considered, it would significantly increase computational cost and complexity. This simplified approach is a key advantage of the methodology. The matrix is discretized with quadratic three-dimensional isoparametric elements (solid elements), each with twenty nodes and three degrees of freedom per node (translations in the x, y, and z directions).

2.4 Modeling of the Interphase Region

The interphase region between the graphene and polymer matrix plays a crucial role in load transfer from the matrix to the graphene, which is key to enhancing composite properties. This region is characterized by its thickness and material properties that transition between those of graphene and the matrix. The interphase region is modeled using three-dimensional solid elements with stiffness values interpolated between the orthotropic properties of graphene and the isotropic properties of the polymer.

The rotational degrees of freedom (DOF) of the graphene's shell elements are coupled with the displacement DOF of the solid interphase region using a Multipoint Constraint (MPC) contact approach. The shell elements represent the carbon atoms of graphene, while the solid elements represent the polymer molecules. Mechanical properties of the interphase region vary, as they depend on the surrounding graphene and matrix properties. A linear interpolation function defines these properties:

$$E_{int(i,j,k)} = E_m + (\overline{x_{int}} - x_m)\frac{\left(E_{g(i,j,k)} - E_m\right)}{\left(x_g - x_m\right)} \tag{4}$$

where $\mathbf{E_{int}}$, $\mathbf{E_g}$ and $\mathbf{E_m}$ denote the moduli for interphase, graphene and matrix respectively. Here, $\overline{\mathbf{x_{int}}}$ represents the mean x-coordinate of the interphase thickness, while $\mathbf{x_g}$ and $\mathbf{x_m}$ correspond to the x coordinates of the graphene and matrix respectively. This formulation creates a heterogeneous interphase environment, indirectly accounting for nanoscopic effects such as van der Waals interactions. Although these effects are not explicitly modeled, the mechanical behavior of the interphase is defined by macroscale parameters.

2.5 Numerical Solution Procedure

The numerical solution integrates the finite element models of the graphene, matrix, and interphase region to solve for the mechanical response of the RVE. A linear static analysis is conducted to predict the elastic mechanical properties. The system of linear equations is expressed as:

$$K \cdot U = F \tag{5}$$

where **K**, **U** and **F** represent the assembled stiffness matrix, displacement vector, and force vector, respectively. This system is solved with appropriate boundary conditions.

Additionally, a series of modal analyses are conducted to study the dynamic behavior of the composite, determining its natural oscillation frequencies and corresponding mode shapes. The governing equation for these analyses is:

$$[M]\{\ddot{u}\} + [K]\{u\} = \{0\} \tag{6}$$

where **[M]** and **[K]** are the mass matrix and the stiffness matrix respectively and $\{ \boldsymbol{u} \}$ and $\{ \ddot{\boldsymbol{u}} \}$ correspond to the acceleration and the displacement vectors respectively. The eigenvalue problem is solved to obtain the natural frequencies of the RVE, which are enhanced by the addition of graphene. These results provide valuable insights into the composite's mechanical behavior.

3 Results and Discussion

3.1 Graphene Mechanical Properties

The mechanical response of graphene is known to exhibit significant size-dependence for small dimensions, as established in the literature [9, 10]. In this study, graphene sheets with a uniform thickness of 0.34 nm [8, 11] were considered. Also, Poisson's ratio of 0.4 is assumed for graphene and interphase region. The geometric and material properties of the graphene models are detailed in Table 1. The macroscopic material properties such as the Young's modulus (E_x), and shear moduli (G_{xy}, G_{xz}, G_{yz}) were used for all three principal directions of the graphene sheets. These moduli were calculated following the procedures outlined by Giannopoulos et al. [10].

Table 1. Geometrical features and elastic properties of the graphene sheets utilized

Model	Dimensions (nm)			Moduli (GPa)					
	Length	Width	thickness	E_x	E_y	E_z	G_{xy}	G_{xz}	G_{yz}
1	3	3	0.34	674	663	58.5	434	3.32	3.35
2	5	5	0.34	730	709	58.5	435	3.32	3.45
3	7	7	0.34	730	736	58.5	436	3.32	3.37

3.2 Composite Mechanical Properties

The mechanical properties of the composite material were evaluated by investigating its Young's modulus and shear moduli. Finite element analyses were conducted on the proposed model under carefully imposed boundary conditions, enabling the determination of elastic moduli. The results highlight the significant influence of interphase stiffness on the composite's mechanical behavior, underscoring the importance of accurately modeling the interphase region. The interaction between graphene and the matrix induces stress concentrations, particularly in regions near the reinforcement, leading to non-convergence in mesh sensitivity analyses. This issue arises due to singularity points and numerical challenges at angular regions of the geometry. To mitigate these effects, critical points in the geometry were smoothed, reducing singularity-induced stress concentrations (Fig. 1a). Additionally, ANSYS tools were used to smooth results at the nodes of the 3D mesh elements, further enhancing convergence and stability. The geometry was fully parameterized to optimize computational performance and ensure repeatability. Figure 2a illustrates the boundary conditions applied for calculating the Young's modulus of the composite. Symmetry planes along the x and y directions and displacements applied at specific boundaries facilitated the separation of the RVE from the surrounding nanocomposite material. These conditions provided the reactions at constrained nodes, enabling the calculation of the composite's longitudinal elastic modulus. Similar procedures were used to determine other elastic properties with appropriate boundary conditions, for example the shear moduli Fig. 2b. The interphase properties were defined using an interpolation equation, Eqs. (4), correlating the stiffness of the interphase with the mechanical properties of the matrix and graphene. This heterogeneous modeling approach allowed for a more realistic representation of the interphase behavior. To validate the proposed model, a simplified sandwich-like RVE model was employed (Fig. 1b). This model, characterized by a simpler geometry, produced results comparable to experimental data. The validation demonstrates the reliability of the proposed methodology.

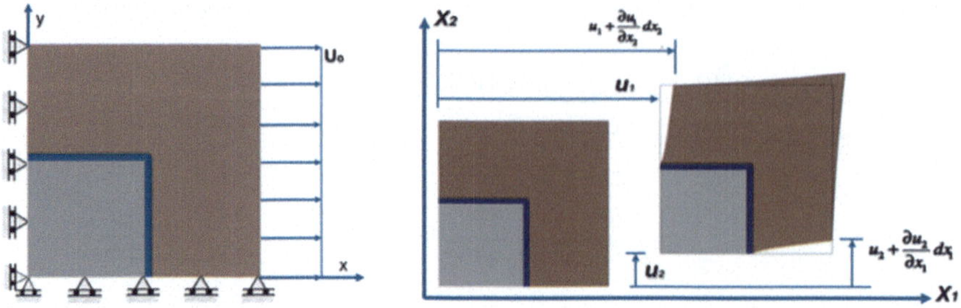

(a) (b)

Fig. 2. (**a**) Embedded RVE cross section with loads and boundary conditions representation, and (**b**) Embedded RVE cross section representation of shear modulus calculation.

3.3 Comparative Analysis of RVEs (Sandwich Model)

Comparisons between the sandwich-like RVE and other theoretical models are presented in Fig. 3. The results include normalized Young's modulus (E_{xc}/E_{xm}) versus graphene volume fraction, highlighting a consistent increase in modulus with growing graphene concentration. Qualitative comparisons with the Halpin–Tsai model and the Rule of Mixtures (ROM) further confirm the trends. The observed differences, particularly the higher predictions of the proposed models, can be attributed to the enhanced representation of the matrix-graphene interphase and the presence of dual interphase sides in the sandwich-like model. Also, this comparison with other numerical analyses like the Spanos et al. [11] shows good agreement with predictions where the graphene and interphase region are represented by discrete structure. Experimental data by Zhao et al. [12] showed deviations due to agglomeration effects, which were not included in the simplified rules-based approaches.

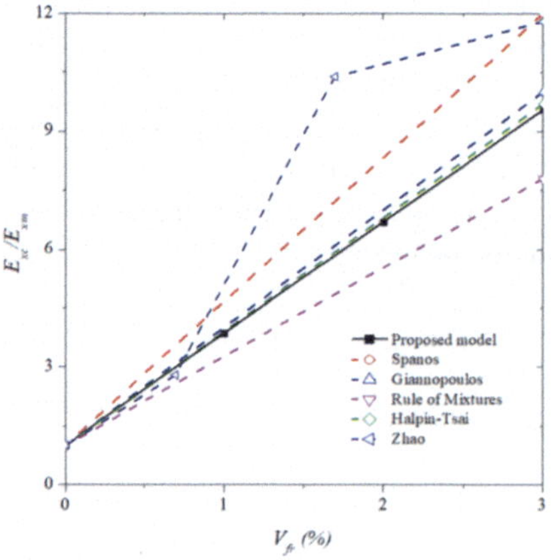

Fig. 3. Normalized Young's Modulus E_{xc}/E_{xm} as a function of the volume fraction in the sandwich model, compared with other published studies and theoretical models (Halpin-Tsai, Rule of Mixtures).

3.4 Parametric Study on Embedded RVE Model

The mechanical behavior of graphene-reinforced nanocomposites was investigated by comparing three different graphene monolayer sizes as reinforcements, analyzed as a function of volume fraction. Assuming a constant Young's modulus for the matrix material, the results presented in Fig. 4a demonstrate a clear increase in the overall composite stiffness with larger graphene reinforcements. This trend aligns with findings reported in similar studies [9, 11]. Notably, the modulus enhancement scales more effectively with higher volume fractions when larger graphene sheets are employed. In Fig. 4b, the influence of interphase stiffness on composite performance was also examined under three scenarios: (a) a perfectly bonded interphase with stiffness equal to graphene, (b) a heterogeneous interphase modeled using the interpolation method and (c) an interphase with stiffness equivalent to the matrix.

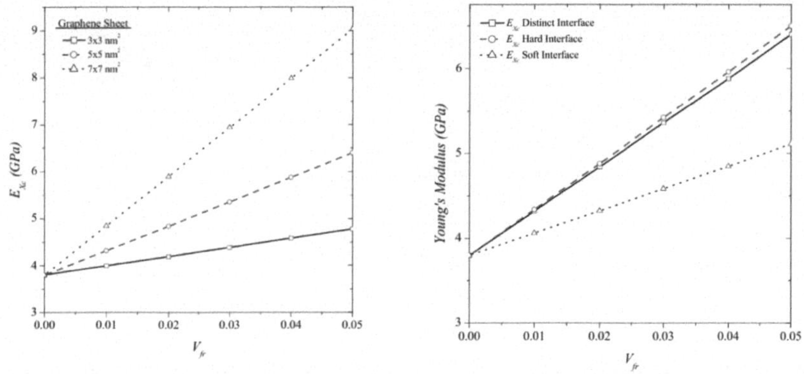

(a) (b)

Fig. 4. (**a**) Normalized Young's modulus in the x direction as a function of graphene volume fraction for different graphene monolayer sizes and, (**b**) presents normalized Young's modulus in the x direction as a function of graphene volume fraction for various interface stiffness conditions.

The heterogeneous interphase model provided a realistic representation, capturing the gradual property transition between graphene and the matrix. The perfect bonding assumption, used as an upper bound, slightly overestimated the composite properties but offered a valuable benchmark for performance evaluation. These findings emphasize the critical role of interphase mechanics in determining the overall performance of graphene-reinforced composites and validate the robustness of the proposed methodology. In Fig. 5a, the shear moduli G_{xyc}, G_{xzc}, and G_{yzc} were evaluated as functions of graphene volume fraction. Results indicate a linear increase in shear moduli with increasing volume fraction. Similarly, the Young's modulus in the x-direction for the embedded model was assessed across volume fractions, presented in Fig. 5b, showing consistent improvements. To further investigate the effect of matrix properties, the normalized Young's modulus of graphene composites was analyzed for different matrix materials as depicted in Fig. 6. This comparison, performed on nanocomposites reinforced with 5×5 nm^2 graphene sheets at the same volume fractions, varied the Young's modulus of the matrix from soft to hard. Results indicate that softer matrices exhibit a more pronounced modulus enhancement due to the interaction with rigid graphene sheets, while harder matrices show a slightly less significant improvement. As expected, increasing the graphene content consistently enhances the composite's stiffness, with the effect being more pronounced in softer matrices. Overall, these findings highlight the dependence of the elastic mechanical properties of graphene-based composites on the graphene sheet size, volume fraction, and interphase region stiffness. The results further underscore that the reinforcement's effectiveness in enhancing stiffness is maximized when paired with softer matrix materials, consistent with observations in prior studies [9, 11].

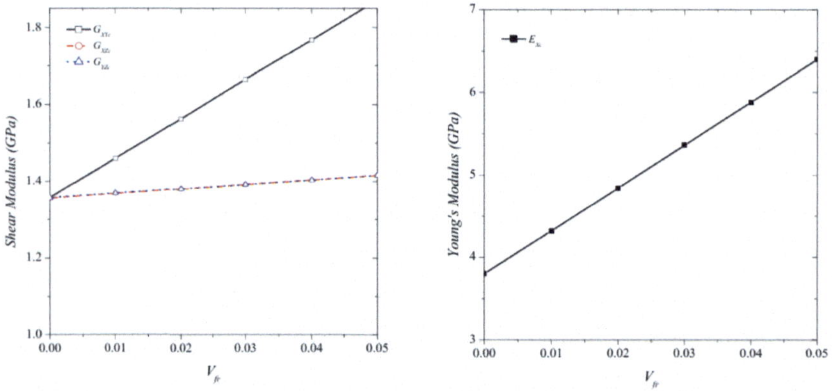

(a) (b)

Fig. 5. (**a**) Shear moduli G_{xyc}, G_{yzc}, G_{zxc} as a function of graphene volume fraction and, (**b**) Normalized Young's modulus in the x direction as a function of graphene volume fraction.

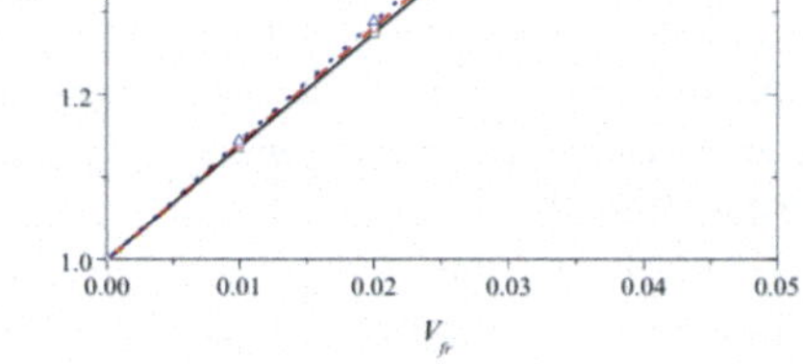

Fig. 6. Normalized Young's Modulus E_{xc}/E_{xm} as a function of the volume fraction for different matrix stiffnesses

3.5 Modal Analysis: Natural Frequencies and Vibrational Behavior

To assess the dynamic behavior of the graphene-reinforced composite, a modal analysis was performed to identify the natural oscillation frequencies and their corresponding mode shapes. This analysis involved solving the eigenvalue problem associated with the RVE, revealing that the inclusion of graphene significantly enhances its natural frequencies. A comparative study was conducted to examine the influence of graphene volume fraction on the fundamental frequency f (Hz) of the composite. The analysis compared the vibration response of an embedded RVE containing varying graphene concentrations (1–5% by volume) with that of a neat resin polymer sample. The results, summarized in Table 2, highlight the impact of graphene on the dynamic performance of the composite. The addition of graphene not only increases the stiffness of the material but also leads to higher natural frequencies, indicating improved resistance to vibration-induced failures. This enhancement underscores the potential of graphene reinforcement for applications requiring superior dynamic stability.

Table 2. Comparison study of the Fundamental frequency *f (Hz)* of embedded RVE nanocomposite

Dimensions (nm)	Vfr (%) (reinforced polymer)	Fundamental Frequency f (Hz) $\times 10^6$	
		Neat polymer	Reinforced polymer
$5 \times 5 \times 2.42$	1	82824	89503
$5 \times 5 \times 1.21$	2	52387	59595
$5 \times 5 \times 0.81$	3	37620	44344
$5 \times 5 \times 0.61$	4	29186	35548
$5 \times 5 \times 0.48$	5	23784	29873

4 Conclusions and Future Work

This study presents an efficient finite element-based micromechanical approach for analyzing graphene-reinforced composites, highlighting their superior mechanical properties, such as increased stiffness, shear strength, and enhanced natural frequencies. These improvements are attributed to effective load transfer facilitated by the interphase region, as modeled within the developed representative volume elements (RVEs). The findings emphasize the critical role of geometric and material parameters, including graphene sheet dimensions, volume fraction, and interphase stiffness, in optimizing composite performance. The implementation of a three-dimensional micromechanical finite element framework demonstrated a balance between computational efficiency and accuracy by avoiding molecular-level complexity. The proposed method effectively modeled the continuum behavior of the matrix, graphene, and interphase, providing a robust and scalable tool for analysing nanocomposite materials. Validation of the framework against a simplified sandwich RVE model confirmed its reliability and underscored its potential applicability to other nanocomposites, such as carbon nanotubes. Future work should focus on incorporating non-linear behaviors and failure mechanisms to enhance the model's predictive capabilities. Experimental validation remains essential to align simulation results with real-world observations, particularly for characterizing the interphase region. Further exploration of interphase mechanics could yield insights into optimizing load transfer, enabling the design of composites with tailored performance. By addressing these aspects, the methodology can be expanded to support broader applications in advanced engineering materials.

References

1. Grujicic, M., Pandurangan, B., Angstadt, D.C., Koudela, K.L., Cheeseman, B.A.: Ballistic-performance optimization of a hybrid carbon-nanotube/E-glass reinforced poly-vinyl-ester-epoxy-matrix composite armor. J. Mater. Sci. **42**(14), 5347–5359 (2007)
2. Soliman, E.M., Sheyka, M.P., Taha, M.R.: Low-velocity impact of thin woven carbon fabric composites incorporating multi-walled carbon nanotubes. Int. J. Impact Eng. **47**, 39–47 (2012)
3. Shahbaz, S.: Istanbul Technical University Graduate School of Science Engineering and Technology. Thesis (2016)
4. Tüzemen, M.Ç., Salamcı, E., Avcı, A.: Enhancing mechanical properties of bolted carbon/epoxy nanocomposites with carbon nanotube, nanoclay, and hybrid loading. Compos. Part B Eng. **128**, 146–154 (2017)
5. Eskizeybek, V.: Fabrication and Investigation of Interlaminar Fracture Behaviour of Woven Glass Fibers/epoxy Nanocomposites Modified With Carbon Nanotubes Chemically Bonded on the Glass Fibers. Unpublished manuscript (2012)
6. Gouda, P.S., Kulkarni, R., Kurbet, S.N., Jawali, D.: Effects of multi-walled carbon nanotubes and graphene on the mechanical properties of hybrid polymer composites. Adv. Mater. Lett. **4**(4), 261–270 (2013)
7. Ciecierska, E., Jurczyk-Kowalska, M., Bazarnik, P., Kulesza, M., Lewandowska, M., Kowalski, M., Krauze, S.: Influence of carbon nanotubes on mechanical properties and structure of rigid polyurethane foam. IOP Conf. Ser.: Mater. Sci. Eng. **64**(1), 012007 (2014) IOP Publishing
8. Chandra, Y., Adhikari, S., Saavedra Flores, E.I., Figiel, Ł.: Advances in finite element modelling of graphene and associated nanostructures. Mater. Sci. Eng. R. Rep. **140**, 100544 (2020)
9. Giannopoulos, G., Kallivokas, I.: Mechanical properties of graphene-based nanocomposites incorporating a hybrid interphase. Finite Elem. Anal. Des. **90**, 31–40 (2014). https://doi.org/10.1016/j.finel.2014.06.008
10. Giannopoulos, G.I., Liosatos, I.A., Moukanidis, A.K.: Parametric study of elastic mechanical properties of graphene nanoribbons by a new structural mechanics approach. Phys. E Low-Dimensional Syst. Nanostruct. **44**, 124–134 (2011)
11. Spanos, K.N., Georgantzinos, S.K., Anifantis, N.K.: Mechanical properties of graphene nanocomposites: a multiscale finite element prediction. Compos. Struct. **132**, 536–544 (2015)
12. Zhao, X., Zhang, Q., Chen, D., Lu, P.: Enhanced mechanical properties of graphene-based poly(vinyl alcohol) composites. Macromolecules. **43**(5), 2357–2363 (2010). https://doi.org/10.1021/ma902862u

13. Georgantzinos, S.K., Giannopoulos, G.I., Markolefas, S.I.: Vibration analysis of carbon fiber-graphene-reinforced hybrid polymer composites using finite element techniques. Materials. **13**(19), 4225 (2020)
14. Georgantzinos, S., Giannopoulos, G., Anifantis, N.: Investigation of stress–strain behavior of single-walled carbon nanotube/rubber composites by a multiscale finite element method. Theor. Appl. Fract. Mech. **52**(3), 158–164 (2009). https://doi.org/10.1016/j.tafmec.2009.09.005

Computational Design, Analysis and Experimental Investigation of the Behavior of 4D Printed Airfoils

Markos G. Tataris(✉), Antonios Fatsis, and Stelios K. Georgantzinos

Aerospace Science and Technology, National and Kapodistrian University of Athens, 34400 Psachna, Greece
mtataris03@gmail.com, sgeor@uoa.com

Abstract. Since the first appearance of additive manufacturing in the late 1980s, significant advancements have been made in this field. Additive manufacturing (AM), also known as 3D printing, has a wide range of applications, including the production of physical prototypes, functional components, and complex biological structures. In recent years, the spotlight has turned towards smart materials, which could change shape or properties in response to external stimuli, leading to the emergence of four-dimensional (4D) printing. This work explores the computational design, analysis, and experimental investigation of 4D printed airfoils made from PLA, a shape memory polymer (SMP). These airfoils can adapt their shape to meet varying aerodynamic conditions and then return to their original form. The study includes an in-depth examination of the potential use of these smart materials in aerospace applications, specifically in the context of sports vehicle aerodynamics. These adaptive airfoils' performance and design were verified by a series of computer simulations using CFD software. Following the calculations, the airfoils were created experimentally using 3D printing technique called Fused Deposition Modeling (FDM). The experimental protocol used the Taguchi De-sign of Experiments (DoE) methodology to examine in detail how different printing factors affect the airfoils' aerodynamic performance and shape memory behavior. To evaluate the viability and efficiency of employing 4D printed airfoils in practical applications, the experimental findings were compared with computer projections. This research demonstrates the potential for significant advancements in the design and functionality of aerodynamic components, paving the way for future innovations in adaptive aerospace structures.

Keywords: 3D printing · 4D printing · smart materials · shape memory materials · airfoils · CFD

1 Introduction

Material Extrusion technology is widely adopted due to its ease of operation and low cost, combined with its ability to produce fully functional components of a product. This technology works by using a filament of thermoplastic polymer material or metal, which is fed into a heated extrusion nozzle that can control the flow rate of the material

N. D. Lagaros et al. (Eds.): ADDOPTML 2024, LNCE 646, pp. 122–133, 2025.
https://doi.org/10.1007/978-3-031-92029-5_12

while simultaneously moving both vertically and horizontally, numerically controlled by Computer-Aided Manufacturing (CAM) software. The nozzle is heated, causing the thermoplastic material to transition into a semi-liquid phase and then be extruded onto the print bed, where it immediately solidifies after extrusion, forming individual layers of the piece from its base to its top. This technology is widely used with two types of polymer filaments: Acrylonitrile Butadiene Styrene (ABS) and Polylactic Acid (PLA), but there is a wide variety of filaments with different properties (e.g., high flexibility, carbon fiber reinforcement, etc.) [1].

Smart materials are defined as materials that have the ability to change their geometry and generally their properties at a later time under the influence of some external stimulus, such as the application of heat or immersion in a water bath. An example of producing three-dimensional printed components from smart materials is self-developing structures and soft robotics systems [2]. The use of smart materials in 3D printing has led to the creation of the term four-dimensional printing (4D printing), meaning the alteration of a component's properties at a future time from that of its three-dimensional printing. These materials, under the influence of external stimuli, exhibit capabilities such as self-assembly, self-healing, geometric memory, and decision-making [3]. Smart materials are classified into different categories depending on the type of stimulation they respond to. One of these categories is thermoresponsive polymers, also known as shape memory polymers (SMPs), which is the focus of this particular work.

Shape Memory Polymers (SMPs) constitute an important part of materials that respond to stimuli and can return to their original shape under the influence of some external stimulus such as pressure, light, or heat. This type of smart material has attracted significant attention in the design and development of smart devices for use in the field of sensors, actuators, and biotechnology. The composition, characterization, application, and behavior mechanisms of SMPs have been thoroughly studied in recent years. Polymers can inherently exhibit shape memory behavior, depending on their rubber elasticity, but with varying rates of deformation recovery.

The behavior of shape memory involves two main steps: shape programming and shape recovery. The programming step for temperature-responsive SMPs begins with heating above a transition temperature (T_t), which is either the glass transition temperature (T_g) for amorphous materials or the melting temperature (T_m) for semi-crystalline materials, followed by a deformation process. With cooling below the transition temperature, the induced shape is maintained while the deformation energy is also stored. The increase in temperature above the transition temperature activates the recovery of the configuration entropy, and in turn, the primary shape is recovered (recovery step). The remarkable advancements in 3D printing technologies allow for the printing of various SMPs, such as polycaprolactone (PCL), polyurethane (PU), PLA, which will be used in this thesis, acrylic-based and epoxy-based resins, etc., for the development of smart structures, such as actuators, robots, and medical devices. Additionally, materials with different transition temperatures can be used in combination with each other to print active composite structures with the ability to remember multiple shapes.

Also, depending on the number of shapes they can memorize, SMPs can be classified into three main categories: (i) one-way or dual SMPs, which have a permanent shape and a temporary shape, (ii) multiple-way SMPs that have a permanent shape and (n

- 1) temporary shapes, and (iii) two-way SMPs that are capable of a single reversible two-way transition between temporary shapes The PLA material that will be used in this thesis belongs to the category of one-way (dual) SMPs [4].

In the present paper, a comprehensive investigation into the computational design, analysis, and experimental evaluation of 4D printed airfoils is presented. These airfoils, made from PLA, are designed to adapt their shape to varying aerodynamic conditions and recover their original geometry when activated. The study focuses on the potential application of these adaptive structures in aerospace engineering, with a specific emphasis on optimizing aerodynamic performance for sports vehicle applications. A combination of computational fluid dynamics (CFD) simulations and experimental testing using the Taguchi Design of Experiments (DoE) methodology is employed to evaluate the influence of 3D printing parameters on the shape memory behavior and aerodynamic efficiency of the airfoils. This work aims to provide a foundation for the development of advanced adaptive aerospace components, contributing to the broader field of smart material integration in additive manufacturing.

2 Methodology

2.1 One-Way SMPs

Unidirectional (or bidirectional) thermo-responsive SMPs exhibit the ability to recover a "permanent" shape (i.e., the shape provided through conventional processing) from a "temporary" shape (i.e., the shape provided through a process called "programming") when heated. The one-way SME phenomenon is associated with a transition temperature (e.g., glass transition, melting, crystallization, or clearing temperature, depending on the type of polymer) and arises from the combination of an applied thermo-mechanical history (i.e., the shape memory cycle) with the polymer's macromolecular architecture. The material is initially shaped into a permanent form through conventional processing. Next, the material undergoes a cycle called "programming" to be shaped into a temporary form (Fig. 1) [5].

The programming step consists of: (i) Heating the material to a temperature $T_1 > T_{trans}$, which leads to an increase in the mobility of the polymer molecular chains, (ii) Deformation of the material, which causes an orientation of the polymer molecular chains and a variation in the macromolecular configuration, accompanied by a change in entropy, leaving the polymer in an unstable high-energy state, (iii) Cooling the material to a temperature $T_2 < T_{trans}$, which leads to the restriction of molecular mobility, trapping of the high-energy state, and "freezing" of the temporary shape, and (iv) Removal of the applied load.

Temperatures T_1 and T_2 are typically 15–40 °C above and below the transition temperature, respectively. In addition, the one-way SME phenomenon is irreversible, which means that after the permanent shape has been restored, cooling cannot change the SMP back to its temporary shape; instead, a new programming step is needed to change it back to the desired temporary shape. The only restriction on the shape memory cycle is the possible slow deterioration of this material characteristic.

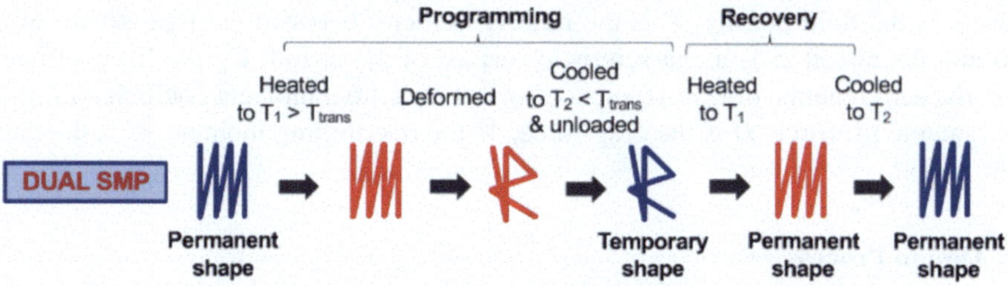

Fig. 1. Representation of a memory cycle shape for one-way or dual SMPs.

2.2 Airfoil Aerodynamics

When an airfoil is in motion within a fluid, the fluid tends to adhere to the surface of the airfoil and move around it, on both its upper and lower surfaces, while at the leading edge, where the flow first impinges, a stagnation point ($V = 0$ m/s) appears. The airfoil is shaped so that the airflow accelerates over its upper surface, and consequently, the pressure decreases. Therefore, the pressure on the upper surface (suction side) is lower than on the lower surface (pressure side), creating a pressure difference between the two surfaces that results in a force directed from the surface with higher pressure to the one with lower pressure, which is called aerodynamic lift (*Lift Force—L*). Also, the airfoil, while moving through the fluid, experiences a force that opposes its motion and is called drag or aerodynamic resistance (*Drag Force—D*). In the case of racing vehicles, the wings are mounted with the suction surface on the underside, so that the lift force produced has an opposite direction and contributes to increasing the vehicles' grip. This force is called a negative aerodynamic lift. (Negative Lift Force). In airfoils, a twisting moment also appears, which is called a pitching moment (*Pitching Moment—M*) and arises from the assumption that aerodynamic forces do not act at the center of pressure, but at the aerodynamic center of the airfoil [6].

The equations that describe drag, aerodynamic lift, and pitching moment are as follows:

$$D = \frac{1}{2}\rho V^2 S C_D \tag{1}$$

$$L = \frac{1}{2}\rho V^2 S C_L \tag{2}$$

$$M = \frac{1}{2}\rho V^2 S C C_M \tag{3}$$

and therefore, the following aerodynamic coefficients arise:

$$C_D = D/\left(1/2\rho V^2 S\right) \tag{4}$$

$$C_L = L/\left(1/2\rho V^2 S\right) \tag{5}$$

$$C_M = M/\left(1/2\rho V^2 S C\right) \tag{6}$$

where ρ is the fluid density, V is the relative velocity between the free stream of the fluid and the airfoil, S is a characteristic surface of the airfoil, C_L the lift coefficient, C_D is the aerodynamic drag coefficient, C_M is the pitching moment coefficient, L is the aerodynamic lift force, D is the drag force, M the overturning moment, C is the chord of the airfoil.

2.3 Design Process

For this work, the 4-digit airfoil NACA 2412 was selected from the NACA database, which is a common choice for the flaps of airfoils in racing vehicles [7]. Initially, a text file was extracted from the database containing the points that describe the surface of the airfoil for a chord length of 1 m. In this database, there are two types of files that contain the coordinates of these points. The first type is the Selig format, where the points of the airfoil start from the trailing edge and move towards the leading edge along the upper surface, and return to the trailing edge along the lower surface. The second type of file is the Lednicer format, where the description of the airfoil for the upper and lower surfaces is done separately, and both start from the leading edge and end at the trailing edge. For this specific model, the Selig format type was chosen. After exporting the specific file, it was processed in Microsoft Excel to ensure that the values matched the desired chord length, which in this case was chosen to be $c = 100$ mm, and it was saved in the form of NACA2412.txt.

For the creation of the 3D model, 3D modeling software was used, in which the NACA2412.txt file was initially imported to create the airfoil curve, as shown in Fig. 2. Then, from this curve, the surface of the airfoil was created, which was offset inward by $t = 2$ mm to create the thickness of the airfoil's wall. Additionally, a depth of $b = 5$ mm was chosen for the airfoil along the z-axis, so that the chord length to depth ratio is sufficiently small ($b/c = 1/20$) and the subsequent computational modeling can be considered two-dimensional.

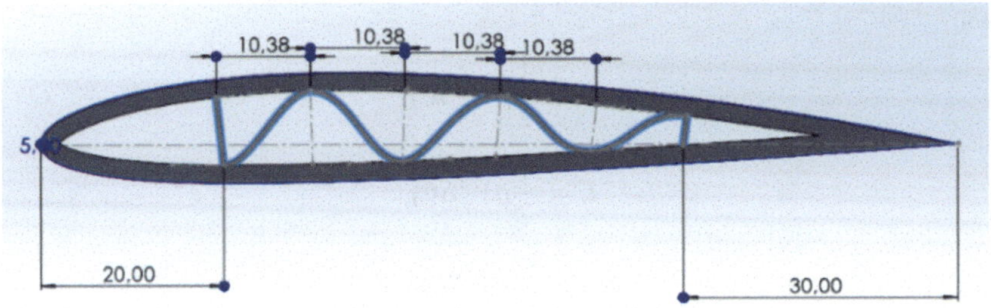

Fig. 2. Final 3D model of the geometry of the examined airfoil NACA 2412.

The next stage of the design involved defining the internal geometry of the airfoil. In the initial stage, test prints were made with the airfoil having no internal geometry. However, after conducting some experiments, it was observed that the absence of geometry inside led to a rapid weakening of the shape memory phenomenon. Based on the literature [8], a damped sinusoidal wave function was chosen as the internal geometry.

The geometry was designed in the region from 20% of the chord to 70% of the chord, and the distance from each peak to each trough is 10.38 mm.

2.4 Computational Simulation Model

For the calculation of the aerodynamic coefficients of the printed airfoils, a two-dimensional computational model must first be created that verifies the predicted values of the aerodynamic coefficients for the NACA 2412 airfoil. For this purpose, a CFD computational simulation software was used. To create the appropriate computational model, a closed surface was initially designed in the 3D modelling software, outlining the selected airfoil, as the internal geometry does not affect the computational model. Then, a STEPAP14 file was exported from the 3D design program to import the airfoil surface geometry into the ANSYS. In the ANSYS, the DesignModeler was used to design the control volume of the computational model. The radius of the semicircle is 7.5 airfoil chords, and the length of the rectangular section on the chord line of the airfoil is 15 airfoil chords (Fig. 3).

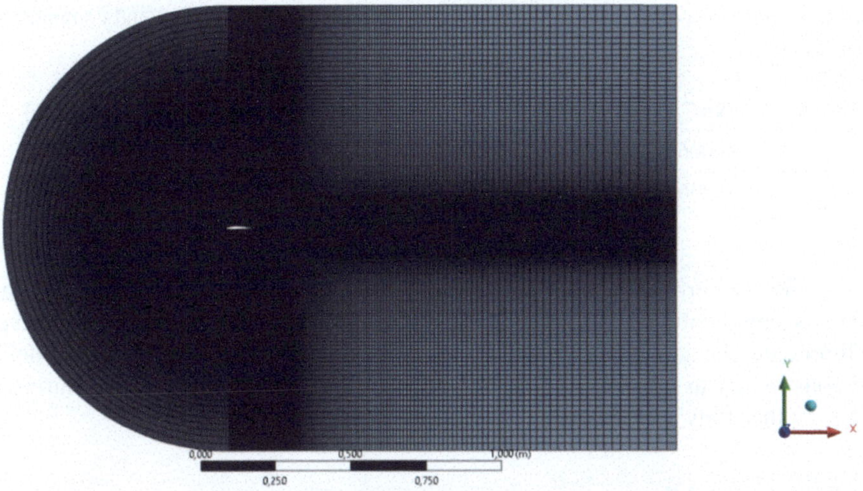

Fig. 3. Final 3D computational model of the geometry of the examined airfoil NACA 2412.

After the creation of the above control volume, the computational grid was constructed. The refinement of the computational grid near the airfoil surface was chosen in such a way that the grid parameter $Y+$ is less than one ($Y+ < 1$).

To determine the required number of computational cells, an iterative process was followed in which the number of cells was continuously adjusted until the aerodynamic coefficients converged to a certain value.

Then, the determination of the solution method to be used in the ANSYS FLUENT, which is a subproduct of ANSYS and intended for use in CFD, followed. From the available models in ANSYS FLUENT, the *SST k-omega* method was chosen for this specific solution.

The computational model was designed to calculate the aerodynamic coefficients of the NACA 2412 airfoil with the characteristics of the incoming airflow as presented in Table 1.

Table 1. Values of the parameters used in the CFD model.

Reynolds number (Re)	Angle of attack (α)	Air density (ρ)	Dynamic viscosity (μ)	Air velocity (V)	Air temperature (T)
5×10^5	0°	1.225 kg/m^3	0.00001802 kg/(m·s)	73.55 m/s	288.16 K

For the specified flow conditions, the aerodynamic coefficients projected by the NACA database are shown in Table 2. The computational model, using a maximum of = 1000 iterations, produced the results listed alongside for comparison.

Table 2. Comparison of aerodynamic coefficients from the NACA database and computational model results.

Parameter	NACA Database value	Computational Model result	Error (%)
C_D	0.00630	0.00655	3.97
C_L	0.2360	0.21108	−10.56

The observed errors are attributed to the limitation in the number of finite elements used in the computational simulation, which impacts the precision of the results. Despite this limitation, the aerodynamic coefficients derived from the computational model are used consistently as a baseline for the comparison of 3D-printed airfoils, ensuring the validity of the study's findings.

2.5 Experimental Process

The experimental setup utilized in this study included a thermostatically controlled water bath (WATER BATH DK-420), enabling precise regulation of immersion temperature during both the deformation and shape recovery stages. Based on the literature, the deformation temperature (T_{deform}) should be at least 5–10 °C above the glass transition temperature (T_g) of the material, while the activation temperature ($T_{\text{activation}}$) should exceed by approximately 10 °C. For the PLA material used here, with $T_g = 60$ °C, three temperature scenarios were established:

- $T_{\text{deform}} = 65$ °C and $T_{\text{activation}} = 75$ °C
- $T_{\text{deform}} = 75$ °C and $T_{\text{activation}} = 85$ °C
- $T_{\text{deform}} = 85$ °C and $T_{\text{activation}} = 95$ °C

The 3D printing parameters investigated in this study to assess their impact on the Shape Memory Effect (SME) include Layer Height, Print Speed, and Nozzle Temperature. These parameters are summarized in Table 3 software to generate the required G-code. All prints were conducted using a Creality CR10 Mini 3D printer. The fixed parameters, including a nozzle diameter of 0.4 mm, a line width of 0.4 mm, a wall thickness of 0.8 mm, 4 top and bottom layers, and an infill density of 100%, were kept constant across all experiments to ensure consistency. Meanwhile, the variable parameters—layer height, print speed, and nozzle temperature—were systematically adjusted to evaluate their effects on the Shape Memory Effect (SME).

Table 3. 3D Printing parameters levels used in the study.

Layer height	Print speed	Nozzle temperature
0.1 mm, 0.2 mm, 0.3 mm	20 mm/s, 40 mm/s, 60 mm/s	190 °C, 200 °C, 210 °C

Given the 4 parameters (*Layer Height, Print Speed, Nozzle Temperature, Deformation—Activation Temperature*) and their respective 3 values for each one, the total number of combinations would be $3^4 = 81$. To reduce the number of experiments, the Taguchi Experimental Design method was applied. (Taguchi DOE). This method allows for the exploration of the parameter spectrum with fewer experiments, while simultaneously maintaining the reliability of the results. For the implementation of the Taguchi DOE, the Minitab software was used, which created an experimental table requiring the execution of only 9 prints instead of 81. The application of the method saves time and resources while ensuring the extraction of conclusions regarding the impact of the parameters on the printing process.

The procedure followed during the execution of the experiments was as follows:

1. Heating the water bath to the desired temperature T_{deform}
2. Immersion of each airfoil for 3 s and its deformation in the first valley of the wave function by applying pressure with the help of a pair of pliers.
3. Removal of the airfoil from the water bath and cool it to room temperature, so that the induced deformation remains.
4. Take a photograph of the deformed airfoil.
5. Heating the water bath to the desired temperature $T_{activation}$
6. Immersion of each airfoil for 1 min to activate the shape memory effect.
7. Removal of the airfoil from the water bath with the help of tweezers and cooling it to room temperature.
8. Taking a photo of the recovered shape of the airfoil in plan view.

For the recalculation of the aerodynamic coefficients of the airfoils after conducting the experiments, the computational model described above was used. Initially, the top view photograph of each airfoil, after shape recovery, was imported into the 3D modeling software to design the outline of the airfoil surface. Then, from this contour, a two-dimensional surface, which was exported as a STEPAP14 file to transfer the geometry to the existing model in the computational simulation software Ansys Fluent, which calculated the drag coefficient C_D and the lift coefficient C_L. The above process was repeated for all the airfoils.

3 Results

Figure 4 provides a visual summary of the experimental procedure applied to one of the 3D-printed airfoils, illustrating the stages of initial 3D printing (A), deformation after heating (B), and shape recovery (C).

Fig. 4. (**a**) 3D printing, (**b**) deformation after heating, and (**c**) shape recovery

Table 4 presents a comparison of the aerodynamic coefficients (C_D and C_L) for each airfoil with the values derived from the computational simulation model. The table also includes the percentage deviations of the experimental coefficients from the baseline computational values.

Table 4. Comparison of aerodynamic coefficients and change calculation.

No.	C_D	C_L	$C_{D,0}$	$C_{L,0}$	C_D Change (%)	C_L Change (%)
1	0.0082248	0.24709	0.0065500	0.21108184	25.57	17.06
2	0.0084435	0.2447	0.0065500	0.21108184	28.91	15.93
3	0.010782	0.2689	0.0065500	0.21108184	64.61	27.39
4	0.0083125	0.21377	0.0065500	0.21108184	26.91	1.27
5	0.0086998	0.2149	0.0065500	0.21108184	32.82	1.81
6	0.0093471	0.26509	0.0065500	0.21108184	42.70	25.59
7	0.0070783	0.25577	0.0065500	0.21108184	8.07	21.17
8	0.0090978	0.21802	0.0065500	0.21108184	38.90	3.29
9	0.0070309	0.22785	0.0065500	0.21108184	7.34	7.94

In Fig. 5, the main effects plots illustrate the influence of the four examined parameters—Layer Height, Print Speed, Nozzle Temperature, and Activation Temperature—on the deviation of the lift coefficient (C_L). The goal is to minimize the error. The left plot shows the mean effect of each parameter, while the right plot presents the corresponding Signal-to-Noise (SN) ratios.

In Fig. 6, similar main effects plots are provided for the aerodynamic drag coefficient (C_D). The left plot highlights the mean effects of the parameters on the C_D deviation, while the right plot demonstrates the SN ratios, where a smaller value is better for minimizing the error.

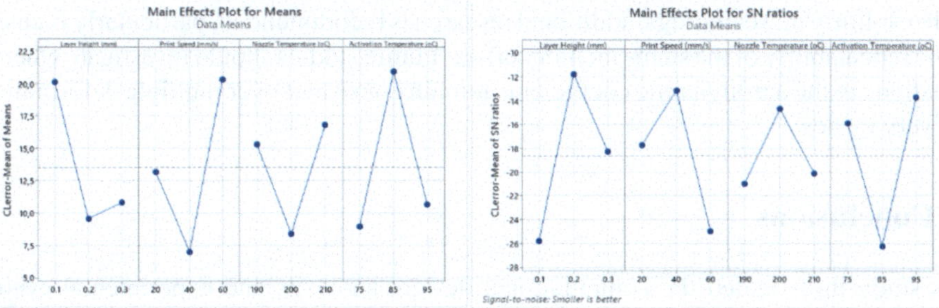

Fig. 5. Effect of the printing parameters on the error of the lift coefficient C_L with the aim of minimizing the error.

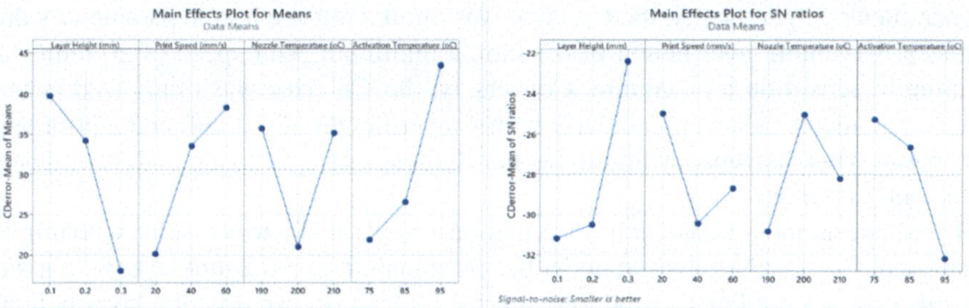

Fig. 6. Effect of the printing parameters on the error of the lift coefficient C_D with the aim of minimizing the error.

The analysis reveals that specific combinations of parameters have differing impacts on C_L and C_D errors, emphasizing the need to optimize printing and activation settings to achieve the desired aerodynamic performance with minimal deviation.

he experimental investigation revealed distinct behaviors for the aerodynamic coefficients C_D and C_L in response to variations in the selected parameters. Specifically, optimization of C_D error was achieved with extreme parameter values—maximum layer height, minimum print speed, and minimum activation temperature. In contrast, the C_L error was minimized at intermediate parameter values, making it challenging to identify a single set of optimal conditions for both coefficients simultaneously.

Notably, the shape memory behavior of PLA demonstrated satisfactory results for the C_L coefficient when using a layer height of 0.2 mm and a print speed of 20 mm/s, achieving a small error of 1–2% compared to the baseline shape. For the C_D coefficient, the lowest error (7–8%) occurred with a layer height of 0.3 mm and the same print speed of 20 mm/s. These findings suggest that a combination of high layer height and low print speed generally favors the process. However, it was also observed that when the C_L error was minimized, the C_D error tended to increase, and vice versa. The only instance where both errors were relatively low (7–8%) occurred in experiment 9.

Based on these results, it can be concluded that the use of PLA for harnessing the shape memory effect in airfoils is effective when the focus is on minimizing error in one of the two aerodynamic coefficients (C_L or C_D) rather than both simultaneously. To further solidify these findings, additional research is recommended, particularly focusing on the repeatability of the shape memory effect. Future studies should investigate whether deviations in the aerodynamic coefficients remain consistent over multiple deformation-recovery cycles.

4 Conclusions

This study investigated the computational design, analysis, and experimental evaluation of 4D-printed airfoils made from PLA, a thermoresponsive shape memory polymer (SMP), to assess their aerodynamic performance following a shape deformation-recovery cycle. The findings revealed distinct behaviors for the aerodynamic coefficients C_D (drag) and C_L (lift) in response to variations in the examined 3D printing and activation parameters. Specifically, the C_D error was minimized at extreme parameter values, such as a maximum layer height of 0.3 mm, a minimum print speed of 20 mm/s, and a minimum activation temperature. Conversely, the C_L error was minimized at intermediate parameter values, particularly with a layer height of 0.2 mm and a print speed of 20 mm/s. This discrepancy highlights the challenge of optimizing both aerodynamic coefficients simultaneously.

The shape memory behavior of PLA demonstrated satisfactory results, with minimal deviation in C_L (1–2%) achieved under the combination of a 0.2 mm layer height and a 20 mm/s print speed. For C_D, the lowest error (7–8%) occurred with a layer height of 0.3 mm and the same print speed. These results suggest that higher layer heights and lower print speeds generally favor the overall process. However, a clear trade-off was observed: when C_L exhibited minimal error, C_D tended to show larger deviations, and vice versa. The only scenario where both coefficients exhibited relatively low errors (7–8%) occurred in experiment 9, indicating a potential balance under specific parameter settings.

Overall, this study demonstrated the feasibility of using PLA for 4D-printed adaptive airfoils, particularly when prioritizing the optimization of one aerodynamic coefficient (C_L or C_D) independently. The results suggest that PLA can be effectively applied in adaptive aerodynamic structures where performance metrics can be selectively prioritized. To further solidify these conclusions, future research should focus on assessing the repeatability of the shape memory effect over multiple deformation-recovery cycles and evaluating the stability of aerodynamic coefficients over time. This work highlights the potential of integrating smart materials like PLA into 4D-printed airfoils, paving the way for advancements in adaptive aerospace components.

References

1. Al-Maliki, J.Q., Ahmed, A., Arya, S., Gupta, V., Furukawa, H., Khosla, A.: The processes and technologies of 3D printing. Int. J. Adv. Comput. Sci. Technol. **4**(10), 161–165 (2015)
2. Ahmed, A., Arya, S., Gupta, V., Furukawa, H., Khosla, A.: 4D printing: fundamentals, materials, applications and challenges. Polymer. **228**(1), 1–25 (2021)
3. Georgantzinos, S.K., Giannopoulos, G.I., Bakalis, P.A.: Additive manufacturing for effective smart structures: the idea of 6D printing. Compos. Sci. **5**(119), 1–11 (2021)

4. Mehrpouya, M., Vahabi, H., Janbaz, S., Darafsheh, A., Mazur, T.R., Ramakrishna, S.: 4D printing of shape memory polylactic acid (PLA). Polymer. **230**(1), 1–17 (2021)
5. Scalet, G.: Two-way and multiple-way shape memory polymers for soft robotics: an overview. Actuators. **9**(10), 1–33 (2020)
6. Rohacs, J.: Goal- and object-oriented models of the aerodynamic coefficients. In: Flight Physics—Models, Techniques and Technologies, pp. 159–191. Intech, London (2018)
7. Azmi, A.R.S., Sapit, A., Mohammed, A.N., Razali, M.A., Sadikin, A., Nordin, N.: Study on airflow characteristics of rear wing of F1. IOP Conf. Ser. Mater. Sci. Eng. **243**(1), 1–13 (2017)
8. Hoa, S., Abdali, M., Jasmin, A., Radeschi, D., Prats, V., Faour, H., Kobaissi, B.: Development of a new flexible wing concept for unmanned aerial vehicle using corrugated core made by 4D printing of composites. Compos. Struct. **290**(1), 1–16 (2022)

Introducing T.O.DAY: A Novel Approach to 3D Printed Facade Design

Dimitrios N. Gonidakis and Nikolaos D. Lagaros(✉)

School of Civil Engineering, Institute of Structural Analysis and Antiseismic Research, National Technical University of Athens, Athens, Greece
nlagaros@central.ntua.gr

Abstract. T.O.DAY (Topology Optimization in DAYlight Analysis) is a novel facade design approach that uses topology optimization techniques to maximize daylight penetration for 3D-printed structures. By strategically rearranging the voids within a facade, T.O.DAY could significantly improve the amount of natural light reaching the interior of a building. Additionally, it could reduce reliance on artificial lighting and energy consumption. All these could improve daylighting by introducing daylight analysis into early design stages, allowing for a more holistic design approach.

T.O.DAY's methodology is based on traditional topology optimization, with a key difference: instead of focusing on structural strength, it optimizes daylight metrics. The methodology involves dividing the facade into finite elements and identifying the optimal locations of voids that enhance daylighting.

T.O.DAY is a design tool that can lead to sustainable facade design in terms of material efficiency, energy consumption, and daylight improvement, respecting the comfort of building users.

Keywords: Topology Optimization · Daylight Analysis · Early-Design Stages · 3D-Printing · Sustainability

1 Introduction

One of the major concerns in building design is sustainability. Architects and Engineers are searching for the optimum layouts that comply with aesthetics and functionality. To achieve that, many new tools have emerged that assist this task. Many software programs that calculate and simulate Natural Daylight, Ventilation, and thermal loads can provide data that will affect the design process [1].

A building's envelope is critical to creating sustainable constructions and crucial to end-user comfort. It is like the "protective skin" that separates a building from the exterior situations. How a façade is designed can improve or worsen the end-user's life and health. Additionally, it could alter a building's CO_2 footprint by minimizing the energy and material that it needs to be constructed and by minimizing the energy that the users spend daily [2].

According to the en.lighten initiative of UNFCCC and IEA, electricity for lighting accounts for approximately 15% of global power consumption. Therefore, it is crucial

N. D. Lagaros et al. (Eds.): ADDOPTML 2024, LNCE 646, pp. 134–144, 2025.
https://doi.org/10.1007/978-3-031-92029-5_13

to minimize the period that we use artificial light by creating well-lit spaces with natural daylight. One of the most complicated parameters that affect the façade's design is daylight. By thoughtfully designing a façade, we could refine a space's natural illumination and significantly reduce the use of artificial lighting, which could benefit both end-user's health and energy reduction [3, 4].

Many different software that calculate and simulate Natural Daylight can provide data that will affect the design process. The only drawback of this procedure is that it normally takes place at the last stages of design by "correcting" the already designed layout [5].

T.O.DAY comes as a positive response to these issues. It is a façade design tool that could customize its layout in the early design stages. Taking topology optimization as a reference, minimizing a geometry's volume by rearranging the solid elements to retain the geometry's structural integrity [6], T.O.DAY's goal is to minimize geometry's volume by rearranging void elements. This method promises to maximize a space's illumination by minimizing material and energy consumption during construction. Moreover, it could facilitate visual comfort and prolong the space's illumination by minimizing energy consumption due to artificial lighting.

This design process could offer customized facades that can only be produced by additive manufacturing tools. By doing that, material waste, energy, and time can be more controlled than by using conventional building techniques.

2 Daylight: Importance, Calculation Methods

Daylight or natural light is the amount of light our eyes perceive during daytime. It is a sum of the sun's light, skylight, reflections from the exterior surroundings, and reflections from the interior surfaces. The last term applies when analyzing interior spaces. A well-designed building envelope that maximizes natural daylight during the daytime, besides reducing artificial light and minimizing CO_2 footprint, is linked to the building's users' health. Various calculation and representation methods have emerged during the last 50 years to approach daylight design, with the most popular being Annual-Based Daylight Analysis. These methods provide insight into how a design will respond to natural daylight [7].

2.1 Daylight and Health

Humans' physical and mental health are connected to Daylight. Sufficient daylight exposure is responsible for visual comfort and circadian rhythm regulation. Well-lit spaces are beneficial for end-users' eye health. Additionally, exposure to natural light helps to develop healthy sleep patterns and mitigate the risk of sleep disorders such as insomnia, which is linked to various health issues, such as depression, anxiety, and cardiovascular diseases. So, incorporating daylight into facade designs apart from an architectural design tool could be a fundamental aspect of creating healthy and sustainable indoor environments [8, 9].

2.2 Calculation Methods

Annual daylight analysis is the most popular method for evaluating daylighting performance in buildings. This method uses climate-based weather data to create a dynamic representation of the sky, taking into consideration variations such as cloud density, sun position, and atmospheric conditions throughout the year [10].

The essential components to perform an annual daylight analysis are a sky model, whose geometry is based on popular sky models such as CIE Standard Overcast Sky, CIE Clear Sky, and the more sophisticated Perez Sky Model [11] that are updated with each region's weather data, a building's envelope geometry with windows and a Horizontal plane (task-plane) (Figs. 1 and 2).

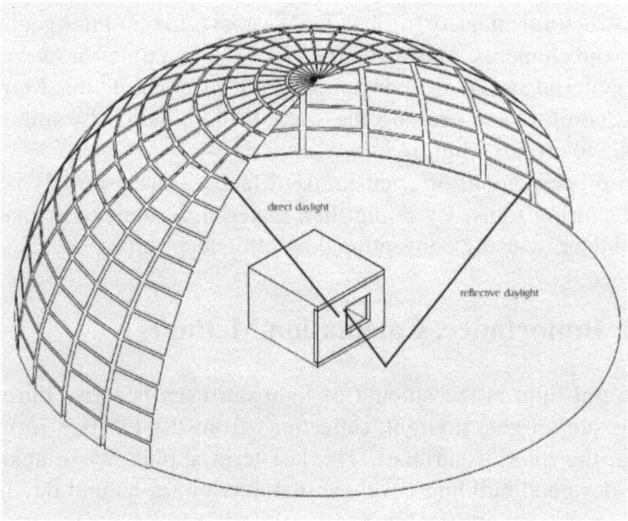

Fig. 1. A Diagram of the computational model representation of climate-based daylight analysis

Once these components are designed, a radiosity-based algorithm will calculate the light distribution over the task plane. This algorithm thoroughly traces the path of light rays interacting with the interior surfaces, considering reflection, absorption, and transmission factors. This complicated process enables the calculation of illuminance levels at various points on the horizontal working plane, providing a detailed understanding of the spatial distribution of daylight [12, 13].

The outcomes of annual daylight analysis are percentages of various annual key metrics. The most popular key metrics are Daylight Autonomy (DA), and Useful Daylight Illuminance (UDI). Daylight Autonomy (DA) is an annual metric that quantifies the percentage of occupied hours during which a specific point on the working plane receives more than 300 lux of daylight without the need for supplementary artificial lighting. A higher DA value indicates a greater reliance on natural light and reduced energy consumption associated with artificial lighting. Useful Daylight Illuminance (UDI): is a metric that defines a desirable range of illuminance levels for visual comfort and productivity. It is typically between a range of illuminance values, such as between 300 lux and

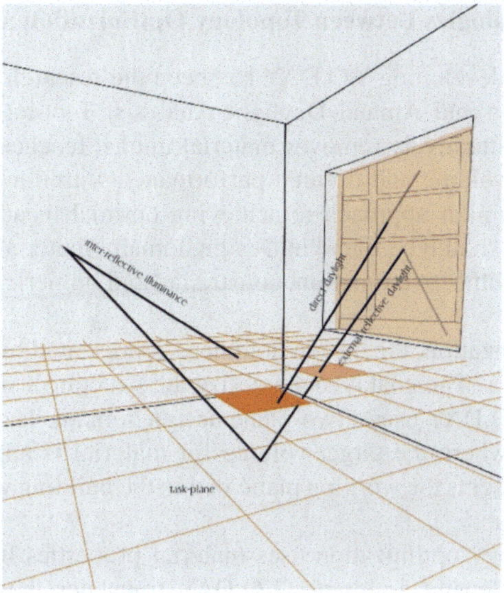

Fig. 2. A diagram of the factors that influence the daylight performance inside a building

2000 lux, where occupants can comfortably perform visual tasks without experiencing glare or discomfort [14].

3 T.O.DAY

T.O.DAY is a topology optimization tool that aims to maximize the annual daylight percentage over a Horizontal task plane A by minimizing the design's domain volume B

$$\text{Max} \int_A UDI\ dA\ or \int_A DA\ dA \tag{1}$$

$$Subjected\ to: V^* - \sum_{i=1}^{N} V_i x_i \tag{2}$$

$$x = 1\ or\ 0\ x \epsilon B \tag{3}$$

Where UDI and DA are Useful Daylight Illumination and Daylight Autonomy, respectively, A is the Horizontal Working task plane where the daylight metrics are calculated, B is a vertical plane that represents a building's envelope wall (in other words a building's façade) and V* is the target Volume that the designer desires to reduce the B plane.

3.1 Finding the Analogies Between Topology Optimization and T.O.DAY

An important step in developing T.O.DAY has been the research of analogies between Topology optimization and Annual Daylight Analysis. Topology Optimization is an iterative process that iteratively removes material until it reaches a desired shape with a targeted minimum volume and optimal performance within a given design domain. T.O.DAY utilizes a similar approach to achieving optimal façade design for daylight access. The analogies should be found in Design domain, material properties, loads and boundary conditions, objective function, constraints and numerical evaluation methods [15].

In topology optimization, the design domain is represented by a single plane where the optimization occurs. The goal is to maximize the structure's stiffness within a target volume. Similarly, T.O.DAY uses a two-plane design domain. The first plane represents the building facade, where the target volume for material is applied, and the second horizontal plane represents the working plane within the building where daylight metrics are evaluated.

Traditional topology optimization uses material properties like Young's modulus, Poisson's ratio, and density as inputs. T.O.DAY translates these properties into the design domain's location and orientation directly influencing daylight penetration into the building.

In topology optimization, loads and boundary conditions are the external forces and constraints applied to the design, such as supports, fixed boundaries, and external loads. T.O.DAY translates these into sky conditions, such as weather data, which significantly impacts the range and availability of daylight throughout the year.

The objective function in topology optimization is a mathematical expression defining the optimization goal, often maximizing stiffness. Correspondingly, T.O.DAY's objective function is to maximize the individual element's DA or UDI of the entire working plane. This ensures sufficient daylight reaches all areas while optimizing overall performance.

Topology optimization employs limitations or requirements, such as stress limits, displacement constraints, or manufacturing limitations. In similar concept, T.O.DAY utilizes a minimum acceptable Daylight Autonomy for each element on the working plane as a constraint. This ensures that all areas receive adequate natural light.

Traditional topology optimization relies on Finite Element Analysis (FEA) to provide information about stresses, strains, displacements, and other relevant parameters. Analogously, T.O.DAY leverages annual daylight analysis to generate information about various annual daylight metrics, such as DA and Useful Daylight Illuminance (UDI).

By drawing these analogies with topology optimization, T.O.DAY establishes a novel framework for optimizing building facade design to maximize daylight penetration while minimizing material usage. This approach can significantly improve energy efficiency and occupant comfort in buildings.

3.2 T.O.DAY Algorithm

Building upon the established principles of daylight analysis and drawing analogies with topology optimization, T.O.DAY presents a novel algorithm for optimizing building facade design. This algorithm iteratively removes facade elements to achieve a balance between maximizing daylight penetration and minimizing material usage.

Core Functionality

- Initial Daylight Analysis:
  - The algorithm commences by conducting an annual daylight analysis on each individual finite element of the facade.
- This analysis generates two crucial metrics for each element:
  - A matrix containing annual daylight metrics, typically Daylight Autonomy (DA) or Useful Daylight Illuminance (UDI) values.
  - A uniformity number is introduced in T.O.DAY which is a division number of the matrix's mean value by its maximum value. A higher uniformity number indicates a more even distribution of daylight across the facade elements.

$$A = \begin{bmatrix} a_{11} & \cdots & a_{1m} \\ \vdots & \ddots & \vdots \\ a_{n1} & \cdots & a_{nm} \end{bmatrix} \tag{4}$$

$$uniformity = \frac{mean(A)}{\max(A)} \tag{5}$$

These metrics provide insights into the amount and distribution of daylight reaching each facade throughout the year.

Initial Element Removal

- To promote a more uniform distribution of daylight within the building, the algorithm strategically removes a small percentage (5–10%) of facade elements during the first iteration.
- These elements are chosen based on their high uniformity values, indicating they contribute less to variations in daylight distribution.

Iterative Optimization

- The algorithm then enters an iterative loop, progressively removing facade elements at a controlled rate (0.05–1%) in each iteration.
- Within each iteration:
  - An annual daylight analysis is performed to identify the darkest spots on the working plane within the building.
  - The algorithm then identifies the facade elements that most effectively illuminate these dark areas.
  - These elements are subsequently removed, and the facade geometry is updated accordingly.

- After each element removal, the algorithm checks if the target objectives have been achieved:
  - Target volume of the facade (minimizing material usage).
  - Mean daylight level on the working plane (ensuring adequate overall illumination).

Convergence and Outcome

- The iterative process continues until the predefined target volume and mean daylight

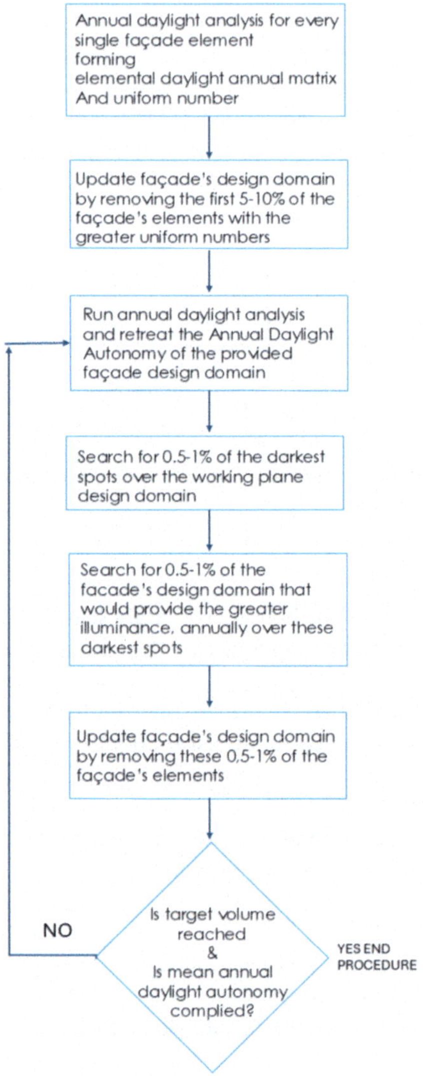

Fig. 3. T.O.DAY's Flow Chart

level are simultaneously met. This convergence criterion ensures an optimal balance between material efficiency and daylight performance.

The flowchart of Fig. 3 renders the Iterative nature of T.O.DAY's Algorithm, which offers a data-driven approach to facade design optimization. By strategically removing facade elements while considering their impact on daylight distribution, T.O.DAY has the potential to create energy-efficient buildings with superior occupant comfort through maximized natural light utilization.

4 Case Study

The case study uses a simple prismatic geometry 3 × 3 × 3 meters where we consider only the south facing wall as vertical plane/window wall which will take place the optimization. This geometry is in Athens therefore the corresponding weather data, from ladybug epw map have been used [15]. To design this geometry the software that has been used is rhinoceros 7.0 educational version and its plug-in grasshopper. For the annual Daylight analysis, we have been using a grasshoppers toggle Honeybee. Honeybee uses Radiance as solver which is widely used for commercial studies in daylight analysis and is a reliable solver.

A grasshopper's configuration has created those exports lists with UDI/DA of the horizontal plane for each facade's element void sate. These lists are used as an input for TODAY's code which is written in python 3. This code uses the annual daylight analysis honeybee's solver as a black-box.

5 Results

By running various target volume scenarios of T.O.DAY we have taken a set of figures (Fig. 4), which illustrate how the algorithm shapes the layout of a south-facing facade in Athens on an unobstructed plot. They also show how the facade design impacts daylight autonomy. As expected, minimizing the facade volume increases daylight autonomy. For example, with a target volume of 70%, some areas on the working plane have a minimum daylight autonomy of 90%, meaning some areas receive less than 300 lux for 36 days a year between 8:00 a.m. and 5:00 p.m. Conversely, with a 40% target volume, the minimum daylight autonomy improves to 95%, reducing the number of days with insufficient light to 18.

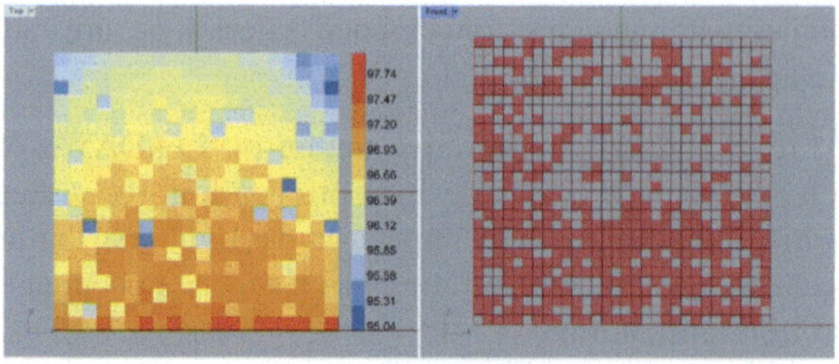

(i) target volume 0.40

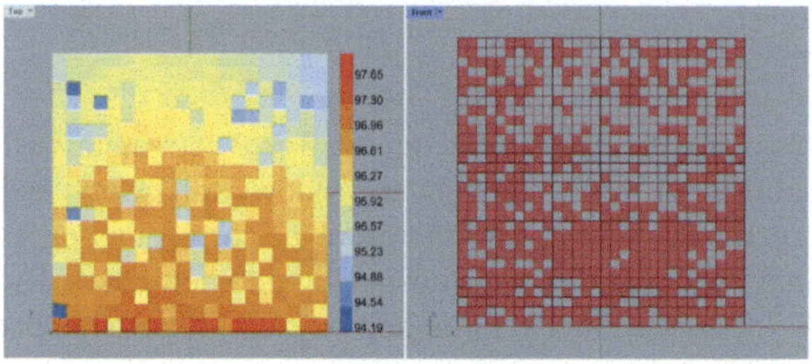

(ii) target volume 0.50

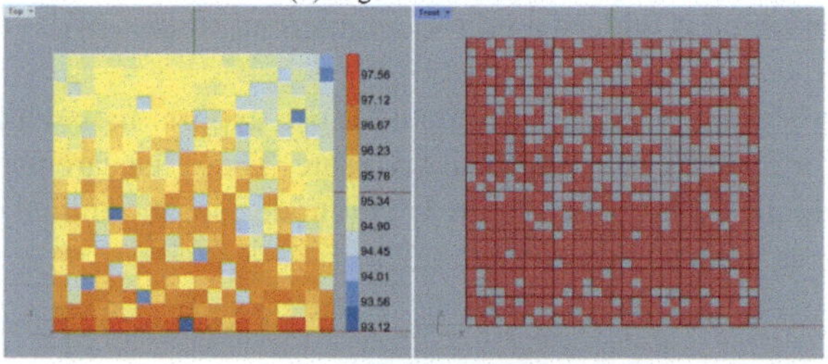

(iii) target volume 0.60

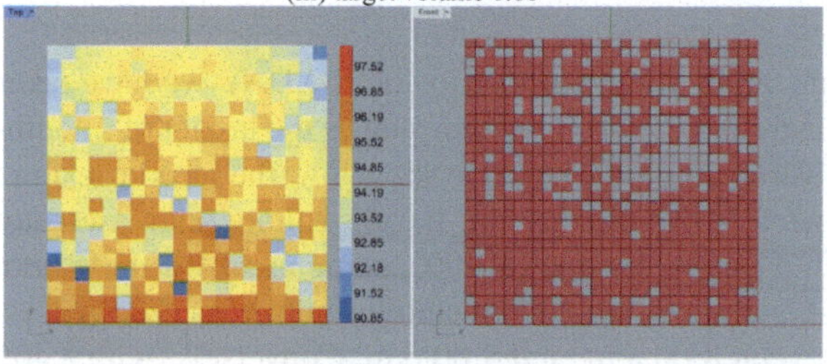

(iv) target volume 0.70

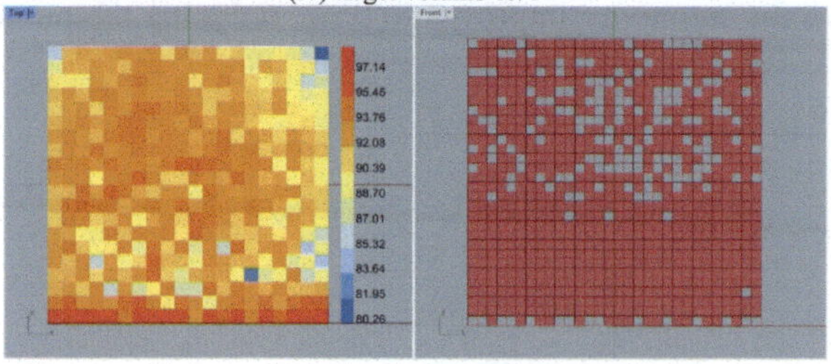

(v) target volume 0.80

Fig. 4. Various T.O.DAY's Results for different target volumes. On the right there is a plot the optimized façade while on the left there are the corresponding daylight autonomy percentages

6 Conclusions

As we saw on the figures on the result section, the results are complied with the algorithm maximization and minimization goals. However, it is obvious that by trying to minimize design domain's volume can lead to a lack of continuity in the facade design. This is the major issue that must be resolved in the future.

At the beginning of this paper 3D printing was mentioned, so the issue which arises is how TODAY is connected to additive manufacturing.

The general idea behind topology-optimized structures is that they are compatible with digital fabrication and additive manufacturing techniques. Early results show that as the volume is minimized, the design domain may lack continuity. Historical examples from structural optimization algorithms demonstrate similar challenges. For instance, Evolutionary Structural Optimization (ESO) often resulted in discontinuities and checkerboard patterns due to its sole subtractive nature. To address these issues, Bidirectional Evolutionary Structural Optimization (BESO) was developed, which adds and subtracts elements to improve continuity. Likewise, T.O.DAY could be further developed to include functions that predict and address continuity and printability challenges, making it more compatible with additive manufacturing processes [16, 17].

Acknowledgement. "The authors acknowledge the support of the Hellenic Foundation for Research and Innovation (H.F.R.I.) under the "2nd Call for H.F.R.I. Research Projects to support Faculty Members and Researchers", AMOSS project: "Additively Manufactured Optimized 3D Printed Steel Structures", (Project Number: 02779)."

References

1. Carmen Llatas, B.S.-V.: Implementing life cycle sustainability assessment during design stages in building information modelling: from systematic literature review to a methodological approach. Build. Environ. **182**, 107164 (2020). https://doi.org/10.1016/j.buildenv.2020.107164
2. Dileep Kumar, M.A.: A critical review for formulation and conceptualization of an ideal building envelope and novel sustainability framework for building applications. Clean. Eng. Technol. **11**, 100555 (2022). https://doi.org/10.1016/j.clet.2022.100555
3. Agency, I. E. Lighting. (2024). https://www.iea.org/energy-system/buildings/lighting#tracking
4. Programme, U. E: Global Status Report for Buildings and Construction 2022. (2024). https://www.unep.org/resources/publication/2022-global-status-report-buildings-and-construction
5. Peiman Pilechiha, M.M.: Multi-objective optimisation framework for designing office windows: quality of view, daylight and energy efficiency. Appl. Energy. **261**, 114356 (2020). https://doi.org/10.1016/j.apenergy.2019.114356
6. Wu, J., Sigmund, O., Groen, J.P.: Topology optimization of multi-scale structures: a review. Struct. Multidisc. Optim. **63**, 1455–1480 (2021). https://doi.org/10.1007/s00158-021-02881-8
7. Ayoub, M.: 100 Years of daylighting: a chronological review of daylight prediction and calculation methods. Sol. Energy. **194**, 360–390 (2019). https://doi.org/10.1016/j.solener.2019.10.072

8. Anna Wirz-Justice, D.J.: The relevance of daylight for humans. Biochem. Pharmacol. **191**, 114304 (2021)
9. Aries, M., Aarts, M., van Hoof, J.: Daylight and health: a review of the evidence and consequences for the built environment. Light. Res. Technol. **47**(1), 6–27 (2013)
10. Brembilla, E., Mardaljevic, J.: Mardaljevic climate-based daylight modelling for compliance verification: benchmarking multiple state-of-the-art methods. Build. Environ. **158**, 151–164 (2019). https://doi.org/10.1016/j.buildenv.2019.04.051
11. García, I., de Blas, M., Torres, J.L.: The sky characterization according to the cie standard general sky: comparative analysis of three classification methods. Sol. Energy. **196**, 468–483 (2020). https://doi.org/10.1016/j.solener.2019.12.039
12. Gonidakis, D.N., Frangedaki, E.I., Lagaros, N.D.: Optimizing daylight performance of digital fabricated adobe walls. Architecture. **4**(3), 515–540 (2024). https://doi.org/10.3390/architecture4030028
13. Ward, G.J., Bueno, B., et al.: Daylight simulation workflows incorporating measured bidirectional scattering distribution functions. Energ. Build. **259**, 111890 (2022). https://doi.org/10.1016/j.enbuild.2022.111890
14. Azza Nabil, J.M.: Useful daylight illuminances: a replacement for daylight factors. Energ. Build. **38**(7), 905–913 (2006). https://doi.org/10.1016/j.enbuild.2006.03.013
15. Tool, L.: EPW Map. (2024). https://www.ladybug.tools/epwmap/. Accessed 8 Sept 2024
16. Prathyusha, A., Raghu Babu, G.: A review on additive manufacturing and topology optimization process for weight reduction studies in various industrial applications. Mater. Today. Proc. **62**(1), 109–117 (2022). https://doi.org/10.1016/j.matpr.2022.02.604
17. George Kazakis, N.D.: A simple matlab code for material design optimization using reduced order models. Materials. **15**(14), 4972 (2022). https://doi.org/10.3390/ma15144972

Analysis, Design and Optimization of Lightweight Lattice Structured Spur Gears Using Finite Element Analysis

Stelios K. Georgantzinos(✉) and Anastasios Athanasiadis

Aerospace Science and Technology, National and Kapodistrian University of Athens, 34400 Psachna, Greece

sgeor@uoa.gr

Abstract. The present study focuses on the behavior analysis of different lattice structures in spur gears. The aim of this study is to compare the mechanical characteristics of various additively manufactured lattice structures when used as the main body of spur gears, compared to a solid gear body. Three different lattice structures were designed, simulated and topologically optimized, in order to generate a spur gear that has similar mechanical characteristics but better strength/weight ratio and others. Characteristics were quantified by measuring the maximum and mean, stress and deformation and calculating their weight ratios. Out of the three lattices, two lattices showed exceptional mechanical characteristics for the application. The Octahedral lattice showed the best stress/weight ratio while in the Double Pyramid lattice, the maximum deformation/weight ratio was observed. This study concludes that lattice structures when used in spur gears can have positive effects in weight reduction but also reveal new characteristics, suited for specific applications. The reported work provides important contributions to the possible applications of lattice structured spur gears. Such results could have huge benefits for a wide range of applications where either the lowest possible stress/weight ratio is required for maximum strength, or a high deformation/weight ratio is required to provide a damping effect for gear transmission applications for which the reduction of vibrations is critical. Furthermore, lattice structured gears showed increased heat dissipation compared to solid gears.

Keywords: Gear · Lattice Structure · Additive Manufacturing · Finite Element Analysis · Mechanical Analysis · Thermal Analysis

1 Introduction

Over the last few years, the technology of 3D printing has developed significantly, and along with it, new technologies based on it are evolving. It is now possible to manufacture gears with lattice structures, which opens up new horizons in the field of power transmission. Additive Manufacturing enables the production of robust, lightweight structures with geometry that cannot be achieved with traditional manufacturing processes, including complex lattice structures. Additive manufactured lattice structures have found to

N. D. Lagaros et al. (Eds.): ADDOPTML 2024, LNCE 646, pp. 145–157, 2025.
https://doi.org/10.1007/978-3-031-92029-5_14

be significantly superior to other cellular structures produced by alternative fabrication methods, primarily due to the greater geometric control and predictability introduced by Additive Manufacturing. A characteristic of lattice structures, in contrast to metal foams, is their structural uniformity, a fact that provides precise control of their collapse point [1]. Lattice structures are topologically optimized, three-dimensional open cell structures that are composed of one or more repeating unit cells. These cells are defined by their dimensions and their lattices are connected in defined nodes. At cell level, lattice structures are considered as structures based on their characteristics and properties, but they behave as homogeneous meta materials. By varying the structural parameters of lattice structures, like the way of connection or the geometry (cell size and lattice dimensions) their behavior can significantly improve and present properties (mechanical or other) that cannot be achieved by their parent materials.

Lattice structures can generally be categorized based on their mechanical response as bending-dominated structures or stretch-dominated structures. Structures dominated by bending can effectively absorb bending moments, while structures dominated by tension have greater stiffness than the former. The cell topology of a lattice structure determines whether it will be bending or tensile dominated. Based on their structure, lattice structures can be classified in:

- Strut-based lattice structures
- Triply periodic minimal surface lattice structures
- Shell lattice structures

Spur gears have external toothing on the circumference of their cylinder and are used to transmit power between parallel axles. During transmission they only impose radial loads on their bearings. A gear is essentially a solid cantilever beam. At the base of the beam, there is tensile stress on the loaded side and compressive stress on the opposite side. Tooth breakage is usually caused by a crack at the base of the tooth on the tensile side. Their ability to resist breaking is commonly referred to as beam flexural strength.

When gears are manufactured with sufficient accuracy that the teeth share the load, the tip-load condition is not the most critical. In almost every gear design, the contact ratio is high enough that two pairs of teeth are engaging, and thus sharing the load, at any given moment. The worst-case loading condition occurs when only one pair of teeth is engaged and receives the full load [2].

The forces acting on the teeth are usually determined by parameters such as power and speed. The values of these forces are calculated theoretically, but in practice the forces can be much higher due to moving masses, manufacturing errors in the teeth, deformation of the teeth causing accelerations and decelerations of short duration, misalignment between axes and due to shock loads. Loads acting in addition to fixed transmission loads are called dynamic loads. Despite the extensive research on the subject by experts such as Earle Buckingham [3], G. Niemann [4] and others, dynamic loads remain one of the least understood topics of gears. The vibrations caused by dynamic loads are so complex that no method has been developed that gives accurate results. The most intense dynamic load phenomena are observed at the moment of disengagement and engagement of the gear pair. Thus, due to the complexity of dynamic loads, their calculation is extremely difficult and not economically viable for cases of simple applications.

Drive systems due to their high required strength have a high mass. Since many of their components are rotating masses, they can put a lot of strain on the drive mechanism as they resist movement. Furthermore, in the case of vehicles, a reduction in mass means less work to move the vehicle. From the reduction of the mass in gears, the rotating mass and also the total mass of the vehicle is reduced. Mass reduction can translate into lower energy consumption or even increased performance (a lighter vehicle develops speed faster). There are several ways to reduce the mass of gears, the most "traditional" being the creation of holes or cavities in their side. Alternative ways to reduce the mass, which seem to have other desirable characteristics such as damping, were proposed by Thales Sardinha Garcia Souza et al. and of course reducing the mass of gears with lattice structures [5]. Traditionally, weight reduction in gears was achieved by creating cavities or holes on their side surface. Additive Manufacturing technology combined with the use of Topology Optimization, has contributed to reducing the weight by replacing holes with lattice structures. It has been experimentally proven that lattice structures in gears can not only reduce weight but also provide damping to reduce vibrations during transmission [5].

Another important advantage that gears with lattice structures can offer is the reduction of vibrations and the level of operating noise, created by the meshing of the teeth. In particular, due to their constantly changing stiffness during engagement, each toothed ring gear turns into an internal vibration generator. In the study by Riad Ramadani et al. fabricated spur gear using SLM printing technology and Ti-6Al-4V alloy. Their aim was to build a lightweight lattice structured gear which would reduce the generated vibrations of the engagement. The lattice structure is expected to be more flexible accommodating to tangential forces, which is expected to have a positive effect on the transmission of vibrations from the ring gear to the hub [6].

In present study, the appropriate computational procedures are developed based on the finite element method for the mechanical analysis, optimization and design of spur gears with lattice structures. Emphasis was placed on the study and interpretation of the forces developed in the gear wheels during contact between them for the purpose of power transfer. The reduction of mass as well as vibrations is an important objective when designing and optimizing the components in the drive system, especially when it concerns high performance vehicles where strong and abrupt forces are encountered, high operating speeds and at the same time it is necessary to minimize the mass in every possible way.

2 Simulation Methodology

2.1 Overview

The methodology aims to analyze and optimize the design of lattice-structured spur gears by adopting a computational framework. The primary objective is to evaluate the mechanical performance of these gears, particularly focusing on weight reduction, stress distribution, and deformation characteristics. This involves defining the gear's geometry, integrating lattice structures for optimization, and applying boundary conditions and operational loads to simulate realistic scenarios. The workflow incorporates a systematic approach to assess structural behavior and optimize the lattice design to enhance the strength-to-weight ratio and vibration-damping properties.

2.2 Geometry, Material Definition, and Simulation Setup

The initial spur gear model was designed with precise geometric parameters, including module, number of teeth, pressure angle, and face width, to replicate standard operational conditions.

Specific material properties and geometric dimensions were chosen to ensure accurate mechanical analysis and optimization. The material properties included an elastic modulus of 105 GPa, Poisson's ratio of 0.3, and a density of 4430 kg/m^3 [7], while the geometric parameters included a module of 4 mm, 26 teeth, a pressure angle of 20°, and a face width of 20 mm. These characteristics, summarized in Table 1, provide the foundation for realistic simulations. For topology optimization, the gear was divided into regions, preserving high-stress areas for structural strength and targeting low-stress regions for material reduction. Boundary conditions involved fixing one gear while applying torque to the shaft of the other. Mesh refinement was applied to critical regions, particularly at the contact points between teeth, to ensure accurate stress and deformation analysis. Worst-case loading scenarios were simulated to identify stress distributions, forming the basis for further design optimization.

Table 1. Geometric parameters of the selected spur gear.

Elastic modulus (E)	Poisson's ratio (v)	Density (ρ)	Module	Number of teeth	Pressure angle	Face width
105 GPa	0.3	4430 kg/m^3	4 mm	26	20°	20 mm

The lattice structure design aimed to balance weight reduction and mechanical performance by integrating optimized lattice configurations. Three lattice types—Cubic, Double Pyramid, and Octahedral—were selected based on their distinct mechanical properties and compatibility with additive manufacturing. The selection criteria included their ability to resist stress, accommodate deformation, and provide vibration damping. The Cubic lattice offered uniformity and simplicity, the Double Pyramid lattice provided flexibility and shock absorption, and the Octahedral lattice demonstrated superior stiffness and strength-to-weight ratio. The geometry of these lattice types is illustrated in Fig. 1.

The integration of lattice structures was guided by the low-stress regions identified during topology optimization, where material removal was permissible without compromising structural integrity. High-stress areas, such as gear teeth and their roots, were preserved. The lattice design process involved defining geometric parameters, including cell size, strut thickness, and nodal connectivity, to ensure mechanical compatibility. The geometric relationship between cell size, strut thickness, and the total volume fraction of the lattice (Vf) was defined as:

$$Vf = \frac{V_{\text{lattice}}}{V_{\text{total}}} \tag{1}$$

Fig. 1. Lattice geometry (**a**) Cubic, (**b**) Double Pyramid, and (**c**) Octahedral.

where $V_{lattice}$ is the material volume in the lattice structure and V_{total} is the total volume of the lattice region.

To ensure a fair comparison between different lattice designs, their masses were controlled by adjusting the density (ρ) and geometric parameters to achieve consistency across configurations. The mass M of the lattice-structured gear was calculated as:

$$M = \rho \cdot V_{lattice} \tag{2}$$

This ensured that observed differences in mechanical performance, such as strength-to-weight and deformation-to-weight ratios, were attributed solely to the inherent mechanical properties of the lattice types.

2.3 Topology Optimization

The primary goal of the topology optimization process was to minimize the weight of the spur gear while ensuring that its mechanical performance, including strength and stiffness, remained robust under operational conditions. This process was formulated as an optimization problem, where the objective was to reduce the total mass of the gear, M, defined by the integral:

$$M = \int \Omega \rho(x) d\Omega \tag{3}$$

Here, $\rho(x)$ represents the material density at a point x, and Ω is the design domain. Constraints were imposed to maintain the mechanical integrity of the gear. The stress constraint ensured that the stress $\sigma(x)$ at any point did not exceed the allowable stress σmax:

$$\sigma(x) \leq \sigma max \tag{4}$$

Similarly, a displacement constraint was included to keep the maximum displacement $u(x)$ within acceptable limits:

$$u(x) \leq umax \tag{5}$$

The gear was divided into two regions based on stress distribution obtained from simulations. High-stress regions, such as the teeth and their roots, were preserved without material removal to ensure structural integrity. Low-stress regions, primarily within the gear's core and intermediate zones, were targeted for material reduction.

A density-based method was employed for the optimization. The design domain was discretized into finite elements, with each element assigned a density variable $\rho(x)$, where $\rho(x) \in [0, 1]$. The algorithm iteratively adjusted $\rho(x)$ values to minimize the objective function while adhering to the constraints. To ensure a clear separation between solid and void regions, a penalization factor p was applied to the density interpolation:

$$E(x) = \rho(x) \cdot E_0 \tag{6}$$

In this equation, $E(x)$ is the effective modulus of elasticity of the element, and E_0 is the modulus of the solid material.

The topology optimization was carried out using advanced computational tools configured for iterative static analysis. Material interpolation models and penalty methods were incorporated to enforce the design constraints effectively. The optimization process concluded when the change in the objective function M between iterations fell below a specified threshold or after reaching a predefined number of iterations. The resulting optimized design provided a lightweight and efficient material layout, suitable for further mechanical analysis and additive manufacturing.

2.4 Static and Heat Dissipation Analysis

The static analysis was conducted to evaluate the mechanical behavior of lattice-structured spur gears under both generalized and realistic operational conditions. Two key setups were employed to provide a comprehensive understanding of their performance. In the general behavior analysis, the configuration involved applying a uniform torque T at the gear shaft while fixing all teeth to simulate an even distribution of load across the gear. The torque applied was calculated using:

$$T = F \cdot r \tag{7}$$

where F is the force applied at the gear teeth and r is the gear's pitch radius. This setup allowed the evaluation of the stress distribution and deformation across the entire gear structure. The stress was assessed using the von Mises criterion [8]:

$$\sigma vm = \sqrt{\frac{1}{2}\left[(\sigma_1 - \sigma_2)^2 + (\sigma_2 - \sigma_3)^2 + (\sigma_3 - \sigma_1)^2\right]} \tag{8}$$

where σ1, σ2, σ3 are the principal stresses. Deformation was evaluated using the strain-displacement relationship:

$$\epsilon = E\sigma \tag{9}$$

where ϵ is the strain, σ is the stress, and E is the Young's modulus of the material. These calculations helped identify weak points and stress concentration regions under uniform loading conditions.

For the operational stress analysis, a more refined setup was employed to simulate realistic gear operation. Torque T was applied at the shaft, while different pairs of teeth were fixed to replicate actual contact points during gear engagement. This approach considered the varying points of force application as the gear teeth engaged and disengaged. The localized stress and deformation patterns were analysed using the same von Mises stress criterion and strain-displacement equations.

The goal of the operational analysis was to capture the dynamic mechanical behavior of the gear under practical conditions. This included identifying high-stress regions, potential deformation zones, and the gear's ability to distribute and absorb loads effectively. The lattice structure's influence on vibration damping and stress distribution was particularly emphasized, providing critical insights into its suitability for specific applications. These findings formed the basis for further design refinement and performance optimization.

The thermal behavior of lattice-structured gears was analysed to assess their ability to dissipate heat, with a focus on the influence of the increased surface area inherent in the lattice design. Effective heat dissipation is critical in preventing overheating during operation, which can degrade material properties and reduce overall performance. The heat dissipation capability was evaluated using the fundamental heat transfer equation:

$$Q = h \cdot A \cdot \Delta T \tag{10}$$

In this equation, Q is the heat transfer rate (in watts), h is the convective heat transfer coefficient (in watts per square meter per kelvin), A is the surface area available for heat dissipation (in square meters), and ΔT is the temperature difference between the gear surface and the ambient environment (in kelvins). The lattice structure increases the surface area A, allowing for greater heat exchange with the environment compared to a solid gear of similar dimensions. To model the internal heat conduction, Fourier's law of heat conduction was applied:

$$\frac{\partial T}{\partial t} = \alpha \nabla^2 T + \frac{\dot{q}}{\rho c_p} \tag{11}$$

In this equation, T is the temperature (in kelvins), t is time (in seconds), α is the thermal diffusivity (in square meters per second), k is the thermal conductivity (in watts per meter per kelvin), ρ is the material density (in kilograms per cubic meter), ρc_p is the specific heat capacity at constant pressure (in joules per kilogram per kelvin), and $\dot{q}$ is the internal heat generation rate per unit volume (in watts per cubic meter). The thermal diffusivity α is calculated as:

$$\alpha = \frac{k}{\rho c_p} \tag{12}$$

The analysis involved applying a uniform heat generation rate within the gear, simulating frictional heating during operation. Boundary conditions were set to model convective heat transfer at the gear surfaces, with the ambient temperature held constant. By increasing the surface area of the gear through the lattice structure, the convective heat transfer was enhanced, resulting in improved thermal management.

Numerical methods were used to solve the heat conduction equation and obtain the steady-state temperature distribution within the gear. Comparisons between lattice-structured and solid gears revealed that the lattice structure reduced the maximum operational temperature, owing to its larger surface area and improved heat dissipation.

The simulations conducted in this study involved several assumptions to simplify the analysis and ensure computational feasibility. One primary assumption was that the material properties of the gear, such as elastic modulus, Poisson's ratio, and thermal conductivity, were uniform and isotropic. Furthermore, the simulations assumed perfect bonding and continuity between lattice elements and solid regions, ignoring possible defects or imperfections that might arise during fabrication. The loading conditions were also idealized to streamline the analysis. For example, the applied torque was considered steady and uniformly distributed, while dynamic effects, such as fluctuating loads or shock impacts during operation, were not explicitly modeled. Contact between gear teeth was assumed to be perfect, neglecting factors such as surface roughness, manufacturing tolerances, and potential wear. Lastly, the iterative nature of the design process was limited by computational resources and time. Convergence criteria, such as changes in objective functions or stress distributions, were set to practical thresholds rather than ideal values.

3 Results

3.1 Topology Optimization Results

Topology Optimization was performed in the low stress areas while high stress areas remained untouched.

After the dimensions of the center ring that the optimization will be applied were defined, the lattice density was determined by running a Static structural simulation on the center ring (Fig. 2a).

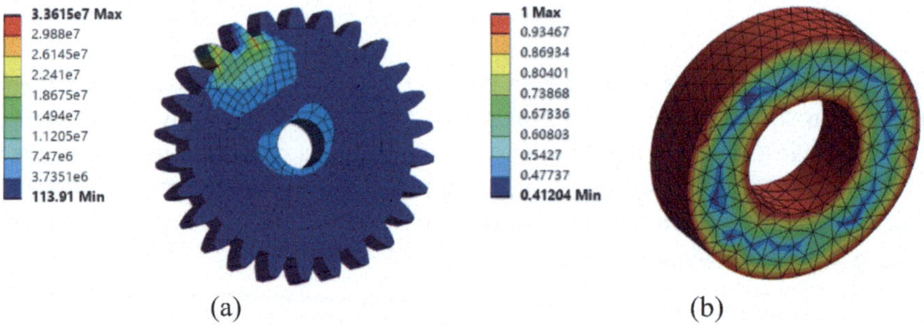

Fig. 2. (**a**) Static structural analysis, indicating the low stress areas (Equivalent von-Mises in Pa) that Topology Optimization will be applied and (**b**) Lattice Density data.

Based on the lattice density data (Fig. 2b), three gears with different lattices were created and subjected to three different sets of simulations, presented in the next chapters.

3.2 Comparison Study of General Behavior Between Different Lattice Structures in Spur Gears

During the analysis of the three different lattice structures of fixed size with equal distribution of the torque on all teeth both the values of the maximum and the average stress and strain were studied.

To understand the properties of a lattice structure in a component, the exclusive study of a single variable value can lead to incorrect conclusions. For example, comparing the maximum stresses of the three lattice structures, it does not necessarily indicate that the part with the highest maximum stress has the lowest strength. The nodes presenting the maximum stresses can be weak points of the structure, but they can also be stress concentration points which are eliminated by local optimization (creating fillet curves on sharp edges or local increase in the thickness of the structure members). Large values of structure deformation can have undesirable effects and create unnatural meshing and thus gear wear. On the contrary, since the deformation is elastic, it can dampen shocks and vibrations (damping effect) of the transmission mechanism (Fig. 3).

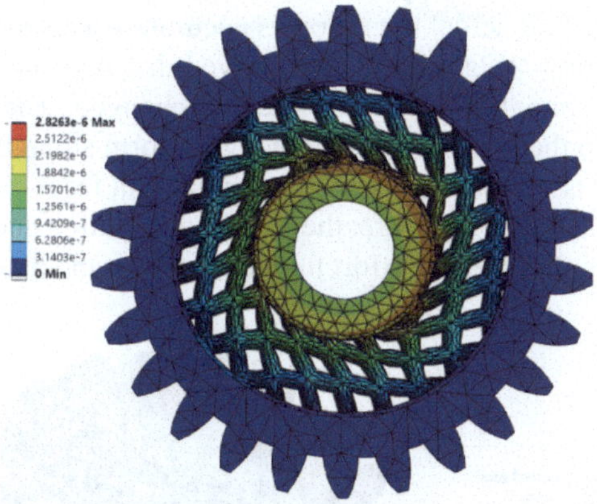

Fig. 3. Exaggerated illustration of lattice structures deformation (in Meters). This deformation gives the gear shock absorbing properties.

It becomes clear that in order to choose the appropriate type of structure, it is necessary to know the application and its requirements. In this section, a comparison study was conducted to determine the most suitable lattice structure per application. For example, if large deformation values are desired to dampen harshness or vibrations caused during power transmission by the engagement of the teeth, and there are no particularly high strength requirements, then the most suitable lattice structure, based on the mean stress to weight ratio, is Double Pyramid as presented in Table 2.

On the contrary, if high strength and/or high precision is required, then the most suitable lattice structure, based on the criterion of mean stress to weight ratio, is the Octahedral structure.

Table 2. Analysis data of spur gears with lattice structures. Considering the boundary conditions used, the given values indicate the general behavior of each structure.

Type	Mass [kg]	Maximum stress [Mpa]	Maximum deformation [μm]	Mean stress to weight ratio [Mpa]	Mean deformation to weight ratio
Solid	0.702	7.399	0.509	1.623	0.177
Cubic	0.487	39.223	2.826	10.561	1.992
Double P.	0.493	47.311	3.230	9.211	2.704
Octahedral	0.507	34.924	2.470	8.040	2.085

3.3 Comparison Study of Strength Between Different Lattice Structures in Spur Gears

In order to find the maximum operating stress of a pair of gear wheels, a new static analysis was performed with the application of the torque on the shaft and two toothed sides, which simulates their actual operation as accurately as possible [9].

During the simulations (Fig. 4), it was observed that by changing the fixed pair of teeth with an adjacent one, the maximum stress is changed. This fact is explained as lattice structures due to their nature do not having uniform strength. Therefore, to draw reliable conclusions it is important to consider stress data from different pairs of teeth. In addition to the results of the software, the standard deviation and average trend of the above values were calculated, in an effort to better understand and interpret them.

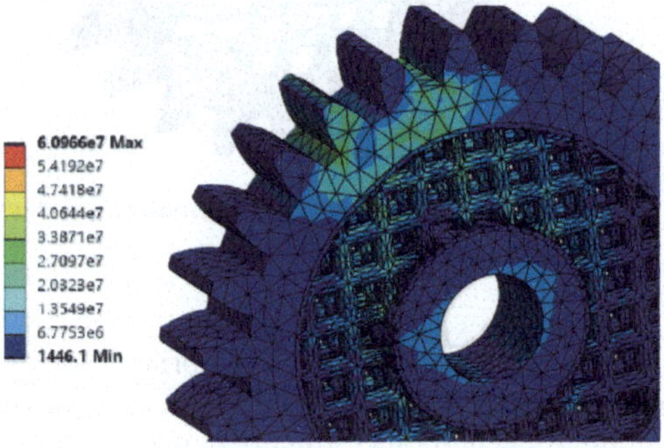

Fig. 4. Random teeth pair (Equivalent von-Mises) stress values in Pa of lattice structured spur gear (Cubic type).

In this section, a comparison study was conducted to determine the most suitable lattice structure when strength is the most important criteria. The boundary conditions of this example focus on finding the actual operating stress values of the spur gears. Although the Cubic and Double Pyramid structures in some of the teeth pairs show lower maximum and/or average stress values, the standard deviation of their three pairs is higher than that of the Octahedral structure (see Table 3). Although high stress values in lattice structures can be eliminated by local optimization, the high standard deviation of the maximum stress, denotes that Cubic and Double Pyramid lattice structures are generally weaker than Octahedral. This fact is confirmed by the previous study of general behavior.

Table 3. Analysis data of spur gears with lattice structures with boundary conditions simulating their actual operation.

Gear type (3 pairs)	Mass [kg]	Standard deviation of maximum stress [Mpa]	Standard deviation of mean stress [Mpa]	Mean value of maximum stress [Mpa]	Mean value of mean stress [Mpa]
Solid structure	0.702	–	–	–	–
Cubic	0.487	42.731	0.229	65.497	7.866
Double P.	0.493	35.354	0.181	95.690	6.563
Octahedral	0.507	3.903	0.459	81.672	6.287

3.4 Comparison Study of Heat Dissipation Between Different Lattice Structures in Spur Gears

During this study, another advantage of lattice structured spur gears was observed. This section examines the effect of the increased surface area of lattices on the Heat dissipation of the gear. Due to the increased surface area that the lattices offered to the ring under analysis, the heat dissipation is expected to increase. The surface area of the solid gears center ring was 14,047 mm^2 compared to 25,529 mm^2 for the Double Pyramid lattice.

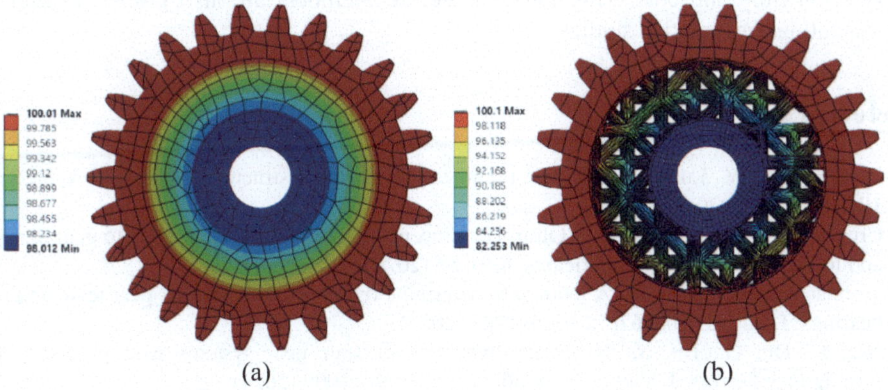

Fig. 5. Heat dissipation illustration of (**a**) solid gear compared to (**b**) Double Pyramid. Plot legend in Celsius degrees.

This fact can be beneficial for the operating conditions of gear transmission systems. Lower maximum temperatures, less thermal load on gear coolant as well as less needs for transmission casing cooling can be achieved.

The simulation comparing the temperatures reaching the shaft between the solid and lattice structured gear, showed a temperature reduction of around 16 °C for all types of lattices (Fig. 5). This number is expected to vary depending on the operating conditions and gear design; however, it is a good indicator of the cooling benefits the lattices can have on the gear.

4 Conclusions

The present simulation methodology and finite element analysis provides a guideline on the applications and mechanical characteristics that each lattice structure can offer to spur gears. Based on the results of the simulations, two of the three lattice structures stand out, each of them being proposed for a different application. The Octahedral structure is recommended for applications where maximum strength is the primary design criterion. By using Octahedral lattice, the mass of the component can be reduced by up to 30% while maintaining the strength at acceptable levels Due to its elasticity, the Double Pyramid structure can be used in applications where there is a very sharp application of forces and shock absorbing properties are required. Also, another proposed area of use is in luxury vehicles or other applications such as medical devices and semiconductor industry, where the reduction of vibrations from power transmission is an important design criterion. Finally, lattice structures increase the external surface area of the part and as a result heat transfer coefficients increase and can be cooled more efficiently, a fact that is particularly useful in high-speed applications. All lattice structured gears showed exceptional heat dissipation characteristics.

Although costs in metal 3D printing are decreasing, they remain higher than conventional gear manufacturing methods. Additionally, 3D printing inherently introduces residual stresses and rough surfaces, requiring extra post-processing, which increases production time and cost. Thin lattice structure members are also prone to fatigue, necessitating preventative studies.

References

1. Pan, C., Han, Y., Lu, J.: Design and optimization of lattice structures: a review. Appl. Sci. **10**(18), 6374 (2020)
2. Croccolo, D., De Agostinis, M., Olmi, G., Vincenzi, N.: A practical approach to gear design and lubrication: A review. Lubricants. **8**(9), 84 (2020)
3. Vedmar, L., Andersson, A.: A method to determine dynamic loads on spur gear teeth and on bearings. J. Sound Vib. **267**(5), 1065–1084 (2003)
4. Xie, S., Du, Q., Hu, Y.: Nonlinear dynamics of spur gear systems with time-varying misalignment errors. J. Vib. Control. **30**(17–18), 3997–4009 (2024)
5. Savoy, J.: Light-weight assembled gears: a green design solution for passenger and commercial vehicles. Gear Technol., 64–71 (2013)

6. Ramadani, R., Pal, S., Kegl, M., Predan, J., Drstvenšek, I., Pehan, S., Belšak, A.: Topology optimization and additive manufacturing in producing lightweight and low vibration gear body. Int. J. Adv. Manuf. Technol. **113**, 3389–3399 (2021)
7. Hasan, R., Mines, R.A.W., Tsopanos, S.: Determination of elastic modulus value for selectively laser melted titanium alloy. J. Mech. Eng. Technol. **2**(2), 17–26 (2010)
8. Georgantzinos, S.K.: A CAD/CAE approach for an efficient analysis of contact mechanics in rolling bearings. J. Appl. Math. Bioinform. **8**(2), 65–82 (2018)
9. Georgantzinos, S.K.: The role of multi-scale finite element methods in analysis and design of advanced mechanical and aerospace structures. Nanotechnology. **9**, 805–808 (2019)

Experimental Investigation of the Mechanical Behavior of PLA Structures Produced by FDM 3D Printing

Panteleimon Bakalis[1], Eleni Papadopoulou[2], and Stelios K. Georgantzinos[2](✉)

[1] General Department, National and Kapodistrian University of Athens, 34400 Psachna, Greece
[2] Aerospace Science and Technology, National and Kapodistrian University of Athens, 34400 Psachna, Greece
sgeor@uoa.gr

Abstract. In the present work, research was conducted to investigate the mechanical properties of PLA (Polylactic Acid) structures fabricated via Fused Deposition Modeling (FDM), a popular 3D printing technique. The main objective of the research was to quantify the influence of key printing parameters, such as printing angle, layer height, layer width and extrusion temperature, on the mechanical properties of 3D printed PLA parts. The specimens which were printed at different angles (0°, 18°, 36°, 54°, 72°, and 90°) were subjected to tensile testing to evaluate their key performances such as tensile strength, elasticity, yield strength and fracture behavior.

The Taguchi method was used to optimize the experimental procedure, this facilitated the identification of optimal combinations of printing parameters, leading to more consistent and reliable results. The study focused on understanding the relationship between printing parameters and the mechanical properties of PLA, providing useful information on how different manufacturing conditions affect the performance of the material.

The results of this study provide valuable information for the additive manufacturing industry by further clarifying the mechanical behavior of PLA under different printing conditions. This information is very important in industries such as automotive and biomedical engineering, where knowledge of the material's properties is crucial to the performance of parts printed using 3D technology. In addition, this work can lay the groundwork for future research aimed at optimizing the design and application of 3D printed components, potentially extending the use of PLA to advanced high-performance packaging.

Keywords: Additive manufacturing · 3D printing · FDM · Printing parameters · Mechanical properties · Design of experiments · Taguchi method

1 Introduction

3D printing, also known as additive manufacturing, has redefined the way manufacturing is done, such that it is now possible to rapidly and economically produce parts with complex geometry. This ability to fabricate complex designs from digitally designed

N. D. Lagaros et al. (Eds.): ADDOPTML 2024, LNCE 646, pp. 158–168, 2025.
https://doi.org/10.1007/978-3-031-92029-5_15

models has reduced the need for molds and cutting tools, making 3D printing a critical tool in industries from aerospace to automotive to biopharmaceutical [1, 2]. The layer-by-layer object manufacturing process reduces waste and energy consumption compared to conventional subtractive manufacturing techniques [3–5].

Among the, PLA is one of the widely used materials in FDM manufacturing, this is due to its biodegradable nature, its low melting temperature and its ability to produce parts with high precision. Pla is produced from renewable sources such as corn starch and sugarcane, for this reason it is a sustainable solution compared to petroleum-based plastics, making it ideal for ecological applications [6–9].

Despite the many advantages of the FMD process, layer-by-layer printing introduces an imbalance in the mechanical properties of the printed parts, making the strength of the material and its stiffness dependent on the orientation of the print, the thickness of the layer and the connection between layers [10, 11]. Previous studies have shown that properties such as ultimate tensile strength (UTS), modulus of elasticity and failure strain are significantly affected by factors such as extrusion temperature, printing speed and raster orientation [12–14]. However, it is still not clear how these parameters can be optimized to improve the performance of standard PLA parts.

To address these difficulties, this study focuses on the mechanical behavior of PLA elements produced by FDM, with particular attention to determining the relationship between key process parameters and material behavior. Tensile tests are performed on PLA samples designed in SolidWorks, varying the printing angles (0°, 18°, 36°, 54°, 72° and 90°), extrusion temperatures and layer heights to systematically evaluate their effect on UTS and elastic properties. The stress-strain curves obtained during these tests provide us with information on how layer bonding and print orientation affect the structural integrity of FDM-printed parts [15]. These results are extremely useful for PLA-based industries for load-bearing applications, where optimizing the strength-to-weight ratio and reducing defects is paramount.

Beyond the mechanical properties of Pla, its biodegradability introduces complexities. While its ability to decompose makes it an environmentally friendly choice, exposure to different temperature changes and humidity levels can accelerate its degradation capacity resulting in long-term alteration of its mechanical behavior [16, 17]. This highlights the importance of evaluating PLA not only under optimal printing conditions but also in environments that simulate real-world use. Such evaluations will help ensure that PLA components maintain their structural integrity over time in applications such as medical devices, packaging and automotive interiors [18].

Furthermore, this study contributes to further understanding of the anisotropic behavior of FDM-issued components. Previous researchers have proposed numerical and theoretical models to expect mechanical structures based mainly on FDM parameters [19]. But, these models typically lack experimental-intellectual validation or fail to account for the effects of intermediate printing angles and variable extrusion states; by combining experimental information with sys-tematic parameter analysis, this study seeks to close these gaps and provide a broader and more comprehensive framework for optimizing FDM methods for PLA [20].

In summary, by investigating the effects of printing angles, extrusion temperatures and layer heights on UTS, elastic modulus and pressure behavior, this observation presents realistic suggestions for optimizing FDM techniques. These insights are highly relevant for industries seeking to leverage the blessings of PLA in lightweight, cost-effective and sustainable packaging, ensuring consistently excellent and overall performance across multiple use cases.

2 Experimental Methodology and Process Optimization

2.1 FDM Technology and Anisotropy

The FMD technique manufactures objects by depositing thermoplastic material layer by layer. As mentioned in the introduction, this process introduces an imbalance in the mechanical properties of the printed objects, due to the strength of the bond between the layers, as well as the strength within the layers themselves. To mitigate the anisotropy, attention was paid to the selection and optimization of printing parameters (Fig. 1).

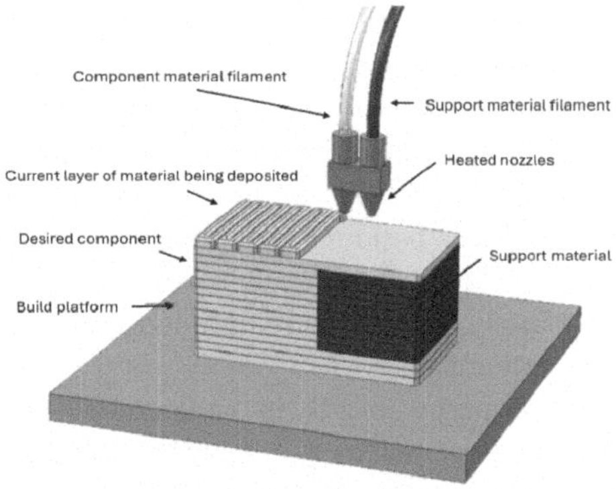

Fig. 1. FDM printing elements.

To reduce the anisotropy and improve the performance of the objects produced by FDM method, some key factors were taken into consideration:

- The orientation of the print: It refers to the orientation of the object during printing. The optimization of the wall thickness aims at balancing the structural integrity with the desired surface finish.
- Printing thickness: Ensuring a uniform print thickness across the object helps to avoid stress concentrations and deformation during the printing process. Thin or uneven layers can lead to weak spots.
- Volume of material: Avoiding excessive deposition of material is essential to avoid distortion due to thermal stresses. Rapid cooling of larger volumes can cause shrinkage, which can lead to distortion of the object.

2.2 Print Parameters and Material Selection

To understand the effect of printing parameters on the mechanical performance of FDM printed objects, the material PLA (polylactic acid) was chosen due to its stable mechanical properties and ease of use. The proposed printing parameters of the PLA used included a filament diameter of 1.75 mm with a tolerance of ±0.05 mm, a proposed printing temperature range of 195 °C to 230 °C and a recommended printing speed between 40 and 90 mm/s. It is worth noting that a heated bed plate (HBP) is not required for this material. The main printing parameters studied are layer thickness, layer height, extrusion temperature, printing angle and printing speed. The parameters and their respective ranges are detailed in Table 1.

Table 1. Experimental parameter levels and ranges.

Code	Control parameters	Unit	Level 1	Level 2	Level 3	Level 4	Level 5	Level 6
A	Printing angle	deg	0	18	36	54	72	90
B	Layer height	mm	0.1	0.2	0.3	–	–	–
C	Layer width	mm	0.3	0.4	0.5	–	–	–
D	Printing Temperature	°C	190	205	220	–	–	–
E	Printing speed	mm/s	30	50	70	–	–	–

These parameters were chosen knowing that they have a direct influence on the mechanical properties of the printed parts such as tensile strength, elastic modulus and deformation, which are essential for its functional performance. Figure 2 illustrates the geometries, printing orientations and dimensions of the specimens used in this study.

In more detail, Fig. 2 illustrates the printing conditions, and the different angles of the 3D printed specimens used for the anisotropy study. Figure 2b shows the standard dimensions of the tensile test specimen, which was designed according to ISO 527 guidelines. Figure 2c illustrates a schematic representation of the printing layers, relative to the tensile direction.

Figure 3 illustrates the relationship between the printed layer height, line width, and the direction of tensile loading.

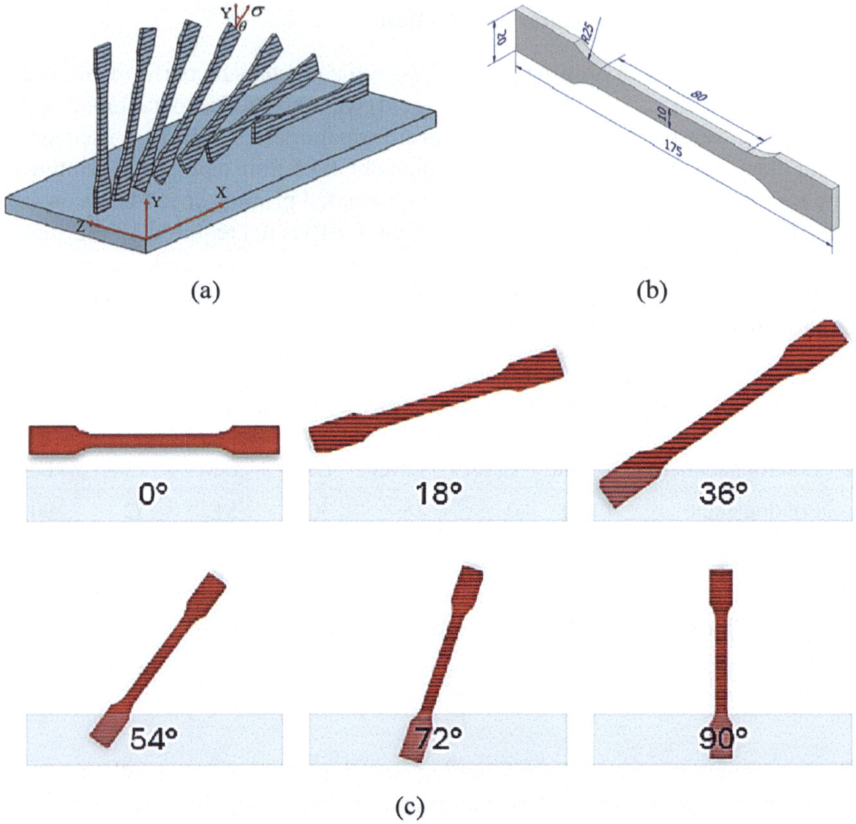

Fig. 2. Specimen geometries, print orientations, and dimensions, (**a**) Specimen geometries, (**b**) specimens dimensions, and (**c**) print orientations.

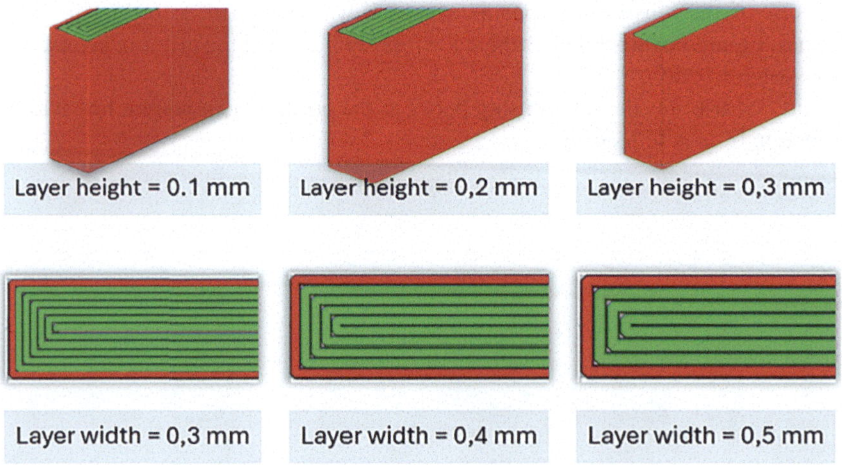

Fig. 3. Schematic representation of layer height and line width relative to the tensile direction.

2.3 Application of the Taguchi Method

The experimental design for four process parameters (Layer height, Wall line width, Infill, Printing temperature, and Printing speed) with three levels (3^4), along with one parameter (Printing angle) with six levels (6^1), was organized using the Taguchi method with an L18 orthogonal array Table 2. The numbers in Table 2 correspond to the parameter levels defined in Table 1, where each control parameter is assigned a specific value for each experiment in the array.

Table 2. L18 orthogonal array for Taguchi optimization.

Ex. No.	Printing angle	Layer height	Layer width	Printing temperature	Printing speed
1	1	1	1	1	1
2	1	2	2	2	2
3	1	3	3	3	3
4	2	1	1	2	2
5	2	2	2	3	3
6	2	3	3	1	1
7	3	1	2	1	3
8	3	2	3	2	1
9	3	3	1	3	2
10	4	1	3	3	2
11	4	2	1	1	3
12	4	3	2	2	1
13	5	1	2	3	1
14	5	2	3	1	2
15	5	3	1	2	3
16	6	1	3	2	3
17	6	2	1	3	1
18	6	3	2	1	2

2.4 Mechanical Testing

The mechanical properties of the specimens were evaluated using a universal testing machine. Figure 4a shows the tensile test setup, including the specimen grips and the testing machine used to evaluate mechanical properties such as tensile strength and strain. Figure 4b illustrates the relationship between the material direction and the load direction for a single polymer layer, where θ represents the angle between the material direction (1) and the load direction (x). This equipment featured high-precision load cells and an extensometer for real-time measurement of stress and strain.

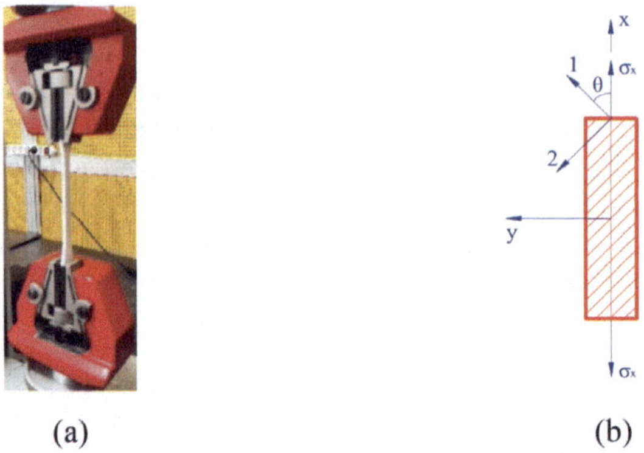

Fig. 4. Tensile test setup and material orientation relative to load direction, (**a**) Tensile test setup, and (**b**) material orientation relative to load direction.

3 Results

The results obtained with the Taguchi method are summarized using comparisons of means for each level of parameters. Figure 5 presents the graphs resulting from the analysis. The mechanical properties analyzed include modulus of elasticity (E), ultimate tensile strength (σ_{ult}), strain at ultimate tensile strength (ϵ_{ult}), yield strength (σ_y), strain at yield (ϵ_y), fracture strength (σ_{fr}), and strain at fracture (ϵ_{fr}).

Figure 5a shows that the modulus of elasticity (E) is strongly affected by the printing angle, with the highest value observed at lower angles (0°) and a decrease at 90°. An average layer height (0.2 mm), larger line width (0.5 mm) and lower extrusion temperature (190 °C) also enhance the elastic modulus, while printing speed has relatively smaller effects with the optimum being at (30 mm/s).

Figure 5b highlights the ultimate tensile strength (σ_{ult}), the best combination values for the highest strength limit were as follows: printing angle 0 degrees, layer height 0.1 mm, line thickness 0.5, printing temperature 190 °C and printing speed 30 mm/s.

Figure 5d demonstrates that strain at ultimate tensile strength (ϵ_{ult}) the best combination values for the largest strain at the ultimate strength limit were: printing angle 0 degrees, layer height 0.1 mm, line thickness 0.4, printing temperature 205 °C and printing speed 30 mm/s.

Figure 5d shows that yield strength (σ_y) best combination values for the higher ratio limit were as follows: printing angle 0 degrees, layer height 0.1 mm, line thickness 0.5, printing temperature 190 °C and printing speed 50 mm/s.

Figure 5e, for strain at yield (ϵ_y) the best combination values for the largest deformation at the ratio limit were as follows: printing angle 0 degrees, layer height 0.1 mm, line thickness 0.5, printing temperature 190 °C and printing speed 50 mm/s.

Figure 5f highlights fracture strength (σ_{fr}), the best combination values for the highest fracture limit were as follows: printing angle 0 degrees, layer height 0.1 mm, line thickness 0.5, printing temperature 190 °C and printing speed 30 mm/s.

Lastly, Fig. 5g reveals that for strain at fracture (ϵ_{fr}) the best combination values for the highest strain at fracture limit were: printing angle 0 degrees, layer height 0.1 mm, line thickness 0.4, printing temperature 205 °C and printing speed 30 mm/s.

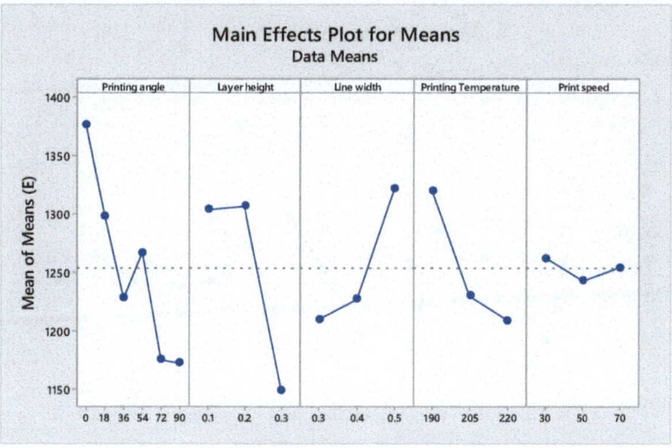

(a)

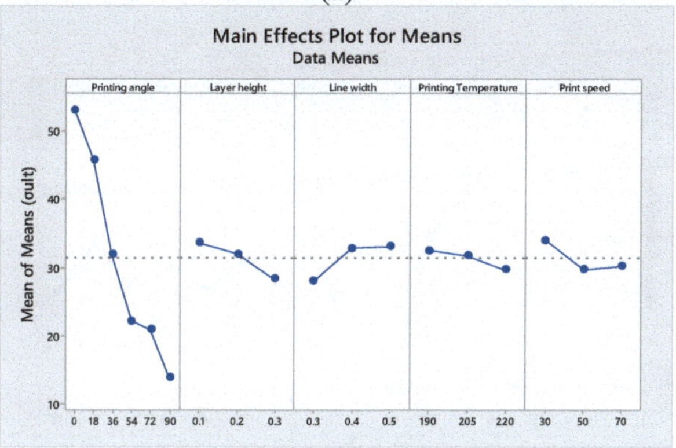

(b)

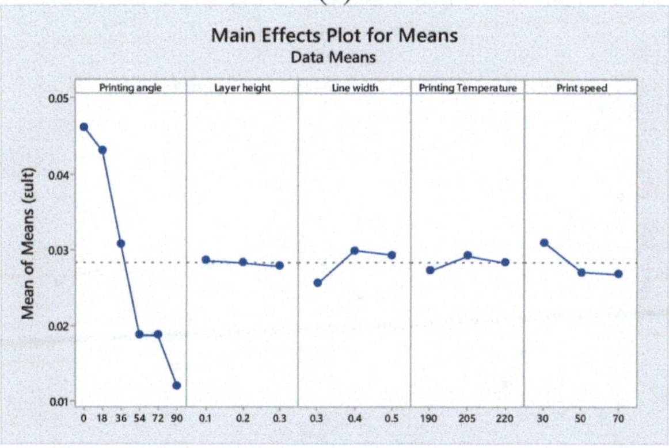

(c)

Fig. 5 Experimental results of quality characteristics with S/N ratio analysis, (**a**) modulus of elasticity (E), (**b**) ultimate tensile strength (σ_{ult}), (**c**) strain at ultimate tensile strength (ϵ_{ult}), (**d**) yield strength (σ_y), (**e**) strain at yield (ϵ_y), (**f**) fracture strength (σ_{fr}), and (**g**) strain at fracture (ϵ_{fr}).

(d)

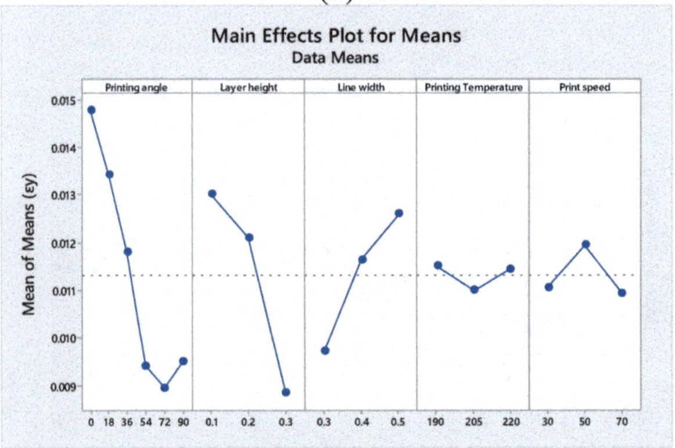

(e)

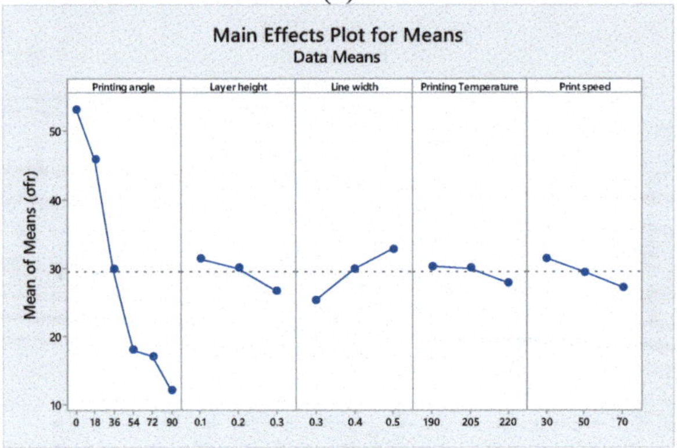

(f)

Fig. 5 (*continued*)

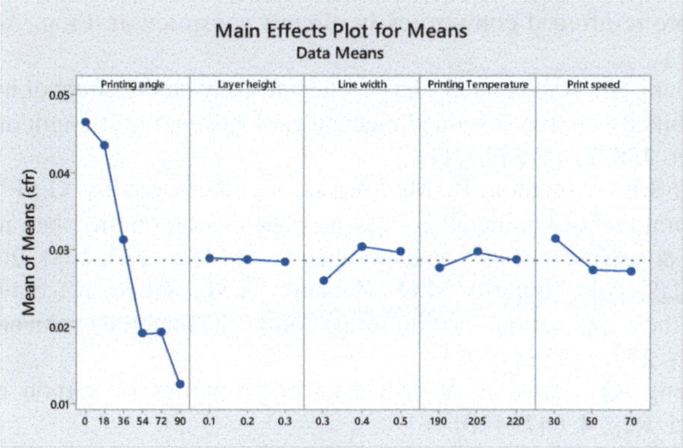

(g)

Fig. 5 *(continued)*

4 Conclusions

The present experiment examined the effects of key FDM printing parameters on the mechanical properties of PLA parts, focusing on optimizing the elastic modulus (E), tensile strength (σ_{ult}), tensile strength strain (ϵ_{ult}), yield strength (σ_y), yield strain (ϵ_y), fracture strength (σ_{fr}) and fracture strain (ϵ_{fr}).

The following conclusions summarize the most significant findings:

- Complex geometries require compromises between minimizing anisotropy and achieving optimal surface finish and mechanical properties.
- For structures requiring specific stiffness, increasing the number of ribs is preferable to increasing their thickness. Thick ribs should have hollow cross-sections to avoid material mass that can cause deformation and increased printing costs.
- Excessive material mass can negatively affect parts by causing distortion or warping during cooling.
- Layer thickness significantly influences surface quality, especially for rounded parts, with thinner layers providing better results.
- The Taguchi method optimizes process parameters by reducing variation and producing high-quality products at a lower cost.

In conclusion, the findings of this study highlight the importance of balancing design complexity with process parameter optimization to achieve mechanically robust and cost-effective PLA components using FDM 3D printing. By leveraging the Taguchi method and tailoring parameters to specific structural and functional requirements, manufacturers can enhance product quality while minimizing material waste and production costs.

References

1. Mallick, P.K.: Fiber-Reinforced Composites: Materials, Manufacturing, and Design, 3rd edn. CRC Press, Boca Raton, FL (2007)

2. Soutis, C.: Fibre reinforced composites in aircraft construction. Prog. Aerosp. Sci. **41**(2), 143–151 (2005)
3. Rafiee, R., Maleki, M., Rahnama, S.: Experimental study on the effect of hygrothermal environments combined with the sustained mechanical loads on the strength of composite rings. Compos. Struct. **258**, 113397 (2021)
4. Georgantzinos, S.K., Antoniou, P., Markolefas, S., Giannopoulos, G.: Finite element predictions on vibrations of laminated composite plates incorporating the random orientation, agglomeration & waviness of carbon nanotubes. Acta Mech. **233**, 2031–2059 (2022)
5. Garg, A., Chalak, H.D., Belarbi, M.O., Zenkour, A.M., Sahoo, R.: Estimation of carbon nanotubes and their applications as reinforcing composite materials—An engineering review. Compos. Struct. **272**, 114234 (2021)
6. Arash, B., Wang, Q., Varadan, V.: Mechanical properties of carbon nanotube/polymer composites. Sci. Rep. **4**, 6479 (2014)
7. Paipetis, A.S., Kostopoulos, V.: Carbon Nanotube Enhanced Aerospace Composite Materials. Solid Mechanics and Its Applications, vol. 188. Springer, Dordrecht (2013)
8. Rafiee, R., Firouzbakht, V.: Multi-scale modeling of carbon nanotube reinforced polymers using irregular tessellation technique. Mech. Mater. **78**, 74–84 (2014)
9. Rafiee, R., Firouzbakht, V.: Stochastic multiscale modeling of CNT/polymer. In: Rafiee, R. (ed.) Carbon Nanotube-Reinforced Polymers, Micro and Nano Technologies, pp. 503–520. Elsevier, Amsterdam (2018)
10. Rafiee, R., Ghorbanhosseini, A.: Stochastic modeling of CNT-grown fibers. In: Rafiee, R. (ed.) Carbon Nanotube-Reinforced Polymers, Micro and Nano Technologies, pp. 521–540. Elsevier, Amsterdam (2018)
11. Rafiee, R., Amohaji, H.: On the application of genetic algorithm for predicting the strength of CNT/ABS filaments using multi-scale modeling. Appl. Phys. A Mater. Sci. Process. **130**, 513 (2024)
12. Rafiee, R., Zehtabzadeh, H., Amini, M.R.: Predicting mechanical properties of 3D printed nanocomposites using multi-scale modeling. Addit. Manuf. **83**, 104055 (2024)
13. Han, J., Kim, C.G.: Low earth orbit space environment simulation and its effects on graphite/epoxy composites. Compos. Struct. **72**(2), 218–226 (2006)
14. Fitriah, S.N., Abdul Majid, M.S., Ridzuan, M.J.M., Daud, R., Gibson, A.G., Assaleh, T.A.: Influence of hydrothermal ageing on the compressive behaviour of glass fibre/epoxy composite pipes. Compos. Struct. **159**, 350–360 (2017)
15. Bisheh, H., Wu, N., Rabczuk, T.: Free vibration analysis of smart laminated carbon nanotube-reinforced composite cylindrical shells with various boundary conditions in hygrothermal environments. Thin-Walled Struct. **149**, 106500 (2020)
16. Mehar, K., Panda, S.K., Sharma, N.: Numerical investigation and experimental verification of thermal frequency of carbon nanotube-reinforced sandwich structure. Eng. Struct. **211**, 110444 (2020)
17. Hassanzadeh-Aghdam, M.K., Jamali, J.: A new form of a Halpin-Tsai micromechanical model for characterizing the mechanical properties of carbon nanotube-reinforced polymer nanocomposites. Bull. Mater. Sci. **42**(117), 1–8 (2019)
18. Thostenson, E.T., Chou, T.W.: On the elastic properties of carbon nanotube-based composites: modelling and characterization. J. Phys. D. Appl. Phys. **47**, 079501 (2014)
19. Chamis, C.C., Lark, R.F., Sinclair, J.H.: Integrated theory for predicting the hygrothermo-mechanical response of advanced composite structural components. In: Vinson, J.R. (ed.) Advanced Composite Materials-Environmental Effects, pp. 160–192. American Society for Testing and Materials, Philadelphia (1978)
20. Avil, E., Kadioglu, F., Kaynak, C.: Contribution of carbon nanotubes to vibration damping behavior of epoxy and its carbon fiber composites. J. Reinf. Plast. Compos. **39**(7–8), 311–323 (2020)

Automated Calibration of the Temperature Parameter in Simulated Annealing for Optimization in Civil Engineering

Athanasios Stamos(✉) and Nikos D. Lagaros

School of Civil Engineering, National Technical University of Athens, Athens, Greece
stamthan@central.ntua.gr

Abstract. Simulated Annealing (SA) is a widely used optimization technique in which the temperature parameter **T** plays a crucial role in controlling the acceptance probability of non-improving trial solutions (negative changes) during the search process. Traditionally, the initial value of **T** is determined by the user, often through trial and error, which can be inefficient and lead to suboptimal performance. In this study, we present an automated method for calibrating the **T** parameter based on the statistical behavior of the objective function. By leveraging the same sequence of changes (both favorable and unfavorable) generated by the SA algorithm, our method dynamically computes an appropriate initial **T** value, eliminating the need for manual tuning.

Extensive experiments on benchmark and real-world optimization problems demonstrate that our approach yields good results. In particular our approach is successfully applied to the topological optimization of discrete structures made of prefabricated or 3D-printed members.

Keywords: Simulated Annealing · temperature · automated

1 Introduction

Simulated Annealing (SA) [3] is a stochastic optimization technique inspired by the annealing process in metallurgy. The process involves heating and then gradually cooling a material to remove defects and achieve a stable state. In optimization, SA mimics this process by starting with an initial state and iteratively making changes to minimize an objective function. It has been extensively used in solving complex engineering problems, including structural optimization, land reclamation, and infrastructure planning.

The key mechanism in SA involves accepting both improving and deteriorating changes in the objective function, with the probability of acceptance controlled by the temperature parameter T. This parameter plays a crucial role in balancing exploration and exploitation in the search space. If T is set too high, the method behaves like a random search, leading to inefficiency. Conversely, if T is too low, SA may prematurely converge to a local minimum, failing to explore the broader solution space effectively.

N. D. Lagaros et al. (Eds.): ADDOPTML 2024, LNCE 646, pp. 169–178, 2025.
https://doi.org/10.1007/978-3-031-92029-5_16

Selecting an appropriate initial temperature is, therefore, crucial for SA's success. Current methods rely on heuristic or empirical approaches, making them dependent on user expertise. While bibliography is rich about SA, most studies do not address the problem of initial temperature (for example [1]), or focus on the procedure of cooling the temperature (for example [4, 6]). [2] suggest to try sequentially a series of initial temperatures, perform the first step of SA method without cooling, and choose the one which yields the best value of the objective function, for the full application of the SA method.

In this study, we present an automated technique to calibrate T, eliminating the need for user-defined values while maintaining efficient convergence. Our method aims to provide a systematic, statistically driven approach to setting T, improving the reliability and efficiency of SA across diverse applications.

2 Simulated Annealing Methodology

SA begins with an arbitrary initial state and iteratively applies changes to improve the objective function. The acceptance of changes follows these rules:

- If the new state decreases the objective function ($\Delta E < 0$), the change is always accepted (favorable change).
- If the new state increases the objective function ($\Delta E > 0$), the change is unfavorable but nevertheless it is accepted with a probability given by the Metropolis criterion [5]: $p = e^{-\Delta E/T}$

ΔE represents the change in objective function value, and T is the current temperature. As T decreases over iterations, fewer unfavorable changes are accepted, leading to convergence to a near-optimal solution [7].

As many other optimization algorithms, SA requires user-defined values for several defined parameters, including:

1. Initial temperature (T), which determines the starting acceptance rate of negative changes.
2. Number N_{tr} of changes per temperature step, which controls the extent of exploration at each stage.
3. Temperature reduction schedule, which dictates how quickly T decreases.
4. Termination criteria, which define when the algorithm stops.

Traditionally, users must experiment with these parameters to achieve optimal results. However, this manual tuning can be time-consuming and inconsistent. The primary focus of this study is to automate the selection of T, eliminating a significant source of uncertainty in SA applications.

3 Automatic Temperature Calibration

Instead of arbitrarily setting T, we propose an approach based on the probability of accepting unfavorable changes. This method dynamically adjusts T to ensure a balance between exploration and convergence.

3.1 Reformulation of the Metropolis Criterion

We redefine the Metropolis formula by introducing a scaling factor k:

$$p = e^{-\Delta E/kT} \tag{1}$$

where k is an unknown factor to be determined, and $T = 100$. To ensure sufficient exploration at the start, we set an initial probability of acceptance p_0 (typically 0.5 or 50%) for unfavorable changes. Solving Eq. (1) for k yields:

$$k = \frac{-\Delta E}{T \cdot \log(p_0)} \tag{2}$$

This formulation allows us to determine an appropriate T value dynamically, ensuring that SA begins with an acceptance probability conducive to effective exploration. Specifically, for random initial state, the chance of a favorable random change is 50% and the same for an unfavorable one. If we accept $p_0 = 50\%$ of unfavorable random changes, the probability of acceptance of a random change, favorable or unfavorable, is high:

$$p = 0.5 + p_0 \cdot 0.5 = 0.75 = 75\%$$

3.2 Estimating ΔE

In order compute k in Eq. (2), ΔE is estimated by making a set of trial changes to a random initial state, and counting the number N_{pos} of unfavorable changes, where $\Delta E > 0$. The number of trial changes is taken to be 10% of the total changes per temperature step (N_{tr}). For example if the number of changes per temperature step is $N_{tr} = 1000$, then we make 100 trial changes.

The $\Delta E_i > 0$ of each unfavorable changes is recorded, and the average of these values is the estimate of ΔE:

$$\Delta E = \frac{1}{N_{pos}} \sum_{i=1}^{N_{pos}} \Delta E_i$$

If fewer than 50% of the trial changes are unfavorable ($N_{pos} < 0.1 \cdot N_{tr}/2$), the process is retried up to five times. Failure to meet this criterion suggests that the objective function may be insensitive to some design variables, or that the initial state is not random at all and in fact it is the worst possible case.

4 Implementation and Parameters

The automatic temperature calibration method is integrated into SA with the following default parameter values:

- After experimentation, the initial probability of acceptance of unfavorable changes is set to: $p_0 \approx 0.4$

- Number of changes per temperature step: $N_{tr} = 200 \cdot N$, where N is dimensionality of the problem – typically the number of design variables [7]
- Temperature reduction schedule: Geometric cooling [7]: $T_i + 1 = 0.7 \cdot Ti$
- Stopping criteria:

  – Maximum of 50 temperature steps.
  – Termination if no accepted changes (favorable or unfavorable) occur in two consecutive temperature steps.

The automatic temperature calibration method, along with the SA, was implemented in the Matlab and the Python programming languages. The former can be downloaded from [9] (function sa) and the latter from [8] (package p_ganneal, class p_ganneal.SimulatedAnnealing).

5 Experimental Validation

To validate the proposed method, we successfully applied it to a range of mathematical and engineering problems:

5.1 Benchmark Functions

- 2D functions with 1 or more minima: Branin, Goldstein-Price, Haupt-1, Haupt-2, Shaffer, Zakharov, Sphere (Fig. 1).
- High-dimensional functions with numerous local minima: Griewank, Ackley, Rastrigin, Schwefel (Fig. 2).

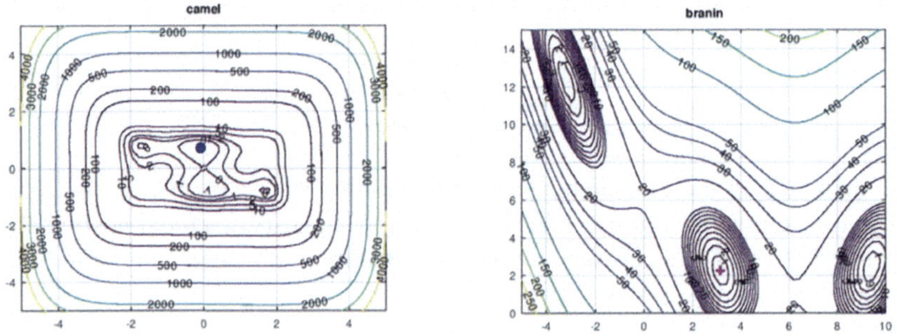

Fig. 1. 2D functions. The results are accurate.

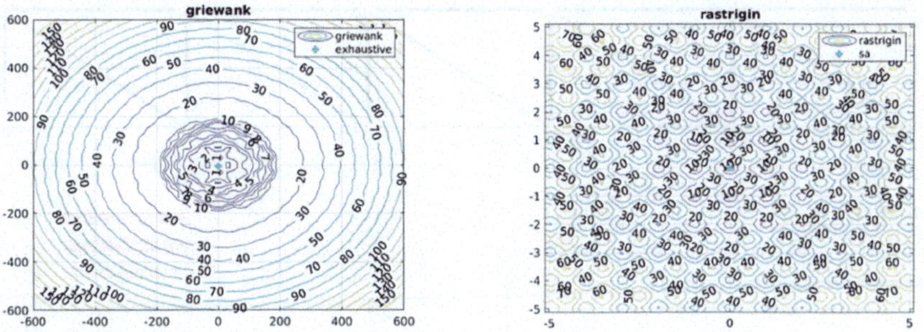

Fig. 2. High-dimensional functions, shown here for 2 dimensions. The results are accurate.

5.2 Civil Engineering Applications

- Size, Shape and Topology Optimization of Trusses (Figs. 3 and 4)
- Optimization of Land Reclamation Layouts (Fig. 5)
- Automatic Floor Planning, Vector and Raster Representations (Fig. 5)
- Stormwater Drainage System Grade Line optimization (Fig. 6)
- Highway Grade Line Optimization (Fig. 7)
- Minimization of plotter paper waste when plotting large drawings (Fig. 8)

SA is a stochastic method and thus each time it runs it produces slightly different results. Typically all the problems were run 5 times and the best solution was chosen. All the problems were also optimized with the Genetic Algorithm Optimization (GA). SA produced analogous or better results. The results from these case studies confirm that the automated calibration of T produces high-quality solutions comparable to, or better than, manually tuned parameters.

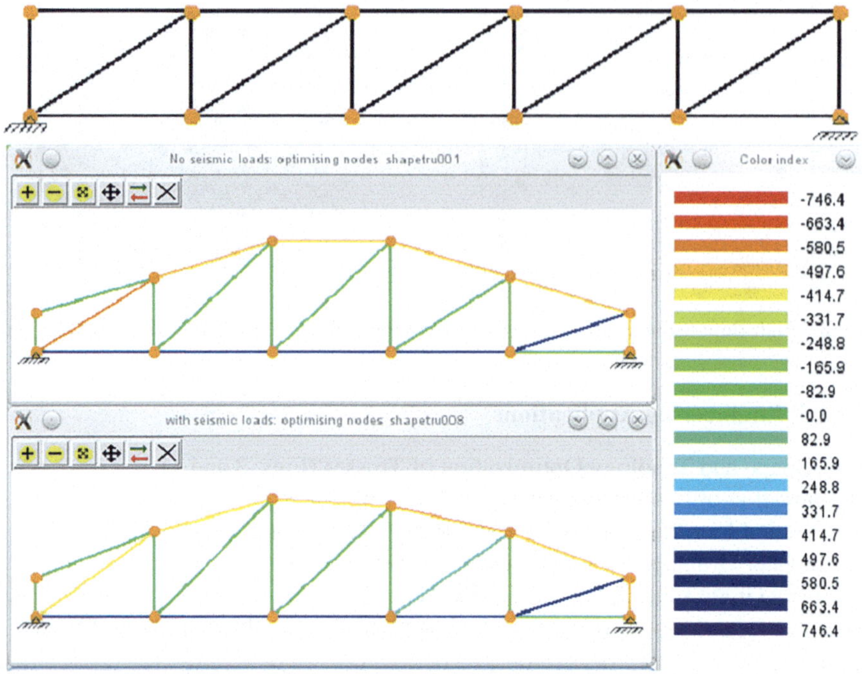

Fig. 3. Shape optimization of truss. Original shape (top), optimized shape under vertical loads (middle), optimized shape under seismic loads (bottom)

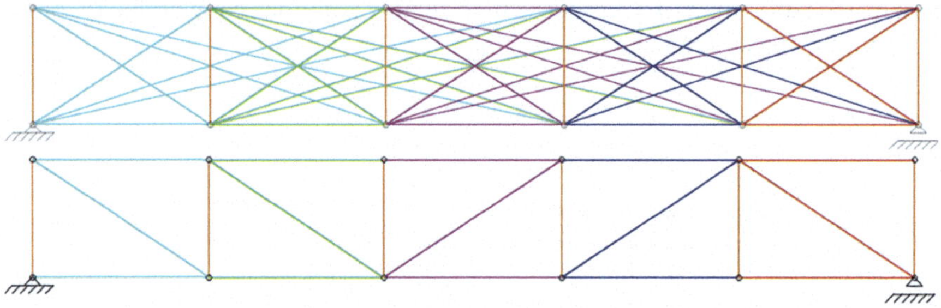

Fig. 4. Topology optimization of truss. Original topology (top), optimized topology (bottom)

Fig. 5. Left: Distribution of 24 unequal land plots to 7 beneficiaries with complex restrictions. Same color land plots are assigned to the same beneficiary. Initial max difference among beneficiaries 200 m^2, optimized difference 48.99 m^2. Right: Optimized floor plan. Bedrooms (green), living room (magenta), kitchen (yellow), bathroom (yellow), corridor (black). Openings are allowed at the top and bottom side

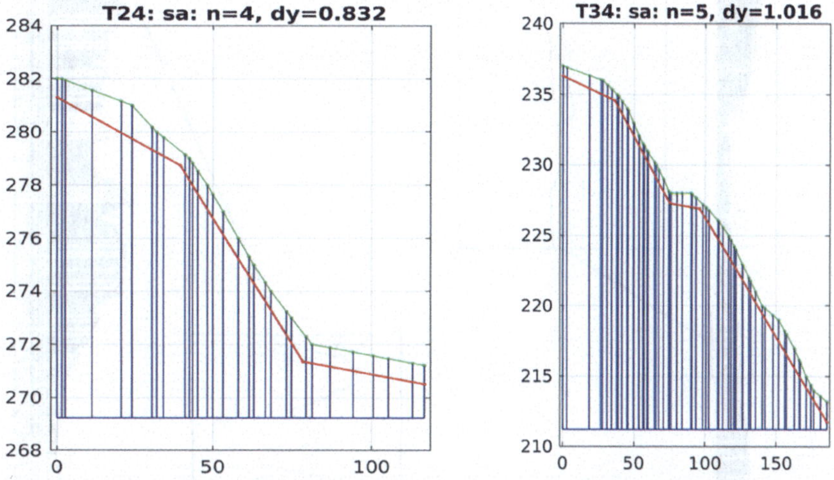

Fig. 6. Stormwater Drainage System Grade Line optimization (2 pipes)

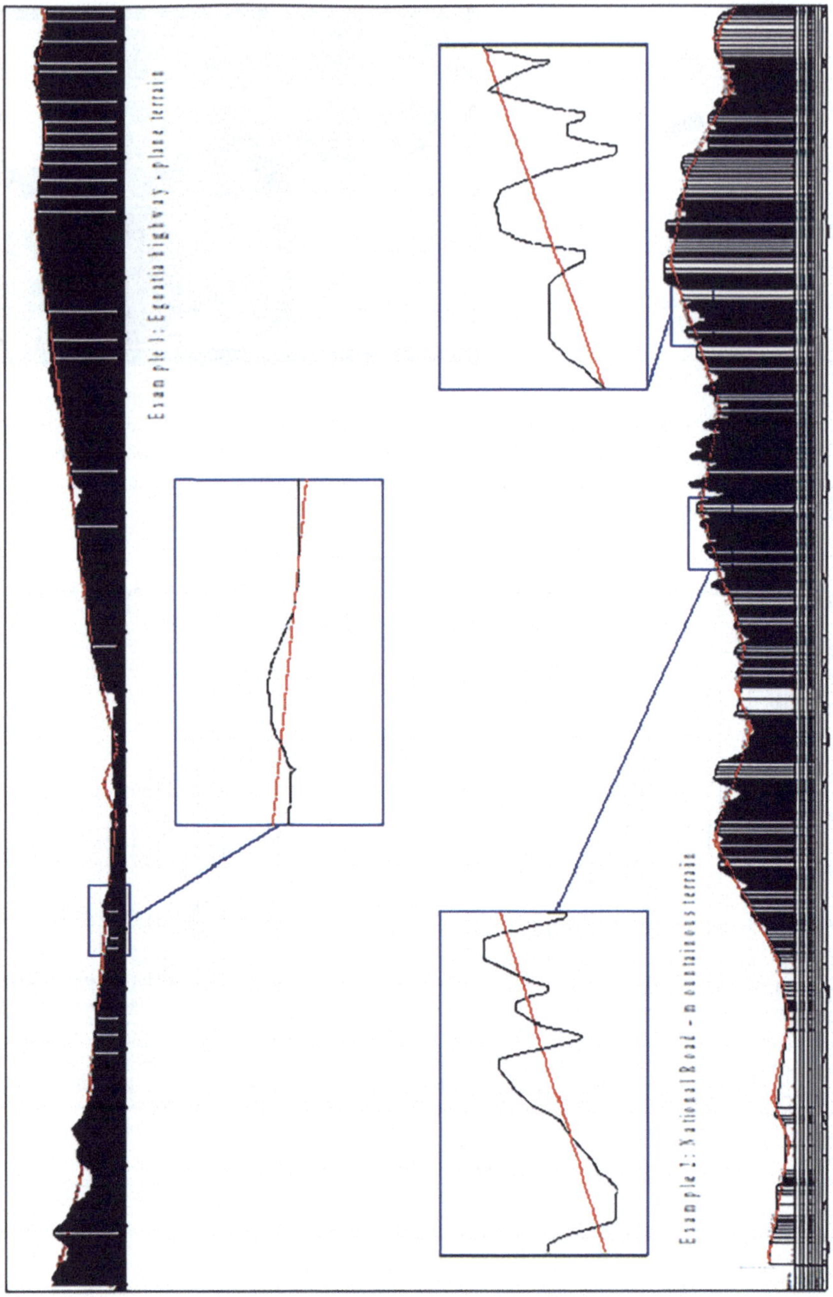

Fig. 7. Highway Grade Line Optimization (2 highways)

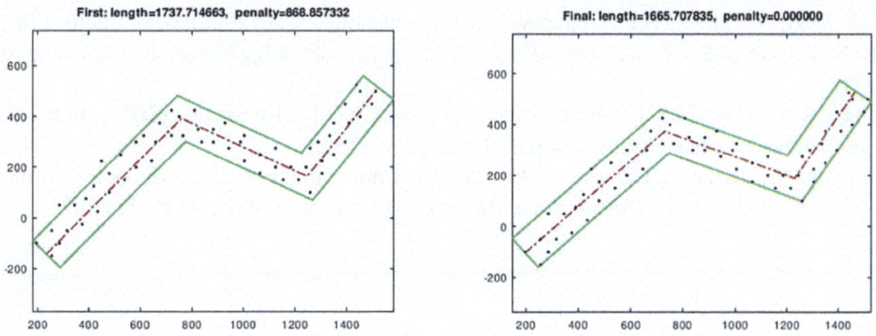

Fig. 8. Minimization of plotter paper waste when plotting large drawings (onyx). Length of original design length 1737 m (left), Length of optimized design 1665 m (right)

6 Conclusion

This study presents an automated method for selecting the initial temperature in Simulated Annealing, significantly improving its efficiency and usability. By leveraging statistical estimates of objective function behavior, our approach dynamically adjusts T to ensure effective optimization. Extensive testing on mathematical and engineering problems confirms that the method is highly effective and eliminates the need for manual tuning. Future work may explore adaptive adjustments for other SA parameters to further enhance its robustness.

Acknowledgments. The authors are grateful for the support of this study through the research program: ADDitively Manufactured OPTimized Structures by means of Machine Learning, MSCA-RISE-2020 Marie Skłodowska-Curie Research and Innovation Staff Exchange (RISE), project number 101007595.

References

1. Černý, V.: Thermodynamical approach to the traveling salesman problem: an efficient simulation algorithm. J. Optim. Theory Appl. **45**, 41–51 (1985). https://doi.org/10.1007/BF00940812
2. Connolly, D.T.: An improved annealing scheme for the QAP. Eur. J. Oper. Res. **46**(1), 93–100 (1990). https://doi.org/10.1016/0377-2217(90)90301-Q
3. Kirkpatrick, S., Gelatt, C.D., Vecchi, M.P.: Optimization by simulated annealing. Science. **220**(4598), 671–680.2 (1983)
4. Laarhoven van, P.J.M., Aarts, E.H.L.: Simulated Annealing: Theory and Applications. Reidel, Dordrecht (1987). https://doi.org/10.1007/978-94-015-7744-1
5. Metropolis, N., Rosenbluth, A.W., Rosenbluth, M.N., Teller, A.H., Teller, E.: Equations of state calculations by fast computing machines. J. Chem. Phys. **21**, 1087–1092 (1953)
6. Nikolaev, A.G., Jacobson, S.H.: Simulated annealing. In: Gendreau, M., Potvin, J.Y. (eds.) Handbook of Metaheuristics. International Series in Operations Research & Management Science, vol. 146. Springer, Boston, MA (2010). https://doi.org/10.1007/978-1-4419-1665-5_1

7. Press, W.H., Flannery, B.P., Teukolsky, S.A., Vetterling, W.T.: Numerical Recipes in FORTRAN: The Art of Scientific Computing, 2nd edn, p. 3. Cambridge University Press, Cambridge (1992)
8. Stamos A.A.: ThanCad: ThanCad, a 2dimensional CAD for Engineers. (2025). https://sourceforge.net/projects/thancad/. Accessed 9 Jan 2025
9. Stamos A.A.: ntuacarlo: Library of Open-Source Implementation of Optimization Algorithms for Matlab. (2025). https://sourceforge.net/projects/ntuacarlo/. Accessed 9 Jan 2025

Experimental Mechanical Characterization of 3D Printed Structures Under Bending Loading

Stelios K. Georgantzinos, Archontoula Koumentakou(✉), Ioannis Sideras, and Grigorios Kostopoulos

Aerospace Science and Technology, National and Kapodistrian University of Athens, 34400 Psachna, Greece

sgeor@uoa.gr, a.koumentakou@yahoo.com

Abstract. In the present work, experiments are conducted to characterize 3D-printed structures under bending loading. Special emphasis is placed on printing with the material PET-G (Polyethylene Terephthalate Glycol), known for its excellent mechanical properties, such as impact and puncture resistance, making it a durable and robust material. The selection of PET-G is considered significant, as it is suitable for applications requiring strong and long-lasting components. A key focus of the study is the investigation of how different printing parameters affect the mechanical performance of the printed structures. The experiments are designed and conducted using the Taguchi method, which enables efficient optimization of multiple parameters. PET-G structures are printed under varying settings, such as printing angle, layer height, line width, printing temperature and printing speed, following Taguchi's orthogonal array, allowing for a systematic analysis of how these factors influence the material's bending behavior. By combining analytical modeling with the experimental results obtained through the Taguchi approach, a deeper understanding is achieved regarding the effects of the 3D printing process on the mechanical properties of PET-G-printed structures. This method provides valuable insights into optimizing printing parameters, leading to improved mechanical performance in real-world applications.

Keywords: 3D printing · Printing parameters · PET-G · Design of experiments · Bending · Taguchi Method

1 Introduction

The history of 3D printing dates back to the 1980s when Chuck Hull invented the first stereolithography (SLA) printer, marking a milestone in additive manufacturing technologies. Throughout the 1990s, advancements such as Selective Laser Sintering (SLS) and Fused Deposition Modeling (FDM) expanded the capabilities of 3D printing, leading to wider industrial adoption. The open-source movement in the 2000s, particularly through the RepRap Project, democratized access to 3D printers, enabling researchers and hobbyists to experiment with the technology. Over the years, 3D printing has become

N. D. Lagaros et al. (Eds.): ADDOPTML 2024, LNCE 646, pp. 179–190, 2025.
https://doi.org/10.1007/978-3-031-92029-5_17

a versatile tool, addressing applications across industries. Its pivotal role during the COVID-19 pandemic, particularly in manufacturing medical equipment, underscored its potential in tackling global challenges [1].

3D printing has redefined the manufacturing process across various industries. It is widely applied in general manufacturing, aerospace, automotive, robotics, and the pharmaceutical industry. In general manufacturing, it is used for rapid prototyping, customized parts, and efficient production. Aerospace benefits from lightweight and durable components, reducing fuel consumption and cost. The automotive industry utilizes it for producing complex, lightweight, and customizable parts, enhancing vehicle performance. In robotics, 3D printing enables the creation of adaptable and durable components. The pharmaceutical sector uses it for personalized medical devices, prosthetics, and surgical models, improving healthcare outcomes. Its versatility drives innovation across diverse fields [2–5].

Among the materials utilized in 3D printing, PET-G (polyethylene terephthalate glycol) stands out due to its ease of printing, impact resistance, and chemical stability. These properties make PET-G suitable for applications requiring mechanical strength and durability, particularly under bending loads.

Recent research has investigated the mechanical behavior of 3D-printed materials under various conditions. Tiantian Li and Lifeng Wang [6] analyzed sandwich composite structures with 3D-printed cores, demonstrating enhanced energy absorption in specific lattice designs. Yi-Tang Kao et al. [7] examined bi-material structures, revealing nonlinear elastic properties through experimental and computational studies. Dalia Calneryte et al. [8] emphasized the role of geometric microstructures in mechanical strength, combining experimental testing with multi-scale finite element modeling.

Zhaobing Liu et al. [9] explored metal-polymer sandwich structures with lattice cores, identifying core cracking as the dominant failure mechanism under bending. Mohammad Reza Khosravani et al. [10] enhanced the strength and stiffness of TPU components by incorporating polyurethane foam. Dorin-Ioan Catana et al. [11] showed that complex modeling techniques better predicted the bending behavior of PLA structures compared to simplified methods.

Kenan Tüfekci et al. [12] highlighted that a $\pm 45°$ fill angle minimizes stress relaxation in PLA objects. Akshay S. Karad et al. [13] demonstrated that increasing infill density improves the tensile and flexural strength of ABS parts. Amin Montazeri et al. [14] investigated foam-filled honeycomb structures, finding that re-entrant honeycomb designs offer superior energy absorption compared to conventional hexagonal designs.

Despite these advancements, further research is needed to comprehensively understand the impact of 3D printing parameters on material behavior. This study aims to address this gap by focusing on the experimental and computational characterization of PET-G structures under bending loads, providing insights for optimizing FDM printing processes.

2 Methodology

2.1 Fused Deposition Modeling

Fused Deposition Modeling (FDM) is a widely used and affordable 3D printing method where a thermoplastic filament is heated and extruded layer by layer to create a three-dimensional object. This technique is ideal for prototypes and small production runs due to its low cost and reliability. The filament is fed through a heated nozzle, melting as it is deposited onto a build surface, usually made of heated materials for better adhesion. Stepper motors control precise movement along the X, Y, and Z axes. After each layer, the print head moves up or the build platform moves down, repeating the process until completion. Post-printing, objects may require finishing, such as surface smoothing or removing support structures.

2.2 Taguchi Method

The first phase of the analysis involves collecting data from experiments conducted using the Taguchi method, a statistical approach developed by Genichi Taguchi to improve processes and product quality by minimizing variability. Each experiment, organized through orthogonal arrays, represents a specific combination of parameters, with results recorded in tables that display parameter values alongside their corresponding performance metrics. The primary goal of the Taguchi method is to produce high-quality products at a lower manufacturing cost. To assess the impact of each variable on the output, the signal-to-noise (SN) ratio is calculated based on the mean and variance of the performance metrics for each experiment as following,

$$SN_i = 10\log\frac{\overline{y_i}^2}{s_i^2} \tag{1}$$

where

$$\overline{y}_i = \frac{1}{N_i}\sum_{u=1}^{N_i} y_{i,u} \tag{2}$$

$$S_i^2 = \frac{1}{N_i - 1}\sum_{u=1}^{N_i} \left(y_{i,u} - \overline{y}_i\right) \tag{3}$$

where i represents the experiment number, while u is the trial number. The total number of trials for experiment i is expressed as N_i. The performance characteristic value for trial u of experiment i is denoted by $y_{i,u}$. The mean value of the performance characteristic for experiment i is represented as $\bar{y}_i$, and the variance of the performance characteristic for experiment i is expressed as s_i^2.

After calculating the *SN* ratio for each experiment, the average *SN* value for each factor and level is determined and presented in a table. The range R (where $R =$ high $SN -$ low SNR) is then calculated for each parameter. A larger R value indicates a greater effect of the variable on the process, as the same change in the signal results in a more significant impact on the measured output variable.

2.3 Experimental Setup

The dimensions of the printed specimens were determined according to ISO 187:2003, with a length (l) of 80 ± 2 mm, a width (b) of 10.0 ± 2 mm, and a thickness (h) of 4.0 ± 2 mm (see Fig. 1).

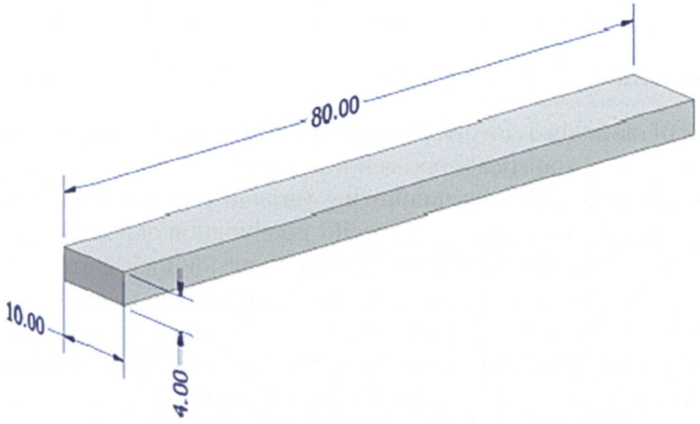

Fig. 1. Dimensions of the specimens

There are 6 printing angles to study their effect on their mechanical properties. The selection of these different print angles seeks to investigate how the direction of the layers affects the strength of the test piece during bending. The angles chosen are 0°, 18°, 36°, 54°, 72° and 90°.

The bending machine used is Testometric M350-5 which is a high precision and control testing device, which includes automated computational control of test methods, offering simplicity in operation. The device is designed for extensive testing applications in materials and products, covering stress, compression, shear, bending and other test types requirements.

3 Experimental Procedure

According to the ISO 178:2003 standard, the experimental procedure begins with measuring the width b of samples to an accuracy of 0.1 mm and the thickness h to 0.01 mm at their centers, followed by calculating the average thickness. The span L is adjusted based on the following equation,

$$L = (16 \pm 1)h_{mean} \tag{4}$$

with L set at 65.5 mm within a range of 60–68 mm. The bending stress must be positive, maintained within $0 \leq \sigma_{f0} \leq 5 \times 10^{-4} E_f$.

The testing speed is established at 2 mm/min. The sample is positioned symmetrically on the supports, and force is applied at the midpoint, with both force and deflection recorded. Finally, relevant stresses and strains are derived from the data collected, while the Taguchi method is used to set the printing parameters for evaluating the bending properties of PET-G.

The specimens were printed by Creality Ender 3 Pro printer, from PET-G filament. During printing, the platform temperature was set to 80 °C and the number of specimens printed for each experiment is 5.

The test was conducted according to the standard, with the specimen's position defined by specific dimensions. The force application cylinder has a radius $R_1 = 10.3$ mm, while the support cylinders have a radius $R_2 = 7.5$ mm. The specimen's thickness is $h = 4$ mm, the support span is $l = 80$ mm, and the distance between the support points is $L = 65.5$ mm (see Fig. 2).

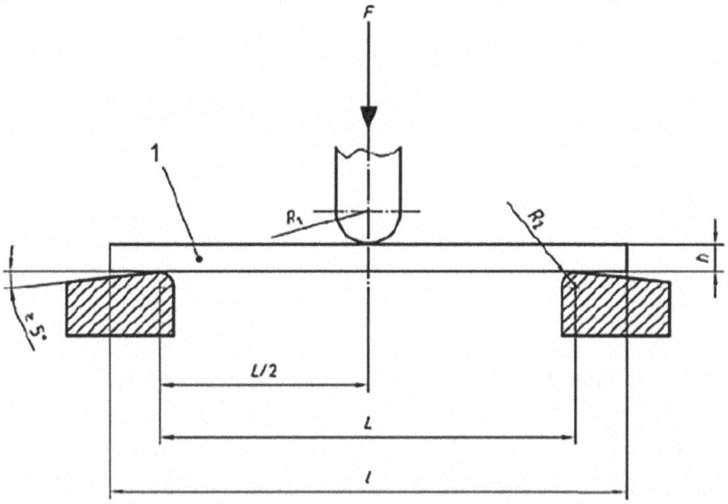

Fig. 2. Position of the specimens on the bending machine according to the ISO 178:2003 standard

To determine the flexural properties of the PET-G material, the printing parameters of each experiment were defined according to the Taguchi Method. To conduct the experiments, we utilized the L18 orthogonal array from Taguchi's method, ensuring an efficient and systematic approach to parameter variation (see Table 1). In the accompanying table, the control variables and their corresponding levels are presented, providing a clear overview of the experimental setup (see Table 2).

Table 1. Taguchi L18

Experiment number	Printing angle	Layer height	Line width	Printing temperature	Printing speed
1	1	1	1	1	1
2	1	2	2	2	2
3	1	3	3	3	3
4	2	1	1	2	2
5	2	2	2	3	3
6	2	3	3	1	1
7	3	1	2	1	3
8	3	2	3	2	1
9	3	3	1	3	2
10	4	1	3	3	2
11	4	2	1	1	3
12	4	3	2	2	1
13	5	1	2	3	1
14	5	2	3	1	2
15	5	3	1	2	3
16	6	1	3	2	3
17	6	2	1	3	1
18	6	3	2	1	2

Table 2. Different control variables and their levels

Code	Parameters	Unit	Level 1	Level 2	Level 3	Level 4	Level 5	Level 6
A	Printing angle	deg	0	18	36	54	72	90
B	Layer height	mm	0.1	0.2	0.3	–	–	–
C	Line width	mm	0.3	0.4	0.5	–	–	–
D	Printing temperature	°C	220	235	250	–	–	–
E	Printing speed	mm/s	30	50	70	–	–	–

4 Results and Discussion

In this study, 18 experiments were conducted, each consisting of 5 specimens. The results of these experiments are graphically presented in the following images (see Fig. 3a–d), illustrating the behavior of the specimens under bending load, where the *x*-axis is the flexural stress (MPa) and in *y*-axis is the flexural strain.

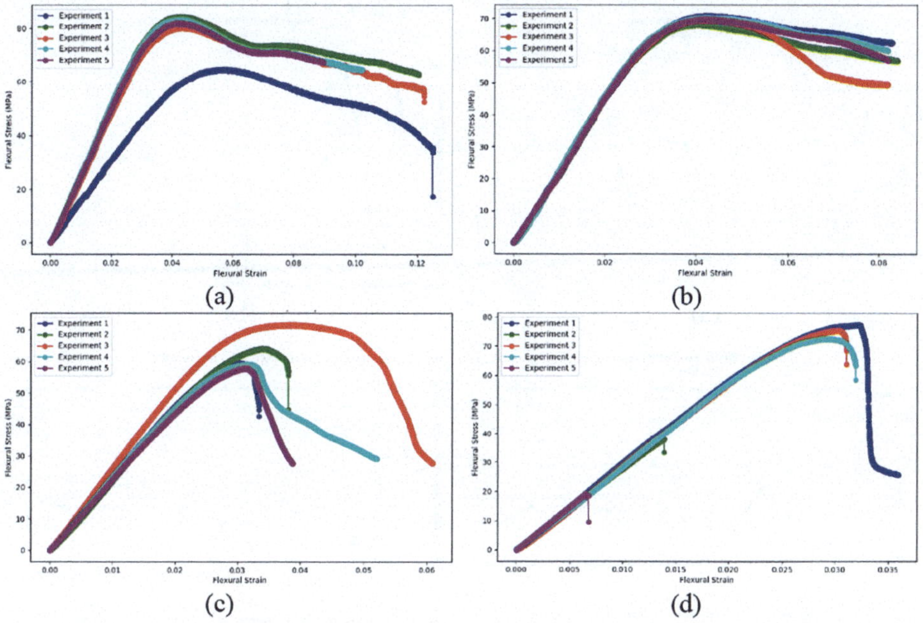

Fig. 3. Flexural stress strain curve for experiment (**a**) 1, (**b**) 6, (**c**) 12, and (**d**) 18

Additionally, this research included a main effects analysis focusing on key parameters such as the flexural modulus, maximum force, maximum deflection, maximum stress, the deflection at maximum stress, with Taguchi method, ANOVA analysis and regression analysis.

The flexural modulus quantifies a material's rigidity under bending, with higher values indicating greater resistance to deformation. Using the Taguchi method, it was observed that the printing angle significantly affects rigidity and material quality, while other parameters, such as temperature and print speed, show minor impacts.

Similarly, max force, which reflects material strength under load, and max stress, indicating load-bearing capacity, are notably reduced by the printing angle, with lesser effects observed from other factors like layer height and print temperature.

Flexibility-related properties analyzed through the Taguchi method, such as max deflection (maximum deformation before failure) and deflection at max stress (displacement under peak load), are significantly influenced by the printing angle, which decreases these values, indicating increased rigidity (Fig. 4).

Elasticity-related metrics, including max strain (maximum deformation before failure) and strain at max stress (deformation under peak load), also show pronounced reductions due to the printing angle, while other parameters contribute smaller but measurable effects. Overall, the Taguchi method highlights the printing angle as the dominant factor across all parameters, with secondary influences from layer height, temperature, and print speed (Fig. 5).

The ANOVA analysis reveals that printing angle is the most significant factor influencing the mechanical properties, such as flexural modulus, max force, max deflection,

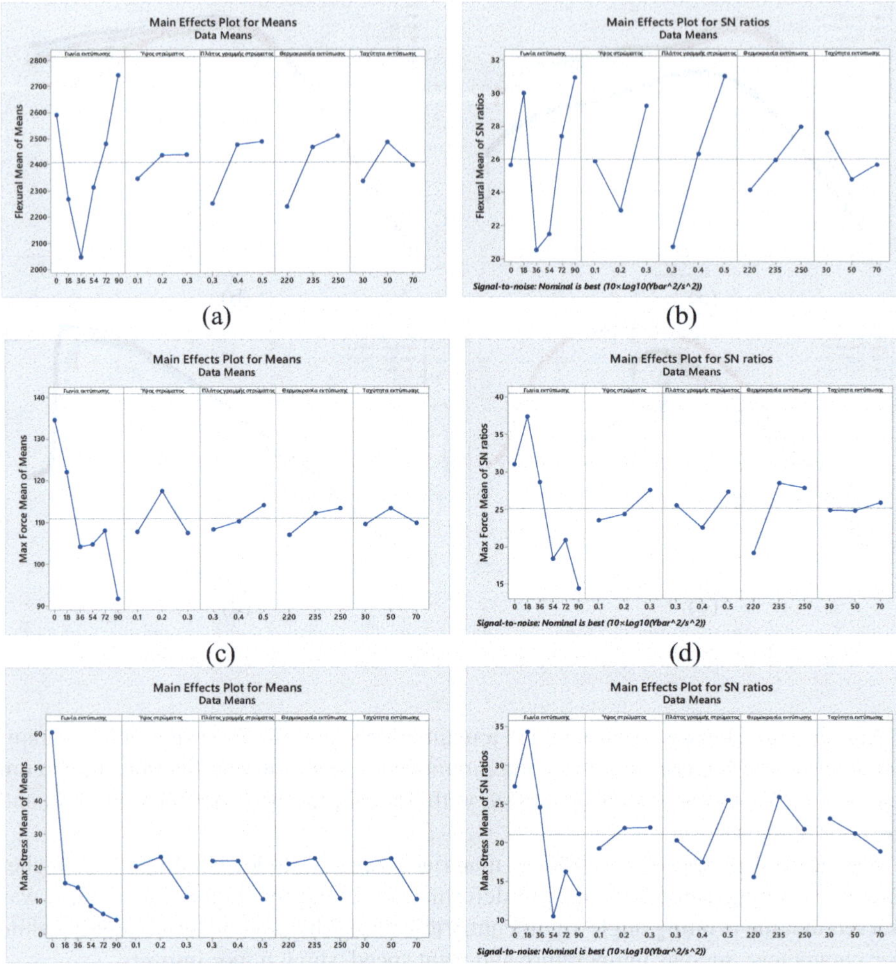

(a) (b)

(c) (d)

(e) (f)

Fig. 4. Main effects plots for means of (**a**) means of flexural modulus, (**b**) *SN* ratios of flexural modulus, (**c**) means of max force, (**d**) *SN* ratios of max force, (**e**) means of max stress, and (**f**) *SN* ratios of max stress

max stress, and max strain. In some cases, it accounts for up to 93% of the total variability (e.g., max deflection, max strain). Layer height and line width also show moderate contributions, typically ranging from 2% to 10%, depending on the property analyzed. On the other hand, printing temperature and printing speed have relatively minor effects, contributing less than 5% in most cases, though their influence is not entirely negligible. Overall, the results emphasize that the orientation of the printed layers, as determined by the printing angle, is the dominant factor in shaping the mechanical behavior of the samples, highlighting its critical importance compared to other parameters. Below, one

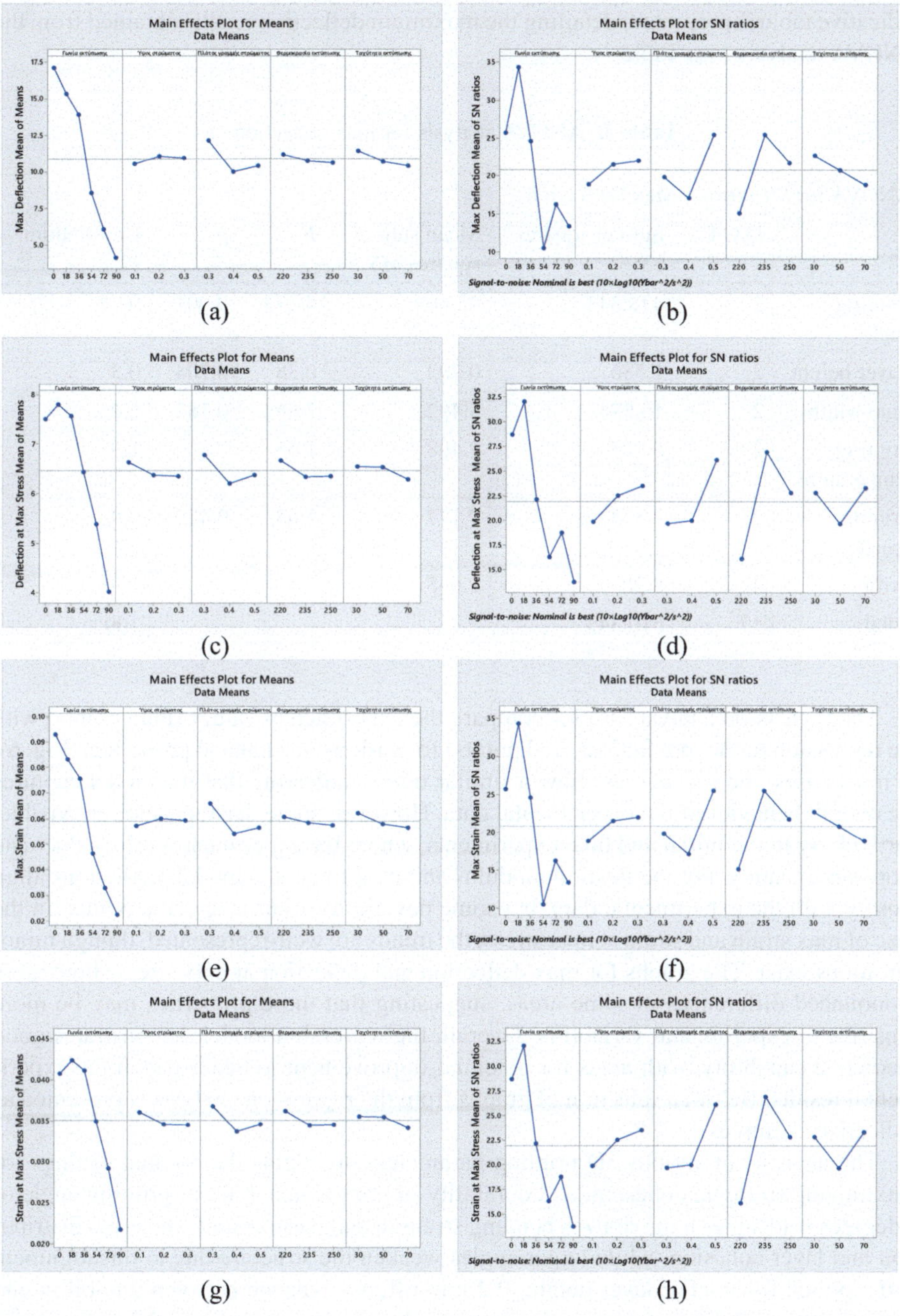

Fig. 5. Main effects plots for means of (**a**) means of max deflection, (**b**) *SN* ratios of max deflection, (**c**) means of deflection at max stress, (**d**) *SN* ratios of deflection at max stress, (**e**) means of max strain, (**f**) *SN* ratios of max strain, (**g**) means of strain at max stress, and (**h**) *SN* ratios of strain at max stress

indicative table is presented, detailing the maximum deflection results obtained from the ANOVA analysis (see Table 3).

Table 3. ANOVA analysis for max deflection

ANOVA for *SN* ratios—Max Deflection						
	D.O.F.	Sum of squares *SS*	Mean sum of squares *MS*	*F*	*p*	Contribution (%)
Printing angle	5	478.343	95.669	46.33	0.001	93.7
Layer height	2	1.586	0.793	0.38	0.704	0.3
Line width	2	10.984	5.492	2.66	0.184	2.2
Printing temperature	2	6.815	3.408	1.65	0.3	1.3
Printing speed	2	4.355	2.177	1.05	0.429	0.9
Error	4	8.26	2.065			1.6
Total	17	510.343				100

The regression analysis graphs compare the experimental values (blue points) with the regression model predictions (red points) for various mechanical properties (Fig. 6). In most cases, the two curves show a similar trend, indicating that the model captures the overall behavior of the experimental data. However, some discrepancies are visible, particularly in the initial and final experiments, where the experimental values fluctuate more significantly. For the flexural modulus and max force, the model predictions align closely with the experimental data, but some deviations occur at specific points. In the case of max strain and strain at max stress, the trends are well-represented, though minor variations exist. The graphs for max deflection and deflection at max stress show more pronounced differences in some areas, suggesting that these properties may be more sensitive to experimental variability. Overall, the regression model demonstrates good predictive capability, with areas for potential improvement to better match the experimental results. Below, a selection of graphs from the regression analysis is presented as indicative examples.

The analysis of various 3D printing parameters highlights the optimal settings for maximizing strength, cohesion, and durability in printed materials. A printing angle of 0 degrees is ideal, as it maximizes bending strength and ensures uniform stress distribution and layer cohesion, while larger angles weaken the structure due to misalignment with applied forces. For layer height, 0.2 mm offers a balance between durability and print quality, as 0.1 mm compromises stability despite finer detail, and 0.3 mm reduces durability due to fewer layers. The optimal wall line width is 0.3 mm, providing a combination of strength and cohesion, whereas wider widths like 0.4 mm and 0.5 mm decrease surface quality and structural strength. A printing temperature of 220 °C is optimal, ensuring the best material flow and layer adhesion, as higher temperatures like

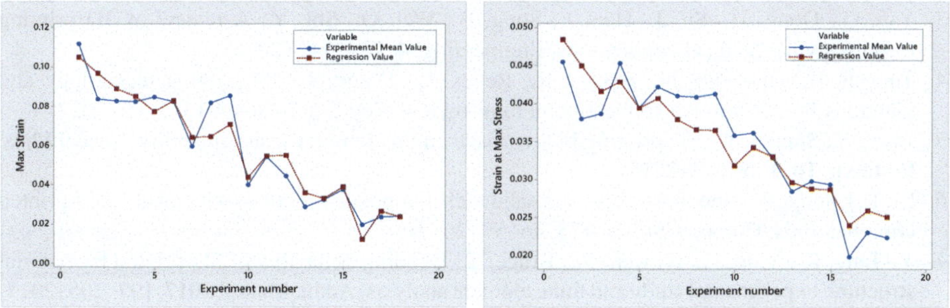

(a) (b)

Fig. 6. Regression analysis of (**a**) max strain, and (**b**) strain at max stress

235 °C and 250 °C result in excessive flow and weaker bonding. Lastly, a printing speed of 30 mm/s is recommended for its positive impact on layer adhesion and durability, as higher speeds, particularly 70 mm/s, reduce print time but lead to poor adhesion and minimal strength (Table 4).

Table 4. Final optimal settings

Optimal settings				
Printing angle	Layer height	Wall line width	Printing temperature	Printing speed
0 degrees	0.2 mm	0.3 mm	220 °C	30 mm/s

5 Conclusions

This study systematically examined the mechanical performance of PET-G 3D-printed structures under bending load, employing the Taguchi method, ANOVA analysis and regression analysis to assess the influence of various printing parameters. The results reveal that factors such as flexural modulus, maximum force, and deflection are significantly affected by the process settings, with printing angle emerging as the most critical determinant of mechanical behavior. These findings underline the importance of optimizing printing parameters to improve the structural performance and durability of 3D-printed components, with the printing angle identified as the dominant factor influencing overall material behavior.

References

1. Su, A., Subhi, J.: 3D Printing Applications in Cardiovascular Medicine. Academic, New York (2018)
2. Hopkinson, N., Hague, R., Dickens, P.: Rapid Manufacturing: An Industrial Revolution for the Digital Age, p. 175. Wiley, Hoboken, NJ (2006)

3. Yan, Q., Dong, H., Su, J., Han, J., Song, B., Wei, Q., Shi, Y.: A review of 3D printing technology for medical applications. Engineering. **4**, 729–742 (2018)
4. Tuazon, B., Custodio, N., Basuel, R., Reyes, L., Dizon, J.: 3D printing technology and materials for automotive application: a mini-review. Key Eng. Mater. **913**, 3–16 (2022)
5. Joshi, S., Sheikh, A.: 3D printing in aerospace and its long-term sustainability. Virtual Phys. Prototyp. **10**, 175–185 (2015)
6. Li, T., Lifeng, W.: Bending behavior of sandwich composite structures with tunable 3D-printed core materials. Compos. Struct. **175**, 46–57 (2017)
7. Yi-Tang, K., Ying, Z., Jyhwen, W., Bruce, T.: Bending behaviors of 3D-printed bi-material structure: experimental study and finite element analysis. Addit. Manuf. **2017**, 197–205 (2017)
8. Dalia, C., Rimantas, B., Daiva, M., Rytis, M., Neciuna, A., Armantas, O., Martynas, P., Andrius, K.: Multi-scale finite element modeling of 3D printed structures subjected to mechanical loads. Rapid Prototyp. J. **24**(3), 177 (2018)
9. Zhaobing, L., Hantang, C., Shuaiqi, X.: Mechanical performances of metal-polymer sandwich structures with 3D-printed lattice cores subjected to bending load. Arch. Civil Mech. Eng. **23**(3), 89 (2020)
10. Khosravani, M., Zolfagharian, A., Jennings, M., Reinicke, T.: Structural performance of 3D-printed composites under various loads and environmental conditions. Polym. Test. **91**, 106770 (2020)
11. Catana, D., Brus, D., Pop, M.: Modelling influence on bending behaviour simulation of the poly(lactic acid) structures, 3D printed. Polymers. **15**(4), 960 (2023)
12. Tüfekci, K., Çakan, B., Küçükakarsu, V.: Stress relaxation of 3D printed PLA of various infill orientations under tensile and bending loadings. J. Appl. Polym. Sci. **140**(39) (2023)
13. Karad, A., Sonawwanay, P., Thakur, M.: Experimental study of effect of infill density on tensile and flexural strength of 3D printed parts. J. Eng. Appl. Sci. **70**, 104 (2023)
14. Montazeri, A., Bahmanpour, E., Safarabadi, M.: Three-point bending behavior of foam-filled conventional and auxetic 3D-printed honeycombs. Adv. Eng. Mater. **25**, 2300273 (2023)

Enhancing Trend-Following Strategies in Financial Markets Using Machine Learning and Time Series Models

Spyros K. Chandrinos(✉) and Nikos D. Lagaros

Department of Structural Engineering, School of Civil Engineering, Institute of Structural Analysis and Antiseismic Research, National Technical University of Athens, 9. Heroon Polytechniou Street, Zografou Campus, 15780 Athens, Greece

Abstract. In the article, ML-based trends following techniques are used to include machine learning (ML) models in trading and investing. Old-school trend-following algorithms like Ichimoku Cloud, get complemented by cutting-edge ML methods such as Naive Bayes, K-Nearest Neighbors, Gradient Boosting, Decision Trees, and so on. They are backtested using currency pairs (2010–2023) data, and their relative performance against parameters such as total return, maximum drawdown, and mean return. Data indicates that Extreme Gradient Boost (XGB) or Naive Bayes algorithms give far better returns at low risk than the standard approach. Additionally, time series models. as TCNs, and Kalman Filters) for accurate predictions. Both models have major upsides on total returns and mean return with lower volatility. The paper describes how machine learning and time series models can be used to improve decision making in the financial markets and increase the profitability of trend-following strategies.

Keywords: Ichimoku Cloud · Trend-following · Trading strategy · Machine learning · Time-series models

1 Introduction

The Financial industry primarily in trading and investment is undergoing an infrastructural change due to machine learning (ML). With ML models entering these areas, decisions and strategies have never been so much easier to analyse data. Traditional approaches mostly based on human intuition and historical statistics are getting supplemented – or in many cases displaced – by advanced algorithms that are able to deal with immense amounts of information identify patterns and predict with faster and better precision than ever before. The ever-increasing volatility of financial markets has also made these models more widely popular allowing traders and investors to compete by being more educated data-driven decisions. With the revolving door of the financial sector. It's important for practitioners as well researchers to know the impact of machine learning on trading and investing.

In recent years, it has become very popular to use machine learning models in trading and investing in the trend following scenarios. Trend-following as presented in classics

N. D. Lagaros et al. (Eds.): ADDOPTML 2024, LNCE 646, pp. 191–201, 2025.
https://doi.org/10.1007/978-3-031-92029-5_18

such as Moskowitz, Ooi and Pedersen's paper [1] on momentum of time series, explains that the assets with strong past returns also tend to return to a high level in the short term which could be applied to future predictive modelling of financial markets. This approach also has historical support, such as the detailed review by Hurst, Ooi and Pedersen [2] that calculates trend-following strategy profitability over time across different asset classes. Chandrinos, Sakkas, and Lagaros [3]: AIRMS is a machine learning-driven risk management decision support system – allowing prediction and management of financial risk.

It's also important to note that the mechanics behind these moves have been investigated in other studies such as Fama and French's study [4] of the permanent and temporal components of stock prices, with consequences for price dynamics and the longevity of market moves. And then there are the feedback traders, where Cutler, Poterba, Summers [5] goes on to show how the movement of prices past could be the basis of the market for the future (an idea at the core of trend-following).

Managed futures, a popular market place asset which uses trend-following, and Hurst, Ooi and Pedersen [6] report a description of these strategies' risk-return profiles. The implementation guide of these techniques can be found in Clenow [7]. Providing an in-depth guide for investors who want to implement trend-following techniques effectively. Chandrinos and Lagaros [8], they propose a currency portfolio-building system using a trend-following optimized investment strategy. This method intelligently allocates funds for highest possible return based on market fundamentals and is a structured approach to managing currency portfolios in volatile markets. These authors in their turn point to the success of trend-following in all kinds of markets making the way for the addition of cutting edge machine learning algorithms to improve upon these tactics even more.

2 Trading Strategy

Our research interests are on trend-following decision-making techniques, a widely used trading method for taking advantage of longer-term price action in markets. This is the essence of trend-following – to recognize and use up or down trend in the price of an asset. These approaches operate on the presumption that, if you start a trend, it's only going to last a period of time so that investors can cash in on the price directionality. We find this decision process is especially applicable to our goal of price direction forecasting which is fundamental to saving on 3D printing expenses.

We have chosen the Ichimoku Cloud as our main trend-following indicator after conducting a lot of correlation and comparison. This multi-purpose, complete technical evaluative instrument was called Ichimoku Kinko Hyo and created by the Japanese journalist Goichi Hosoda in the late 1930s. Ichimoku Cloud gives you a full-body picture of the price direction using various lines and parts to provide information about the market, support/resistance, entry and exit points. We can simply say that "Ichimoku Kinko Hyo" meaning "one glance equilibrium chart" in Japanese perfectly reflects its ability to give you a complete market picture at a glance. The Ichimoku Cloud consists of several components, including:

- Tenkan-Sen (Conversion Line): It is the short-term force of the market.
- Kijun-Sen (Base Line): It is the market's future pattern.

- Senkou Span A (Leading Span A): It's the first edge of the cloud, as well as one of the clouds borders.
- Senkou Span B (Leading Span B): It is the cloud's second line and another boundary.
- Chikou Span (Lagging Span): It allows traders to know the intensity of the trend based on historical price movements.
- Kumo (Cloud): The cloud acts as a live support/resistance line and can be used to help traders see whether the trend reverses or continues.

3 Machine Learning Models

The use of machine learning algorithms for trading is on the rise. Proving to be much more effective than old approaches. For example, Krauss, Do, and Huck [9] tried deep neural networks. Gradient-boosted trees. & random forests for statistical arbitrage rules on the S&P 500, demonstrating that these models are capable of significantly improving trading performance. The same goes for Abe, Nakayama, and Matsui [10] predicting stock returns using statistical learning models including support vector machines and random forests, particularly with limited data sets, and found that such models successfully represent elaborate financial cycles.

Deep learning techniques were also shown to provide further machine learning advances. Fischer and Krauss [11], who were using Long Short-Term Memory (LSTM) networks to predict stock prices, showed that the capabilities of deep learning models to identify temporal dependencies in financial data improves the accuracy of predictions and trade execution. Takeuchi and Lee [12] even employed convolutional neural networks for enhancing momentum trading pointing to how deep learning can help to increase the profitability of these legacy methods.

Moreover, the application of machine learning to a dynamic trading approach, was used with reinforcement learning (Zhang and Zhou [13]). They simulated investor behavior under prospect theory to optimize consumption and portfolio selection, showing how machine learning can be used to improve financial market decision-making.

In our application of machine learning algorithms, we've mostly used them as a classification scheme to scrub generated signals to get at the ones that are most likely to make money and minimize the risk involved instead of directly boosting the overall system's performance. The models that we have used are: Naive Bayes Classifier, K-Nearest Neighbors (KNN, Gradient Boosting, Decision Trees, Random Forest, K-Means Clustering, Gaussian Mixture Models (GMM) and Extreme Gradient Boost (XGBoost).

3.1 Back-Testing Results

In our research, we have used currency pair data. Their microstructure is similar to commodity data (such as oil, steel, copper etc). We have backtested our trend following decision-making strategy from 2010 up until 2023 by applying all these machine learning models. Below we present the performance results (total return, maximum drawdown, mean return, standard deviation, number of trades) from our research (Tables 1, 2, 3, 4 and 5).

Table 1. Total returns (%) from backtesting period 2010–2023 by applying machine learning models

Total returns (%)					
	eurusd	usdjpy	audusd	gbpusd	nzdusd
Strategy	71.64	29.00	40.92	75.92	42.18
XGB	97.22	36.79	45.51	82.06	55.89
GMM	26.13	62.04	21.28	66.35	50.96
Kmeans	64.81	28.67	38.87	31.28	26.59
RF	70.39	17.92	22.94	35.57	57.11
DT	38.02	48.54	10.98	28.81	37.51
GBM	55.36	15.44	12.21	63.96	37.56
kNN	41.79	14.99	33.00	49.27	19.93
NB	88.62	43.26	49.92	96.56	62.64

Table 2. Number of trades from backtesting period 2010–2023 by applying machine learning models

Number of trades					
	eurusd	usdjpy	audusd	gbpusd	nzdusd
Strategy	1986	2070	2037	2108	2050
XGB	972	1009	913	1076	962
GMM	1149	1155	1153	1263	1301
Kmeans	1192	1172	1236	1295	1215
RF	949	931	936	1005	958
DT	995	1047	899	1041	965
GBM	951	939	909	1027	981
kNN	702	1006	692	786	688
NB	1187	1138	1238	1307	1302

Table 3. Maximum drawdown (%) from backtesting period 2010–2023 by applying machine learning models

Maximum drawdown (%)					
	eurusd	usdjpy	audusd	gbpusd	nzdusd
Strategy	16.21	25.91	15.46	12.08	11.18
XGB	6.56	16.03	12.41	9.34	7.03
GMM	20.56	15.42	13.74	14.16	8.70
Kmeans	10.19	21.71	18.60	17.19	16.50

(continued)

Table 3. (*continued*)

Maximum drawdown (%)					
	eurusd	usdjpy	audusd	gbpusd	nzdusd
RF	8.54	26.84	16.71	9.53	7.89
DT	15.31	20.41	16.65	13.96	7.51
GBM	7.50	23.45	16.92	7.18	10.02
kNN	10.72	25.22	12.61	13.26	13.71
NB	9.44	17.53	14.76	9.98	9.60

Table 4. Mean return (%) from backtesting period 2010–2023 by applying machine learning models

Mean return (%)					
	eurusd	usdjpy	audusd	gbpusd	nzdusd
Strategy	0.037	0.015	0.020	0.039	0.021
XGB	0.100	0.037	0.050	0.076	0.058
GMM	0.022	0.054	0.018	0.052	0.039
Kmeans	0.054	0.024	0.032	0.024	0.022
RF	0.074	0.019	0.025	0.035	0.059
DT	0.038	0.046	0.012	0.028	0.038
GBM	0.057	0.016	0.013	0.062	0.038
kNN	0.059	0.015	0.047	0.062	0.027
NB	0.075	0.038	0.040	0.074	0.048

Table 5. Standard deviation (%) from backtesting period 2010–2023 by applying machine learning models

Standard deviation (%)					
	eurusd	usdjpy	audusd	gbpusd	nzdusd
Strategy	0.78	0.81	0.73	0.78	0.70
XGB	0.73	0.78	0.68	0.74	0.66
GMM	0.75	0.90	0.78	0.78	0.75
Kmeans	0.74	0.91	0.79	0.76	0.73
RF	0.73	0.80	0.78	0.80	0.72
DT	0.76	0.92	0.75	0.77	0.73
GBM	0.76	0.85	0.75	0.78	0.73
kNN	0.80	0.83	0.83	0.94	0.75
NB	0.73	0.81	0.73	0.74	0.69

3.2 A Subsection Sample

The results presented above demonstrate that the application of the Extreme Gradient Boost (XGB) model leads to an increase in total returns across all assets when compared to the benchmark basic strategy. A similar improvement is observed with the Naive Bayes (NB) classifier model. Additionally, both XGB and NB models exhibit lower maximum drawdowns across all assets relative to the original strategy indicating a reduction in risk. However, the K-Nearest Neighbors (kNN) model consistently underperforms the original Ichimoku strategy across all currency pairs. The remaining models do not show consistent behavior across all assets. Another key finding is that models such as XGB. Random Forest (RF). Decision Trees (DT). Gradient Boosting Machines (GBM) and kNN significantly reduce the number of trades which is a positive outcome in terms of lowering trading costs. Despite kNN generating fewer trades and producing lower total returns than the original strategy, it yields a higher mean return, likely due to the reduced trading frequency.

4 Time Series Models

The prediction of financial time series has long been a central challenge in the field of quantitative finance given the inherent complexity, volatility, and non-stationarity of financial markets. Among the various approaches to financial forecasting time series models have emerged as a fundamental tool, offering powerful methods for analyzing and predicting future price movements based on historical data.

The time series models are primarily built to extract temporal dependences and structure of sequential financial data. Standardised models (like autoregressive integrated moving average (ARIMA) or exponential smoothing) have been widely applied because they are easy to interpret. But as high-frequency data became more available and computation power grew, newer models have emerged, such as state-space models, Generalized Autoregressive Conditional Heteroskedasticity (GARCH) models and more advanced machine learning techniques such as Long Short-Term Memory (LSTM) networks, Temporal Convolutional Network (TCN) and other deep learning systems.

Financial indexes have also been predicted using multivariate time series models, such as Alonso and Pea [14], who applied the models to build trading strategies based on forecasted index moves. Hybrids that combine traditional models with new techniques, like the ARIMA/ neural network combination that Zhang [15] has developed, have further improved prediction accuracy for the financial markets provides the best of the latest trading strategy. Engle [16] has also been influential, with his GARCH model, a powerful tool to simulate financial time series with a fluctuating volatility that helps us to predict risk and volatility in the markets. Tsay [17] is the leading work on financial time series and describes the use of models such as GARCH to estimate stock prices, volatility, and returns and it is a must-read for understanding and using these techniques in financial forecasting.

We didn't just categorize signals, we added time series models to the Ichimoku Cloud strategy to increase the precision of its original signals. With time series analysis and forecasting by the Python program "Darts", we tried to model the Ichimoku components to have a better idea of their near-future trajectory. In this way, we are able to build a more accurate system with less false signals. In particular, we used two models for modeling the Ichimoku pieces namely the TCN model and the Kalman Forecaster.

4.1 TCN Model

Bai, Kolter, and Koltun [18] give an overall empirical review of TCNs as compared to recurrent networks for sequence modeler tasks with emphasis on the performance and cost-efficiency gains of TCNs. which is an argument in favour of them for time series prediction. Lea et al. [19] first introduced TCNs in a more general manner where they have shown their ability to represent sequential data and apply that directly to time series models in finance. Another example by Zhou, Li and Gao [20] is a joint application of TCNs and LSTM networks to stock market forecasting: TCNs identify temporal dependences in financial data.

Temporal Convolutional Network (TCN) is a deep learning architecture for sequence modelling like time series prediction. It also well combats shortcomings of traditional RNNs and LSTM networks (vanishing gradients and slow training rates). The long-range dependencies in time series data can be identified using TCNs that use dilated causal convolutions and are without need of recurrent layers. This is the capability of TCNs to model non-linear and non-linear relationships in financial time series that traditional models such as ARIMA or GARCH are not able to handle. By inferring the behavior from the past, TCNs can also be used in trading applications to predict future price movements and volatility with a stronger and more adaptive prediction framework. Trading strategies can include TCNs to generate more accurate prediction, which will ultimately result in better informed trading decisions and higher performance across various financial assets.

A few basic elements to the TCN architecture:

1. Convolutional Layers: They are 1D convolutions with the given filters, kernels, and dilation. The convolutions are causal meaning they don't take future data into the present prediction.
2. ReLU activation: A ReLU activation function is a non-linear operation on the model.
3. Normalization: Weight normalization/batch normalization is used to lock in the training speed.
4. Dropout: Dropout is used to avoid overfitting.
5. Residual Blocks: Each convolutional layer belongs to a residual block which passes the block input into the output so as to preserve the gradient when it's backpropagated.

4.2 Kalman Forecaster

In his pioneering article, Kalman [21], he developed the Kalman filter that is now being used extensively in many different disciplines such as finance. Harvey [22] gives a clear and extensive discussion of Kalman filter's contribution to the predictive and structural time series modelling, providing valuable guidance for its use in financial markets. Kim and Nelson [23] describe how the Kalman filter is applied to state-space models in regime-switching settings, a feature extremely relevant to financial time series analysis. Durbin and Koopman [24] provide an exhaustive treatment of state-space models and the Kalman filter in the context of financial time series analysis, such as asset price models and risk management. Kalman and Bucy [25] take the classic Kalman filter theory to a whole new level, bringing theoretical innovations that have practical applications in financial markets such as algorithmic trading and portfolio management. In addition, Mitter and Newton [26] discuss the Kalman filter in the context of feedback control systems (which applies to financial trading systems that need to be continuously recalculated and optimized).

The Kalman Forecaster is based on the Kalman Filter, an advanced state-space model known for forecasting the state of a dynamic system from a stream of noisy measurements. The Kalman Filter is a powerful algorithm used in control systems, signal processing and economics that excels at recursively estimating process states (i.e., the actual price of an asset) and removing noise from the data.

The Kalman Filter is a critical part of time series analysis and trading strategies in the financial industry. Its capacity to deal with uncertainty and volatility also lends itself perfectly well in situations where the data is noisy and highly dynamic. As the filter is recursive, its estimations are constantly being updated with new information, which is really useful when dynamically tuning model parameters, detecting patterns, and performing predictive tasks in real-time. This feature is very helpful when tracking moving averages and applying mean-reversion strategies. We find that Kalman Filter's process of improving the prediction over time helps traders to be more dynamic and responsive, making it more accurate to make predictions and make trading more efficient in highly volatile markets.

4.3 Back-Testing Results

See Tables 6, 7, 8 and 9.

Table 6. Total returns (%) from backtesting period 2010–2023 by applying time series models

Total return (%)					
	eurusd	usdjpy	audusd	gbpusd	nzdusd
Strategy	71.64	29.00	40.92	75.92	42.18
TCN	153.99	117.71	145.71	116.79	139.24
Kalman filter	169.28	119.84	183.67	158.34	155.98

Table 7. Standard deviation (%) from backtesting period 2010–2023 by applying time series models

Standard deviation (%)					
	eurusd	usdjpy	audusd	gbpusd	nzdusd
Strategy	0.78	0.81	0.73	0.78	0.7
TCN	0.77	0.82	0.729	0.73	0.7
Kalman filter	0.76	0.82	0.731	0.78	0.7

Table 8. Mean return (%) from backtesting period 2010–2023 by applying time series models

Mean return (%)					
	eurusd	usdjpy	audusd	gbpusd	nzdusd
Strategy	0.037	0.015	0.02	0.039	0.021
TCN	0.078	0.06	0.07	0.059	0.071
Kalman filter	0.085	0.06	0.09	0.079	0.078

Table 9. Maximum drawdown (%) from backtesting period 2010–2023 by applying time series models

Maximum drawdown (%)					
	eurusd	usdjpy	audusd	gbpusd	nzdusd
Strategy	23.84	31.6	20.95	16.80	16.92
TCN	14.57	16.25	15.87	14.62	11.87
Kalman filter	13.62	19.96	12.70	22.75	12.6

It is clearly evident from these results that the introduction of time series models into the original approach improves its performance dramatically. The total returns of the strategies based on time series models consistently outperform the original strategy across all assets. Among these the Kalman filter-based strategy yields higher total returns than the Temporal Convolutional Network (TCN) strategy in all currency pairs. However, regarding the standard deviation of returns time series models do not introduce any meaningful differences. Another noteworthy finding is that the time series-based strategies effectively reduce the maximum drawdown. This. combined with higher total returns, makes the TCN and Kalman filter strategies considerably more effective in terms of risk-return performance compared to the original strategy.

5 Conclusion

This study explored the integration of machine learning models with a trend-following trading strategy. Specifically using the Ichimoku Cloud indicator to enhance decision-making in predicting market trends. By leveraging the Ichimoku Cloud's comprehensive view of market dynamics, we aimed to filter and optimize trading signals with advanced ML techniques to improve accuracy and profitability in trend-following strategies.

Our result shows that Machine learning models and especially ensemble approaches like Gradient Boosting, Random Forest can be applied in the right direction to improve classic trend-following approaches. These models outperformed less complicated classifiers such as Naive Bayes and K-Nearest Neighbors in the recognition of profitable signals and showed that they can identify the subtleties of financial markets. Specifically, Gradient Boosting and Extreme Gradient Boost (XGBoost) always gave better prediction performance and reduced risk over Ichimoku generated signals.

Furthermore, we test the theory that when you combine machine learning with technical signals like the Ichimoku Cloud, you get better and more profitable trading strategies,

Machine learning models can predict and target trading signals with the least number of false positives which makes it very superior to using just rule based models. The combination of classic financial theory and contemporary computation techniques in this hybrid solution is effective for trend-following in highly volatile and ambiguous markets.

Forecasts for movements in the financial markets have been a success using time series models with some real advantages over standard technical analysis. We will see how using the newer time series models (TCN, Kalman Forecaster etc.) in trading algorithms with Ichimoku Cloud indicator works well. The results demonstrate emphatically that combining these models significantly improves the predictive and risk-adjusted performance of trend-following methods. In particular, TCN model which accounts for long-range dependencies and non-linear relationships in financial time series improves total returns and drawdown with substantial reduction in multiple currency pairs.

The Kalman Forecaster, instead, uses recursive filtering to adjust forecasts as data change which is especially useful in volatile markets. Kalman Filter Strategy Outperformed TCN on all pairs of currency with better total returns and R/R ratios. Furthermore, because it kept on updating its estimates of market direction and volatility, it was able to take smarter, more responsive trades. Machine Learning models (XGBoost, Naive Bayes and other) are good at identifying the trades that can be profitable, but time series like TCN and Kalman Forecaster provides high total returns and lower risk. Machine learning models can filter signals better and make the trade cheaper but time series models can better identify trends and volatility in time and hence are more appropriate for long-term trading.

The research could be built upon in the future using deep learning models like Long Short-Term Memory (LSTM) networks or CNNs to achieve still better signal classification and model efficiency. Furthermore, if this hybrid method could be applied to other asset classes and market environments, then there would be a greater picture on the generalizability and scalability of these strategies. All in all, machine learning and time series models applied to the classical trading strategies holds enormous promise to improve the algorithmic trading space and give traders better tools to leverage on a growing array of increasingly complex financial markets.

Acknowledgements. The support of the Hellenic Foundation for Research and Innovation (H.F.R.I.) under the "2nd Call for H.F.R.I. Research Projects to support Faculty Members and Researchers", AMOSS project: "Additively Manufactured Optimized 3D Printed Steel Structures", (Project Number: 02779) is highly acknowledged.

References

1. Moskowitz, T.J., Ooi, Y.H., Pedersen, L.H.: Time series momentum. J. Financ. Econ. **104**(2), 228–250 (2012)
2. Hurst, B., Ooi, Y.H., Pedersen, L.H.: Demystifying managed futures. J. Invest. Manag. **11**(3), 42–58 (2013)
3. Chandrinos, S.K., Sakkas, G., Lagaros, N.D.: AIRMS: a risk management tool using machine learning. J. Financ. Eng. Risk Manag. **12**(3), 45–62 (2023)

4. Fama, E.F., French, K.R.: Permanent and temporary components of stock prices. J. Polit. Econ. **96**(2), 246–273 (1988)
5. Cutler, D.M., Poterba, J.M., Summers, L.H.: Speculative dynamics and the role of feedback traders. In: NBER Working Paper No. 3243 (1991)
6. Hurst, B., Ooi, Y.H., Pedersen, L.H.: A century of evidence on trend-following investing. J. Portf. Manag. **44**(1), 15–29 (2017)
7. Clenow, A.F.: Following the Trend: Diversified Managed Futures Trading. Wiley, Hoboken, NJ (2013)
8. Chandrinos, S.K., Lagaros, N.D.: Construction of currency portfolios by means of an optimized investment strategy. J. Financ. Optim. **15**(2), 123–138 (2023)
9. Krauss, C., Do, X.A., Huck, N.: Deep neural networks, gradient-boosted trees, random forests: statistical arbitrage on the S&P 500. Eur. J. Oper. Res. **259**(2), 689–702 (2017)
10. Abe, T., Nakayama, H., Matsui, Y.: Forecasting stock returns using statistical learning models with limited dataset. J. Econ. Dyn. Control. **87**, 66–78 (2018)
11. Fischer, T., Krauss, C.: Deep learning with long short-term memory networks for financial market predictions. Eur. J. Oper. Res. **270**(2), 654–669 (2018)
12. Takeuchi, L., Lee, Y.-Y.: Applying deep learning to enhance momentum trading strategies in stocks. In: Proceedings of the 5th Workshop on Machine Learning and Data Mining for Sports Analytics, pp. 47–56 (2013)
13. Zhang, Y., Zhou, X.Y.: Optimal consumption and portfolio choices with a prospect theory investor. Math. Financ. **19**(2), 267–291 (2009)
14. Alonso, A.M., Peña, D.: Multivariate statistical modeling and forecasting of financial indexes. J. Bus. Econ. Stat. **23**(4), 394–411 (2005)
15. Zhang, G.P.: Time series forecasting using a hybrid ARIMA and neural network model. Neurocomputing. **50**, 159–175 (2003)
16. Engle, R.F.: Autoregressive conditional heteroscedasticity with estimates of the variance of United Kingdom inflation. Econometrica. **50**(4), 987–1007 (1982)
17. Tsay, R.S.: Analysis of Financial Time Series. Wiley-Interscience, Hoboken, NJ (2005)
18. Bai, S., Kolter, J. Z., Koltun, V.: An empirical evaluation of generic convolutional and recurrent networks for sequence modeling. arXiv preprint arXiv:1803.01271 (2018)
19. Lea, C., Flynn, M.D., Vidal, R., Reiter, A., Hager, G.D.: Temporal convolutional networks for action segmentation and detection. In: Proceedings of the IEEE Conference on Computer Vision and Pattern Recognition, pp. 156–165 (2017)
20. Zhou, W., Li, L., Gao, Z.: Stock market prediction based on a TCN-LSTM model. IEEE Access. **8**, 150546–150554 (2020)
21. Kalman, R.E.: A new approach to linear filtering and prediction problems. J. Basic Eng. **82**(1), 35–45 (1960)
22. Harvey, A.C.: Forecasting Structural Time Series Models and the Kalman Filter. Cambridge University Press, Cambridge (1989)
23. Kim, C.J., Nelson, C.R.: State-Space Models with Regime Switching: Classical and Gibbs-Sampling Approaches with Applications. MIT Press, Cambridge, MA (1999)
24. Durbin, J., Koopman, S.J.: Time Series Analysis by State Space Methods. Oxford Statistical Science Series (2012)
25. Kalman, R.E., Bucy, R.S.: New results in linear filtering and prediction theory. J. Basic Eng. **83**(1), 95–108 (1961)
26. Mitter, S.K., Newton, N.J.: Kalman filtering with information-based feedback: stability and optimality. IEEE Trans. Autom. Control. **48**(9), 1615–1627 (2003)

Composite Beams with Partially Encased Steel Section Using Circular Web Openings: Flange and Web Thickness Effect

Rajai Z. Al-Rousan and Bara'a R. Alnemrawi(✉)

Department of Civil Engineering, Faculty of Engineering, Jordan University of Science and Technology, P.O. Box 3030, Irbid 22110, Jordan
bralnemrawi19@eng.just.edu.jo

Abstract. Structural systems with many components are hard to execute, costly in terms of time and money, and require high caution levels to ensure their composite interaction and optimized performance. Therefore, there is a need to find a simplified and powerful method with reduced construction work and improved actions. This study introduces a modified composite beam construction with a partially encased cellular section and eliminated shear connection parts. The steel sections have wide flanges, cellular webs with systematically arranged circular openings, and encased concrete parts. ABAQUS software was used to numerically assess the extent of the cellular section properties on the flexural behavior of the partially encased cellular beams (PECBs) using different flange thicknesses (3, 6, and 9) mm, and web thickness values (3, 4.5, and 6) mm. The interaction between the two welded horizontal and vertical parts was studied and their effect on the PECBs was introduced. It was found that the new system is innovative with superior properties to ensure favorable strength, lightweight, and ductility characteristics.

Keywords: Steel sections · Web openings · Encasement · Composite beams · Web thickness · flange thickness

1 Introduction

The utilization of simply supported composite beams made of concrete and steel sections can get the full benefit from the material properties where the concrete slab is subjected to compression stress and the steel beam is under tension stress [1, 2]. Partially encased beams are structural members where a steel section is encased by concrete where different components are connected using shear studs, hooks, or shaped steel to ensure the full connection between the various parts [3–5]. This type of composite beam is proposed to provide improved fire and corrosion resistance. Research on this type of composite has increased and continued in the last years and they were preferred due to the improved stiffness and strength properties resulting from the encasement of the steel section [6].

The method of partially encased composite beams has preferable properties compared to conventional reinforced concrete beams [7]. Partially encased beams of similar

N. D. Lagaros et al. (Eds.): ADDOPTML 2024, LNCE 646, pp. 202–213, 2025.
https://doi.org/10.1007/978-3-031-92029-5_19

depth and width values with traditional beams show higher shear and flexural resistance along with the effective contribution of the wide-flange part [8]. Additionally, significant improvements were observed in the beam's bending and flexural rigidities with enhanced contribution for the encasing concrete material. Moreover, the presence of encasing concrete plays a major role in preventing the inward local buckling within the wide flange components and the outward local buckling of the steel web [9]. The simplicity of the erection of the partially encased beams by welding or bolting makes it a suitable and easy choice that upgrades the construction process quality [10]. Finally, corrosion and fire protection were also enhanced by the presence of the encasing concrete which ends up with a reduced life-cycle cost for the composite section compared with the conventional one. Consequently, the efficiency of the wide flange members is revealed as an improved and applicable structural system in the civil engineering field.

Various experimental and analytical investigations were previously conducted in the literature and demonstrated the advantages of the encased beams over the conventional ones [11–13]. In addition, design standards are now available for this type of promising structural system which enables their easy utilization as basic components. In contrast, some major problems limit the utilization of the partially encased composite beams, including (i) the placement of stirrups that hold and stabilize the flexural reinforcement in their position, and (ii) the necessity for shear connectors to transfer the interfacial shear loads between the encased wide-flange and the encasing concrete parts [14]. Therefore, there might be congestion regions in the beam reinforcement which result in compromising the consolidation of the concrete material while increasing the possibility of premature failure. However, the congestion might be avoided using wider or deeper encased members which in turn increases the weight of the structure and consumes more material during the construction process [15].

The method of steel perforated beams plays a major role in reducing the depth of the structural member when behaving compositely with other system components, such as slabs. Moreover, down stands composite beams and conventional composite floor beams have increased flange thickness resulting in increasing the span of the structural system which in turn makes the steel section heavier than needed [8]. Finally, it could provide an efficient solution to the steel congestion and the load transferring problems along with their preferable and comparable properties. The method of partially encased composite beams with web holes called partially encased cellular beams (PECB) has become widespread in recent years.

Studying the behavior of large-scale structural systems requires a lot of time and cost if it is done by experimental testing procedure [16–18]. Therefore, the utilization of a feasible, simple, and accurate method became a must, such as finite element analysis modeling. In this study, ABAQUS software [19] was adopted to address the flexural behavior of PECB under the effect of various parameters after a well-validation was made using experimental data from the literature [20]. The behavior of the beams was examined using different geometrical properties of the cellular steel section, including the flange and the web parts. The thickness effect was highlighted when different flange thicknesses (3, 6, and 9) mm, and web thickness values (3, 4.5, and 6) mm were adopted and the interaction between them was introduced. The flexural performance was introduced in detail in this study.

2 Experimental Work Review

The flexural behavior of partially encased cellular beam (PECB) was addressed by Wang et al. [20] using experimental testing of a total of five specimens. The CB specimen was introduced with similar geometrical dimensions for that wide-flange cellular section (300 mm depth, 150 mm width, and 3000 mm length) to compare the result of the cellular steel section only and highlight the resulted improvement due to the concrete encasement in the PECBs, as appears in Fig. 1a. The flange and web thickness were designed as 4.5 mm and 3.2 mm, respectively, with 235 MPa nominal yield strength. The cellular beam is composed of nine circular holes of 200 mm diameter with the steel web spaced at 250 mm center-to-center distance.

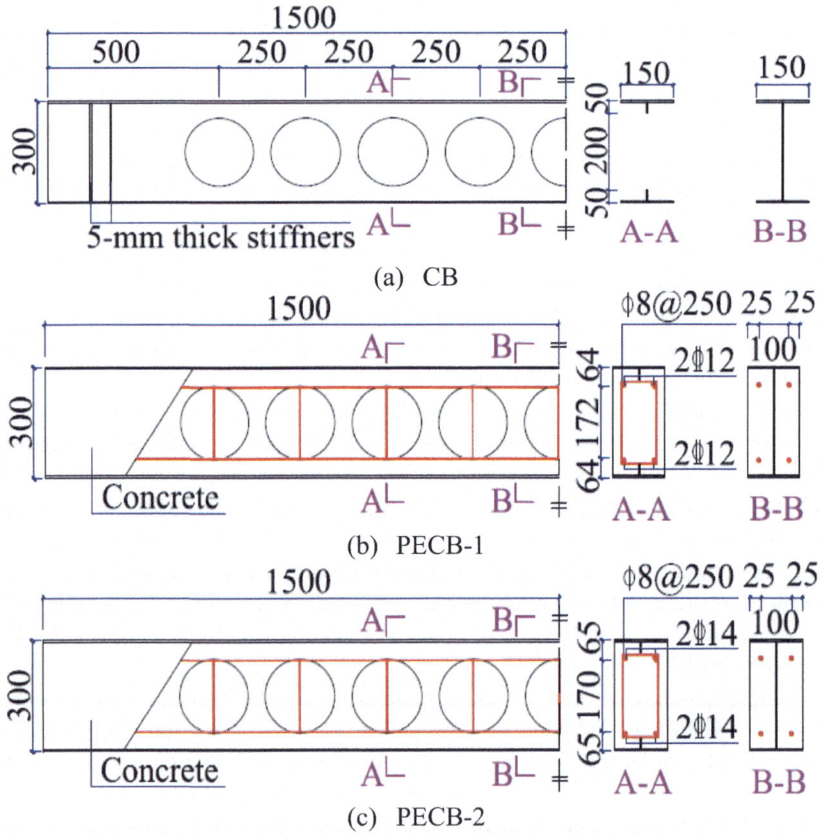

Fig. 1. Specimens geometrical and reinforcement details [20]

The PECB specimens (PECB-1 to PECB-4) were cast using the same concrete patch and the same steel reinforcement properties (longitudinal and transverse). Stirrups were closed with 8 mm diameter and 300 MPa nominal yield strength and located at each one of the circular holes, passing from one side into another and surrounded by concrete that fills the hole's gap. The PECB-1 specimen was reinforced with 400 MPa nominal yield strength rebars of 12 mm diameter where it was reinforced with two rebars at the top and two at the bottom with the absence of shear connectors, as illustrated in Fig. 1b. The shear loads were basically transferred using the stirrups and the concrete materials that pass through the circular holes where a composite action was assumed to be generated between the various components. In addition, the PECB-2 specimen was similar to the material and geometrical properties except that the diameter of the flexural reinforcement (top and bottom) was 14 mm (Fig. 1c). The PECB-1 and PECB-2 specimens were tested to address the effect of the flexural reinforcement on the system behavior and whether it is proper without increasing the depth of the beam.

The construction of the PECB specimen is initiated by the placing of the cellular steel beam with the required shear studs (if any) followed by the positioning of the stirrups at the web hole's location with the longitudinal reinforcements placed at the top and the bottom of the stirrups. Finally, the formworks were erected, and the concrete was cast at the two sides of the composite beam with a wide flange left without embedment inside the concrete. The material properties used in this study were tested according to the Chinese national standards. Concrete was tested using 150 mm cubes and resulted in 19.3 MPa compressive strength. In addition, the steel composite, flexural reinforcement, and shear stud properties were also evaluated and reported. Loading was applied using a four-point loading procedure with load-controlled and displacement-controlled procedures being applied during the pre-cracking and post-cracking stages, respectively.

3 Numerical Modeling

3.1 Material Properties and Element Types

Simulation was accomplished in this study using ABAQUS software [19] where the flexural behavior of PECBs was examined and the different components were modeled, including the concrete, steel section, longitudinal and transverse reinforcements. Concrete was modeled using a three-dimensional solid element with reduced integration, having the ability to translate and rotate in all directions at their eight nodes (C3D8R). The compressive behavior was captured using the stress-strain expression proposed by Carriera and Chu [21] where the behavior was linear elastic before the concrete cracking appeared followed by a nonlinear behavior until the maximum compressive strength was reached, and finally a softened descending branch appeared (Fig. 2a). In addition, the tensile behavior of the concrete material was captured using the stress-strain behavior proposed by Hsu [22], as shown in Fig. 2b. The behavior is initiated by a linear elastic relationship until the maximum tensile strength is attained followed by a nonlinear descending branch with 0.85 power value.

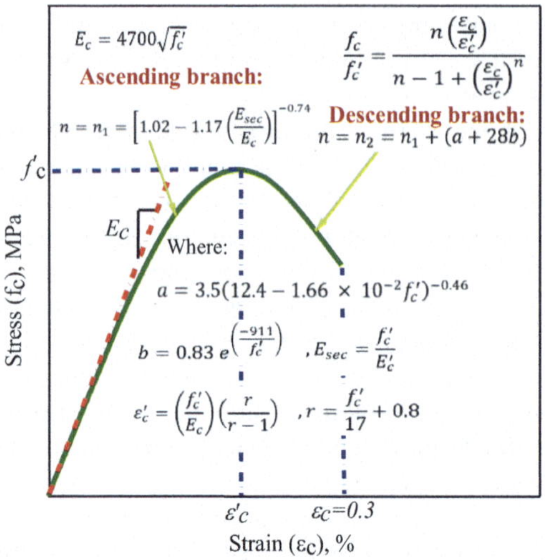

(a) Concrete under compression [21]

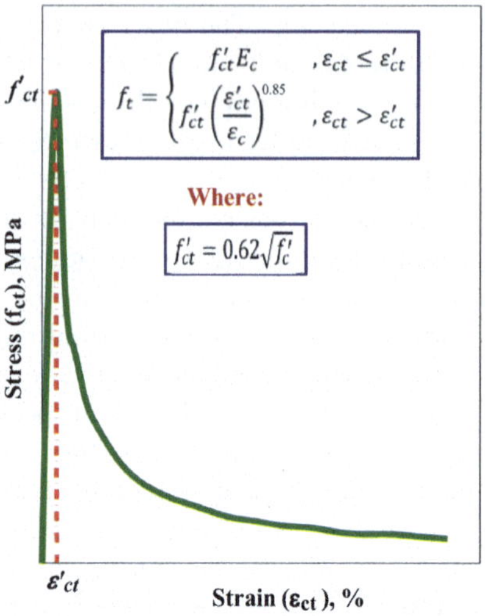

(b) concrete under tension [22]

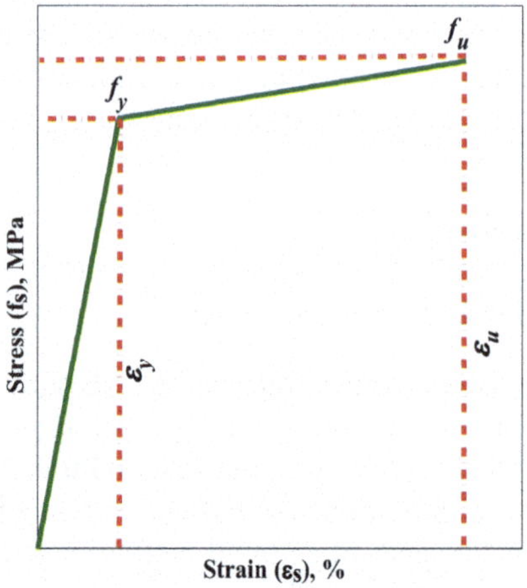

(c) Steel

Fig. 2. Concrete and steel stress-strain behavior

Steel was modeled using a three-dimensional truss element having two nodes that could carry the axial compression/tension loads. Steel components were all modeled using a bilinear strain hardening stress-strain behavior, as shown in Fig. 2c. The behavior is initiated with a linear elastic curve connecting the yield strength and zero points while the second line is the strain hardening region connecting the ultimate and yield strength points where strain is significantly increased compared to the lower increased rate for the strength.

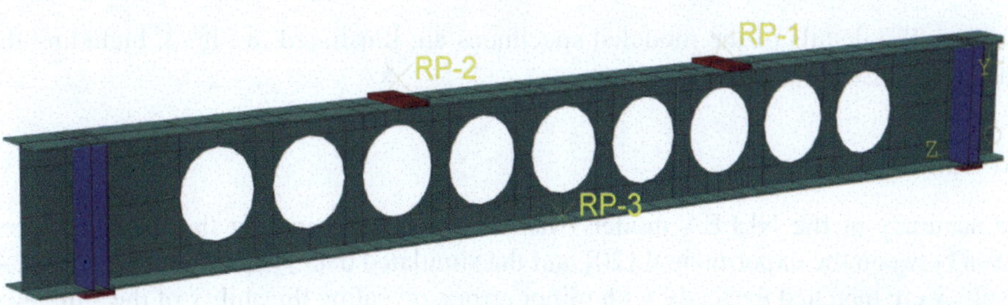

(a) CB model

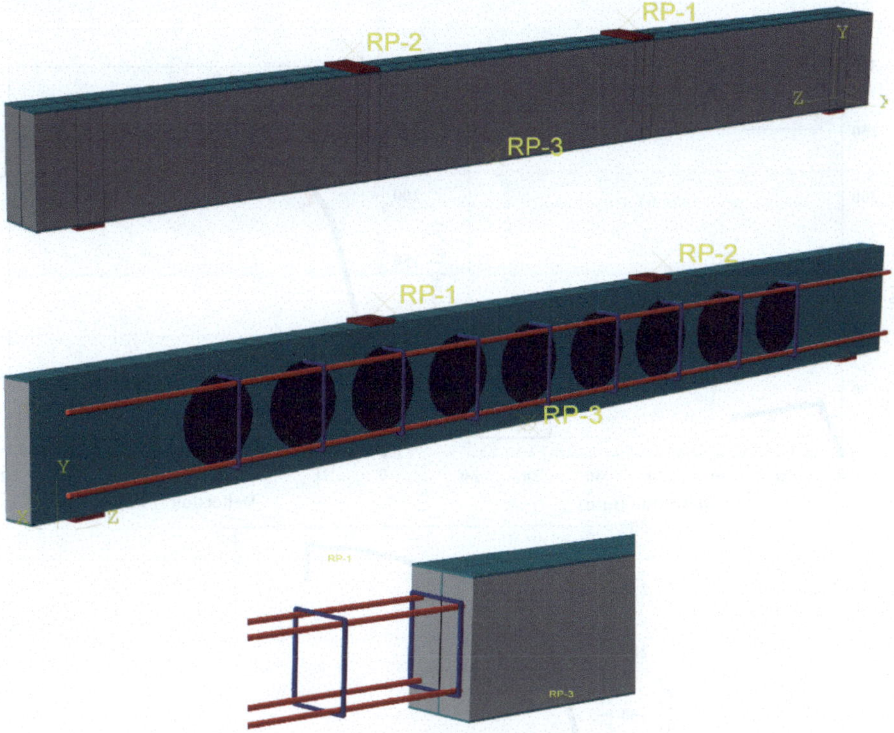

(b) PECB models

Fig. 3. Simulated models

3.2 Loading and Boundary Conditions

The flexural behavior was addressed using the four-point testing procedure where beams are supported at two points and loaded using two other points. Steel plates were provided at the supporting and loading locations and have tie constraints with the beam surface. Loading was applied using two reference points coupled in the top of the steel plates and loaded using a displacement-controlled procedure. In addition, steel reinforcement was positioned inside the concrete using the embedded reinforcement. Finally, the steel section interacted with the concrete parts using tie constraint, revealing the absence of slippage. The details of the modeled specimens are illustrated in Fig. 3, including the control and PECB specimens.

3.3 Validation

The accuracy of the NLFEA models was checked by comparing the load-deflection curves between the experimental [20] and the simulated data [19], as appears in Fig. 4. Results were matched perfectly with minor errors, revealing the ability of the simulated models to address the flexural behavior of PECBs efficiently.

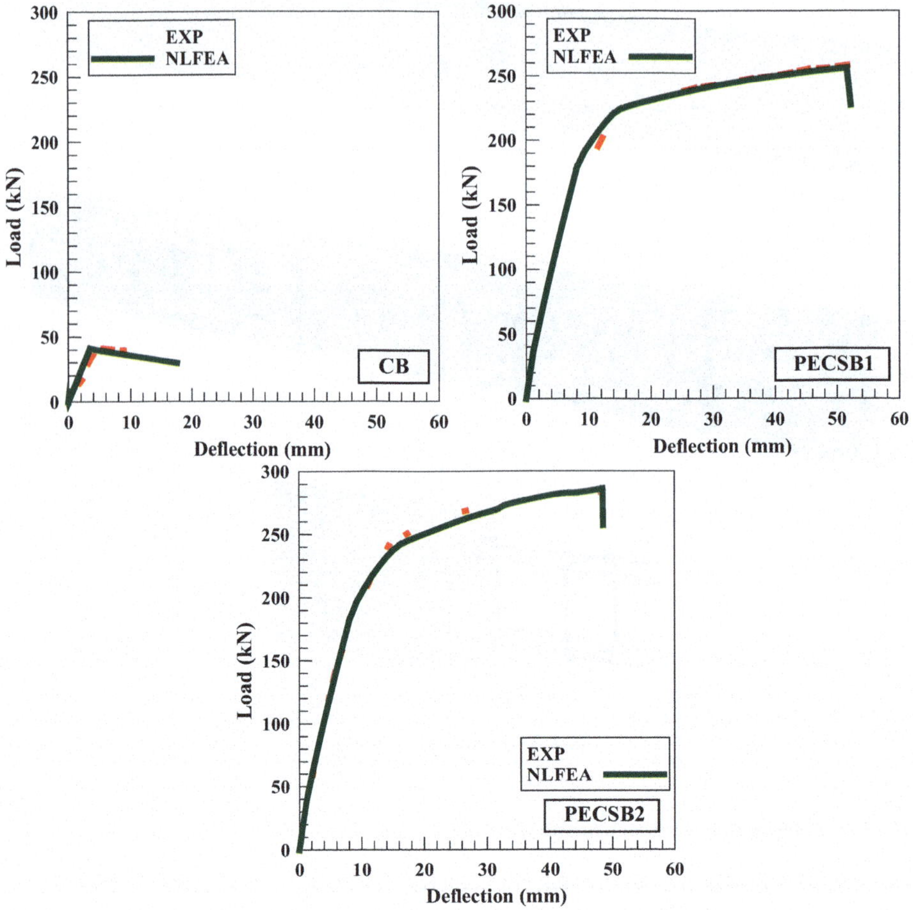

Fig. 4. Load vs. deflection curves validation [19, 20]

4 Results and Discussion

4.1 Overview

The flexural behavior of the beams was examined and compared between the control beams (steel section only with two 5 mm stiffeners at the supporting location) and the PECBs. The investigated parameters in this study include the effect of different flange thicknesses (3, 6, and 9) mm, and web thickness values (3, 4.5, and 6) mm. The results of this study are presented in three main pillars: the first one presents the behavior of the control beams (CB), the second presents the behavior of the PECBs, and the third compares the behavior of the CB and the PECBs and highlights the differences between the two systems. The study is composed of eighteen NLFEA models having the designation illustrated in Fig. 5.

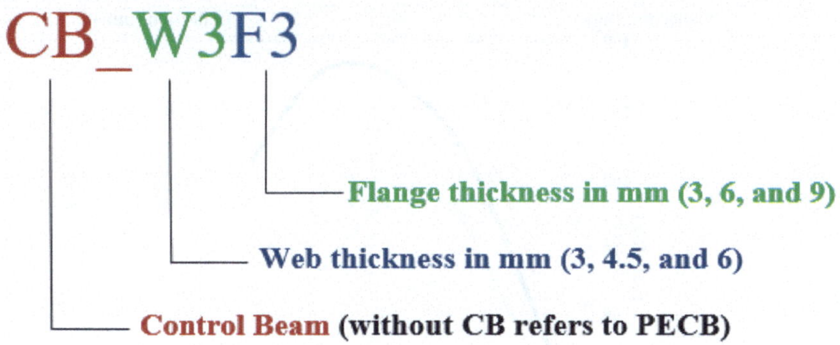

Fig. 5. Specimens' designation

4.2 Control Sections' Behavior

The behavior of the control sections is plotted in terms of their load vs. deflection curves where a total of nine sections were simulated, as presented in Fig. 6. It is clear that the behavior of the steel sections is highly dependent on the thickness of the web part that plays a major role in controlling the local buckling phenomenon that limits the behavior of the steel sections, including the strength and serviceability behavior.

It was observed that the relative behavior depends on the ratio between the web and flange thickness where the best performance was attained when the thickness values are equal. For example, the higher load capacity for the control beams in Fig. 6 was in the first and last plots once the web and flange thicknesses were similar (i.e. CB_W3F3 and CB_W6F6). However, it could be stated that the behavior depends on the local buckling situation within the web and the flange parts where higher differences between the thicknesses of the two parts resulted in scattered performance, especially if the flange thickness was higher, compared to more stability when the web thickness is higher associated with larger deflection values.

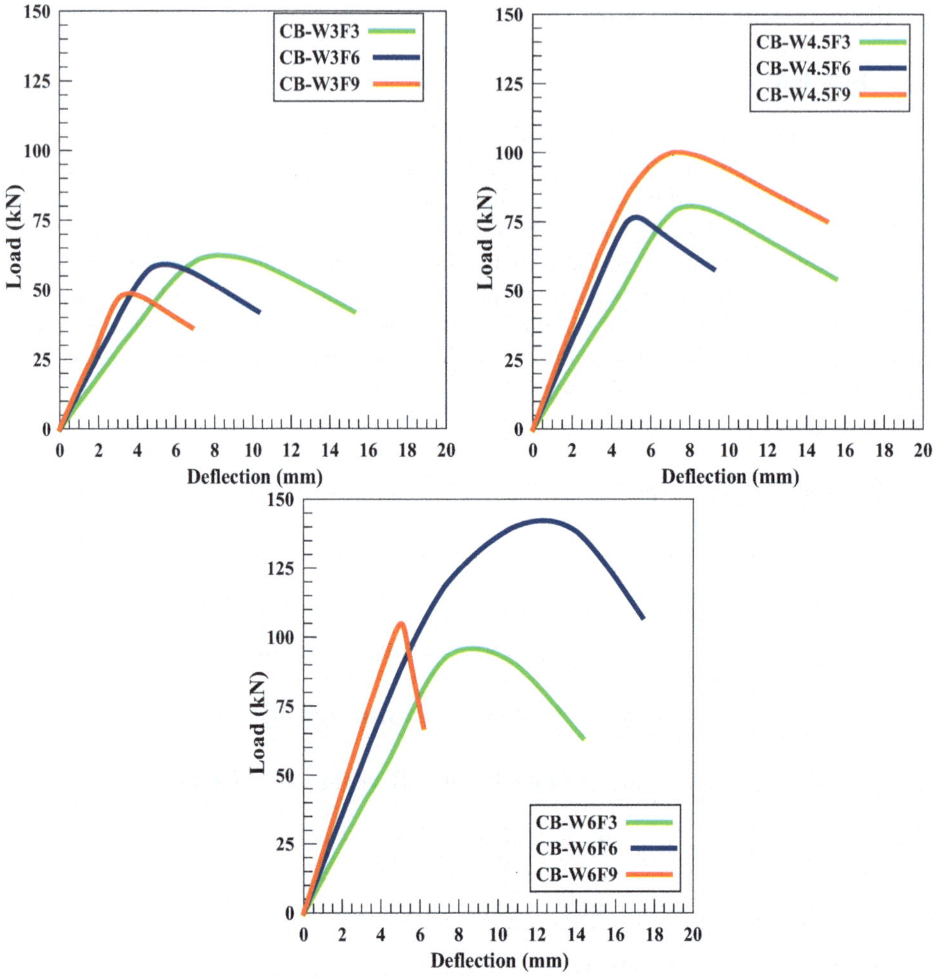

Fig. 6. Control beams load vs. deflection curves

4.3 PECB Sections' Behavior

The load vs. deflection behavior of the PECBs is presented in Fig. 7 for a total of nine specimens. The behavior was ductile and uniform under the effect of various web and flange thicknesses. Generally, increasing web thickness has a major effect on the load capacity compared to a minor effect on the deflection values. In addition, the thickness of the flange is the major contributor to the flexural strength of the beams where beams with higher flange thickness sustained more loads compared to lower deflections. Increasing the flange thickness from 3 mm to 6 mm and 9 mm increases the ultimate loads of the beam by (45.6% and 82.3%), respectively, for beams with 3 mm web thickness, compared to (44.7% and 80.4%) for beams with 4.5 mm web thickness, and (44.7% and 75.7%) for beams with 6 mm web thickness.

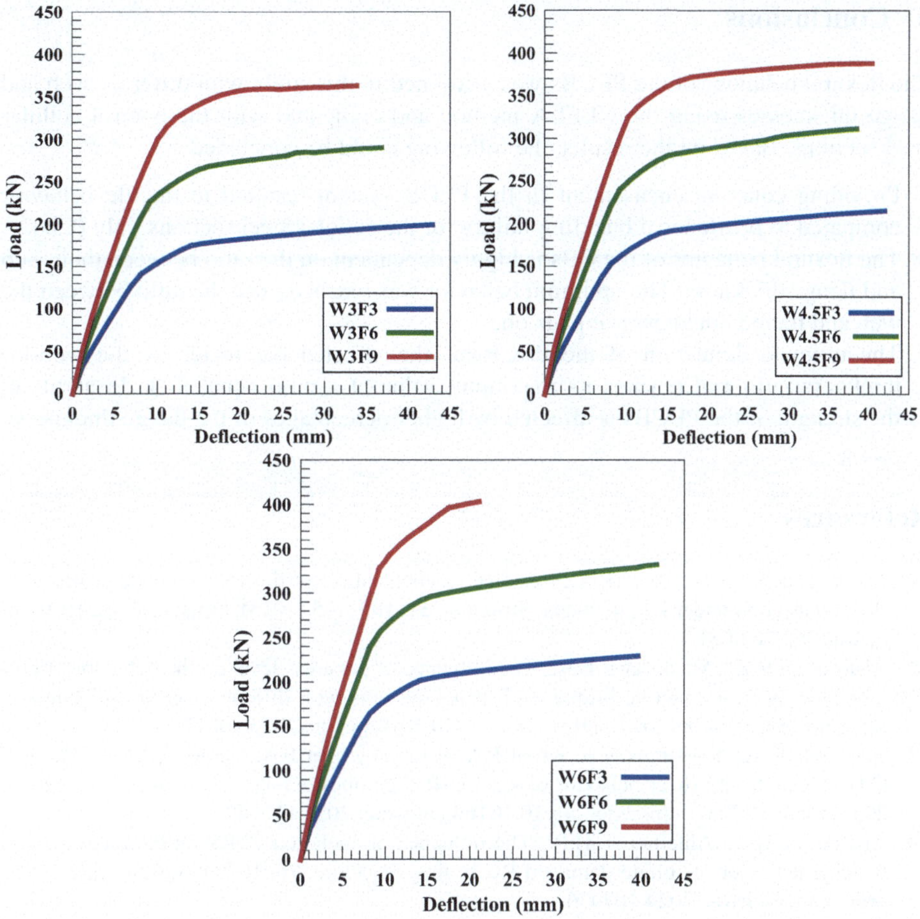

Fig. 7. PECBs load vs. deflection curves

4.4 Comparison

Comparing the behavior of the control steel section and the PECBs shows major differences between the behavior of the two types and reveals the importance of the concrete encasement provided in the PECBs. Generally, the total behavior of the section is different where control beams show brittle behavior compared to ductile ones for the PECBs that could be related to the dominant web buckling failure observed for the control beams. In addition, there is a scatter in the behavior of the control beams where the ratio between the thicknesses of the web and flange parts were dominant and optimized the behavior when it equals one. In contrast, the ratio affects the deflection values only for the PECBs.

5 Conclusions

The flexural behavior of the PECBs was examined in this study with different web and flange thicknesses using the NLFEA method and compared with the control cellular steel sections. Based on the results, the following could be concluded:

1. Providing concrete encasement in the PECB system resulted in ductile behavior compared to brittle local buckling failure for the cellular steel sections only (CB).
2. The flexural behavior of the CBs is highly dependent on the ratio between their web and flange thickness. The optimum behavior was reached once the ratio between the web and flange thicknesses equals one.
3. The ultimate deflection of the PECBs is the affected parameter by the web-to-thickness ratio and reaches its maximum value at a ratio equal to 1. In addition, the strength of the PECBs is affected by higher percentages of the flange thickness.

References

1. Xu, Y., Liu, X., Wang, B., Yi, X.: Residual mechanical properties of composite beams with PBL connectors under fatigue loads. Structure. **51**, 185–195 (2023). https://doi.org/10.1016/j.istruc.2023.03.051
2. Al-Rousan, R.Z., Alnemrawi, B.R.: The influence of prestress level on the behavior of prefabricated precast concrete bridge deck panel systems with different overlay thicknesses. Structure. **58**, 105478 (2023). https://doi.org/10.1016/j.istruc.2023.105478
3. Benedetty, C.A., dos Santos, V.B., Krahl, P.A., Rossi, A., Silva, F.A., Cardoso, D.C.T., Martins, C.H.: Flexural and shear behavior of steel-UHPC composite beams: a review. Eng. Struct. **293**, 116649 (2023). https://doi.org/10.1016/j.engstruct.2023.116649
4. Al-Rousan, R.Z., Alnemrawi, B.R.: The influence of anchored CFRP on the torsional and bending behavior of sulfate damaged RC beams. Structure. **70**, 107936 (2024). https://doi.org/10.1016/j.istruc.2024.107936
5. Al-Rousan, R.Z., Alnemrawi, B.R.: Influence of overlay strength degradation on bond stresses of bridge deck system. Heliyon. **10**(10), e31037 (2024). https://doi.org/10.1016/j.heliyon.2024.e31037
6. Zou, Y., Yu, K., Heng, J., Zhang, Z., Peng, H., Wu, C., Wang, X.: Feasibility study of new GFRP grid web—concrete composite beam. Compos. Struct. **305**, 116527 (2023). https://doi.org/10.1016/j.compstruct.2022.116527
7. Jiang, Y., Hu, X., Zheng, H., Shuai, H.: Flexural behavior of partially encased composite beams with a large tensile reinforcement ratio. Buildings. **14**(6), 1606 (2024). https://doi.org/10.3390/buildings14061606
8. Huang, H., Chen, Z., Lu, Y., Li, M., Zhang, W.: Experimental study on seismic performance of frame with steel-hollow core partially encased composite spliced beam. J. Constr. Steel Res. **212**, 108239 (2024). https://doi.org/10.1016/j.jcsr.2023.108239
9. Mansour, W., Tayeh, B.A., Tam, L.-h.: Finite element analysis of shear performance of UHPFRC-encased steel composite beams: parametric study. Eng. Struct. **271**, 114940 (2022). https://doi.org/10.1016/j.engstruct.2022.114940
10. Yan, Q., Zhang, Z., Yan, J., Laflamme, S.: Analysis of flexural capacity of a novel straight-side U-shaped steel-encased concrete composite beam. Eng. Struct. **242**, 112447 (2021). https://doi.org/10.1016/j.engstruct.2021.112447

11. Al-Rousan, R.Z., Abdalla, K.M., Alnemrawi, B.R.: The behavior of heat-damaged rc beams reinforced internally with CFRP strips. In: Ungureanu, V., Bragança, L., Baniotopoulos, C., Abdalla, K.M. (eds.) 4th International Conference "Coordinating Engineering for Sustainability and Resilience" & Midterm Conference of CircularB "Implementation of Circular Economy in the Built Environment", pp. 165–174. Springer Nature Switzerland, Cham (2024). https://doi.org/10.1007/978-3-031-57800-7_15
12. Liang, J., Zou, W., Wang, C., Li, W.: Flexural performance of partially encased steel-concrete composite beams with web opening. Structure. **64**, 106666 (2024). https://doi.org/10.1016/j.istruc.2024.106666
13. Kang, L., Hong, S., Liu, X.: Shear behaviour and strength design of cellular beams with circular or elongated openings. Thin-Walled Struct. **160**, 107353 (2021). https://doi.org/10.1016/j.tws.2020.107353
14. Azad, S., Mirghaderi, S.R., Epackachi, S.: Numerical investigation of steel and composite beam-to-encased composite column connection via a through-plate. Structure. **31**, 14–28 (2021). https://doi.org/10.1016/j.istruc.2021.01.040
15. Ahmad, S., Masri, A., Abou, S.Z.: Analytical and experimental investigation on the flexural behavior of partially encased composite beams. Alex. Eng. J. **57**(3), 1693–1712 (2018). https://doi.org/10.1016/j.aej.2017.03.035
16. Alnemrawi, B.R., Al-Rousan, R.Z.: Finite-element based-prediction of the cracking-torsional behavior of prestressed concrete beams with hollow cross-sections. Results Eng. **25**, 103712 (2025). https://doi.org/10.1016/j.rineng.2024.103712
17. Al-Rousan, R.Z., Alhassan, M., Al-wadi, R.: Nonlinear finite element analysis of full-scale concrete bridge deck slabs reinforced with FRP bars. Structure. **27**, 1820–1831 (2020). https://doi.org/10.1016/j.istruc.2020.08.024
18. Alrousan, R.Z., Alnemrawi, B.R.: The behavior of alkali-silica reaction-damaged full-scale concrete bridge deck slabs reinforced with CFRP bars. Results Eng. **16**, 100651 (2022). https://doi.org/10.1016/j.rineng.2022.100651
19. ABAQUS Dassault Systemes Simulia Corp. P: ABAQUS Analysis User's Guide (6.17), 17th edn, (2017)
20. Wang, N., Hou, H., Wang, Y., Qu, B., Zeng, X., Fang, H., Yan, H., Gao, M., Xiong, F.: Flexural behavior of partially encased cellular beams: tests and design implications. Eng. Struct. **293**, 116631 (2023). https://doi.org/10.1016/j.engstruct.2023.116631
21. Carreira, D.J., Chu, K.-H.: Stress-strain relationship for reinforced concrete in tension. J. Proc. **1**, 21–28 (1986). https://doi.org/10.14359/1756
22. Hsu, T.T.: Unified theory of reinforced concrete—a summary. Struct. Eng. Mech. **2**(1), 1–16 (1994)

Effects of Process Parameters on Characteristics of Wire Arc Additively Manufactured Steel Elements

V. Panagiotopoulos[1(✉)], N. Hadjipantelis[2], C. J. Gantes[1], and N. Lagaros[1]

[1] School of Civil Engineering, Institute of Steel Structures, National Technical University of Athens, 9 Heroon Polytechniou Street, 15780 Zografou, Greece
vasilispanagiotop@mail.ntua.gr

[2] Department of Civil and Environmental Engineering, University of Cyprus, 1 Panepistimiou Avenue, Nicosia, Cyprus

Abstract. Wire Arc Additive Manufacturing (WAAM) is a metal 3D printing method that enables the manufacture of geometrically complex metal components with lower cost and higher deposition rates relative to other metal 3D printing methods. Although numerous studies have recently been conducted on the mechanical properties and physical characteristics of WAAM steel elements, there are still uncertainties regarding the selection of the appropriate process parameters when certain mechanical or geometric characteristics are required. To address this shortcoming, a comprehensive experimental programme has been conducted to investigate the influence of the key process parameters on the weld quality and geometric properties of WAAM wall elements. Following an investigation matrix approach, the present study was carried out in two stages. Fifty-six combinations of wire feed speeds and travel speeds covering a big range of heat inputs were used to produce short WAAM walls. The walls were classified according to the weld quality and their average thickness and average layer height were determined. Subsequently, to investigate the sensitivity of the average layer height to the number of printed layers, fifteen taller WAAM walls were printed based on the high quality printing strategies from the initial short wall investigation. From this investigation useful conclusions are obtained regarding the influence of heat input during the WAAM process on the geometric and mechanical properties of the 3D-printed components.

Keywords: Geometric accuracy · Heat input · Weld quality · Wire Arc Additive Manufacturing

1 Introduction

Additive manufacturing (AM) is a 3D printing technology used to manufacture components through successive layer deposition [1]. AM is widely applied in the aerospace and, automotive industries while, recently, structural engineering applications are also explored [2]. The capabilities of AM, including the ability to fabricate customized structural components obtained from topology optimisation and to carry out in-situ repairs

N. D. Lagaros et al. (Eds.): ADDOPTML 2024, LNCE 646, pp. 214–224, 2025.
https://doi.org/10.1007/978-3-031-92029-5_20

and strengthening, are expected to have a significant impact in the construction industry [3].

Wire arc additive manufacturing (WAAM) is a direct energy deposition technique that can be used to economically produce large-scale metal components [4]. Most WAAM systems use a 6-axis robotic arm and welding equipment [5]. The WAAM process takes advantage of conventional welding techniques, in which the power source generates an electric arc to melt the feedstock welding wire, and the material is deposited layer-by-layer to obtain the desired geometry [6]. The welding area is protected from atmospheric contamination by means of shielding gas [7].

Due to the layer by layer deposition, it is impossible to completely avoid variations in the geometric and material characteristics, which affect the dimensional and mechanical properties of WAAM fabricated parts [8]. Careful calibration of the WAAM process parameters, such as the arc current and voltage, the torch travel speed, and the wire feed speed, is necessary in order to minimize such variations [9, 10]. Additional factors influencing WAAM production have been also investigated, including the contact tip-to-work distance, the torch inclination, the wire type and diameter, the interlayer temperature and shielding gas properties such as composition and flow rate [11].

In order to standardise the WAAM process, most researchers study individual process parameters to isolate their effect and ensure sufficient accuracy. Ma et al. (2015) [12] studied the effect of the interlayer temperature on the quality of the deposited material. Gokhale et al. (2019) [13] investigated the effect of the torch inclination in the geometric dimensions of obtained layers. Evans et al. (2024) [14] explored the effects of the deposition rate on the mechanical characteristics of WAAM fabricated equal-leg angle sections. Huang et al. (2024) studied the influence of temperatures on the mechanical properties of WAAM components. Huang et al. (2022) [15] researched the influence of the angle between loading direction and deposition direction on the mechanical characteristics of as-built and machined WAAM coupons subjected to tension. The tests showed isotropic response of the machined coupons, while the as-built coupons exhibited some anisotropy. Kyvelou et al. (2020) [16] and Hadjipantelis et al. (2022) [17] reported that WAAM stainless steel is more anisotropic than WAAM carbon steel.

Several studies have focused on the geometric characteristics of WAAM elements. Among others, Yildiz et al. (2020) [18] investigated the influence of the wire feed speed to travel speed ratio on the bead geometry, concluding that both the layer thickness and height increase with increasing heat input. Bharat et al. (2020) [19] studied the effect of process parameters on the layer geometry, showing that the weld bead dimensions can be directly correlated with the voltage and the wire feed speed, but they are inversely proportional to the travel speed.

In the present paper a comprehensive experimental programme is described, in which fifty-six combinations of wire feed speeds (between 4 and 16 m/min) and travel speeds (between 5 and 12 mm/s) were employed to produce short WAAM walls using heat inputs ranging between 81 J/mm and 1032 J/mm. Correlations are established between the heat input and the weld quality and geometric characteristics of the WAAM walls, including wall thickness and average layer height. Subsequently, fifteen taller walls were printed to determine the sensitivity of the average layer height to the number of printed layers. Thus, the paper contributes to more reliably characterising the weld quality and geometry of WAAM components, towards standardizing the printing process to obtain the desired mechanical properties.

2 Experimental Setup

The WAAM system employed for the purposes of the present paper is shown in Fig. 1. The system comprises an Oerlikon CITOWAVE II 520 welding power source and an ABB IRB 2600-12/1.85 6-axis robotic arm. Normal-strength ER70S-6 carbon steel wire with a diameter of 0.80 mm was used as feedstock; its composition is given in Table 1. Shielding gas comprising 82% Argon and 18% CO_2 was used at a flow rate of 16 L/min. The WAAM walls were printed using short-circuit welding mode transfer, utilising a bidirectional deposition, while the contact-to-tip work distance (CTWD) was kept constant at 11 mm (see Fig. 1). Furthermore, by utilising a pyrometer, a maximum interlayer temperature of 270 °C was ensured.

Table 1. Characteristic composition of the ER70S-6 feedstock wire (in % by weight) as provided by the manufacturer

C	Mn	Si	Ni	Cr	Mo	S	P	Fe
0.08	1.40	0.85	–	–	–	<0.03	<0.025	bal.

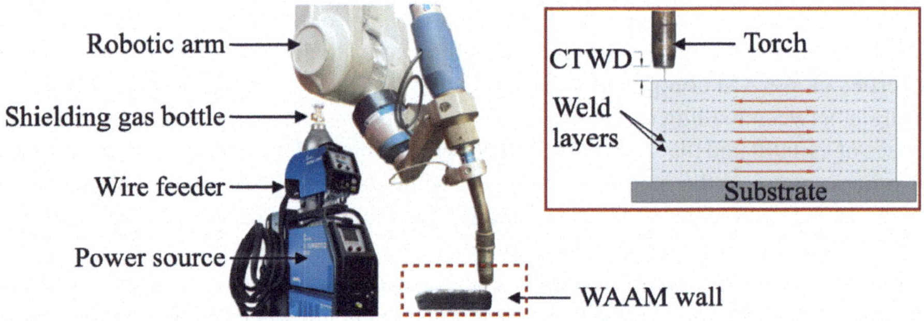

Fig. 1. Experimental setup system of the WAAM process

To investigate the influence of the process parameters on the weld quality and the geometry of WAAM wall elements, seven wire feed speeds (v_t) and eight travel speeds (v_w) were employed to produce short WAAM walls using fifty-six different deposition strategies. The nominal current (I_{nom}) and voltage (V_{nom}) are pre-programmed in the welding power source relative to the wire feed speed. The nominal heat input (Q_{nom}), expressed in J/mm, is given by Eq. (1):

$$Q_{nom} = \eta \frac{I_{nom} V_{nom}}{v_t} \tag{1}$$

where η is the thermal efficiency coefficient, taken as equal to 0.8 [20]. The printing parameters and average geometric characteristics of the short walls are given in Table 2, where t_{av} and h_{av} are the average wall thickness and average layer height, respectively.

Table 2. Process parameters and geometric characteristics of the short WAAM walls

Strategy	v_w (m/min)	v_t (mm/s)	I_{nom} (A)	V_{nom} (V)	Q_{nom} (J/mm)	t_{av} (mm)	h_{av} (mm)	Quality class
W-4-5	4	5	74	16.4	194	4.60	1.377	A
W-4-6	4	6	74	16.4	162	4.73	1.274	A
W-4-7	4	7	74	16.4	139	4.33	1.221	B
W-4-8	4	8	74	16.4	121	4.07	1.091	A
W-4-9	4	9	74	16.4	108	3.80	1.056	B
W-4-10	4	10	74	16.4	97	3.50	1.048	C
W-4-11	4	11	74	16.4	88	3.60	0.992	D
W-4-12	4	12	74	16.4	81	3.50	1.000	C
W-6-5	6	5	106	18.6	315	7.00	1.511	A
W-6-6	6	6	106	18.6	263	6.23	1.408	A
W-6-7	6	7	106	18.6	225	5.67	1.341	A
W-6-8	6	8	106	18.6	197	4.80	1.290	B
W-6-9	6	9	106	18.6	175	4.57	1.207	D
W-6-10	6	10	106	18.6	158	4.63	1.155	C
W-6-11	6	11	106	18.6	143	4.27	1.079	D
W-6-12	6	12	106	18.6	131	4.40	1.078	D
W-8-5	8	5	130	18.8	391	8.10	1.703	B
W-8-6	8	6	130	18.8	326	7.30	1.617	B
W-8-7	8	7	130	18.8	279	6.17	1.483	A
W-8-8	8	8	130	18.8	244	5.97	1.418	C
W-8-9	8	9	130	18.8	217	5.60	1.395	C
W-8-10	8	10	130	18.8	196	5.17	1.315	B
W-8-11	8	11	130	18.8	178	5.03	1.248	D
W-8-12	8	12	130	18.8	163	4.67	1.181	D
W-10-5	10	5	156	20.4	509	9.30	1.836	A
W-10-6	10	6	156	20.4	424	7.53	1.682	A
W-10-7	10	7	156	20.4	364	8.57	1.628	B
W-10-8	10	8	156	20.4	318	7.00	1.556	C
W-10-9	10	10	156	20.4	255	5.27	1.419	D
W-10-10	10	11	156	20.4	231	6.43	1.374	B
W-10-11	10	9	156	20.4	283	5.93	1.403	D

(continued)

Table 2. (*continued*)

Strategy	v_w (m/min)	v_t (mm/s)	I_{nom} (A)	V_{nom} (V)	Q_{nom} (J/mm)	t_{av} (mm)	h_{av} (mm)	Quality class
W-10-12	10	12	156	20.4	212	5.53	1.323	D
W-12-5	12	5	168	21.8	586	10.57	1.839	A
W-12-6	12	6	168	21.8	488	9.30	1.700	A
W-12-7	12	7	168	21.8	419	8.23	1.673	C
W-12-8	12	8	168	21.8	366	7.80	1.544	B
W-12-9	12	9	168	21.8	326	7.30	1.553	B
W-12-10	12	10	168	21.8	293	6.50	1.458	B
W-12-11	12	11	168	21.8	266	6.07	1.392	C
W-12-12	12	12	168	21.8	244	5.90	1.31	C
W-14-5	14	5	191	27.6	843	13.03	1.727	A
W-14-6	14	6	191	27.6	703	11.30	1.627	A
W-14-7	14	7	191	27.6	602	10.70	1.488	B
W-14-8	14	8	191	27.6	527	9.30	1.418	B
W-14-9	14	9	191	27.6	469	8.50	1.364	B
W-14-10	14	10	191	27.6	422	8.33	1.330	C
W-14-11	14	11	191	27.6	383	6.97	1.294	C
W-14-12	14	12	191	27.6	351	6.83	1.217	B
W-16-5	16	5	215	30.0	1032	14.50	1.723	A
W-16-6	16	6	215	30.0	860	13.27	1.610	A
W-16-7	16	7	215	30.0	737	11.30	1.463	B
W-16-8	16	8	215	30.0	645	10.80	1.393	B
W-16-9	16	9	215	30.0	573	9.60	1.388	C
W-16-10	16	10	215	30.0	516	10.13	1.288	D
W-16-11	16	11	215	30.0	469	8.13	1.212	D
W-16-12	16	12	215	30.0	430	7.90	1.264	D

3 Results

3.1 Geometric Analysis of Short WAAM Walls

The short WAAM walls were ten to sixteen layers tall and 60 mm long, as shown in Fig. 2. The walls were printed in a matrix format, where the columns represent the seven wire feed speeds, ranging from 4 m/min to 16 m/min, and the rows represent the eight travel speeds, ranging from 5 mm/s to 12 mm/s – see Fig. 2a. The thickness and overall height were measured at the quarter of the walls, as shown in Fig. 2b, using a

Vernier caliber. Furthermore, to evaluate the weld quality, the walls were categorised in Class A (highest quality) to Class D (lowest quality) by means of visual inspection that considered criteria such as the surface roughness and the presence of defects.

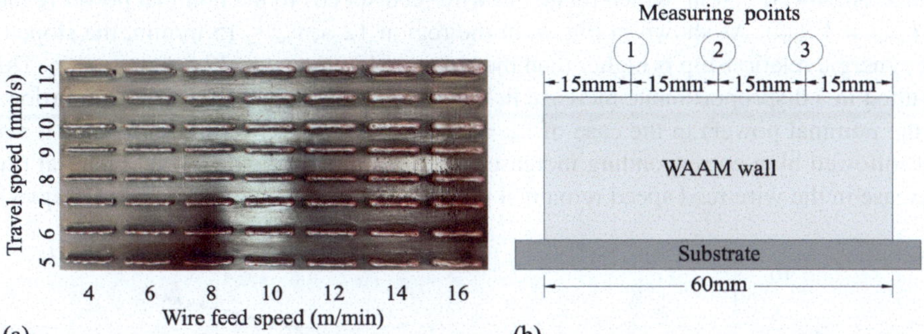

Fig. 2. (**a**) Short WAMM wall investigation matrix; (**b**) location of measurement points

The resulting average thickness and average layer height values are listed in Table 2. The variation of the average thickness and the average layer height with the nominal heat input for each strategy are shown in Fig. 3.

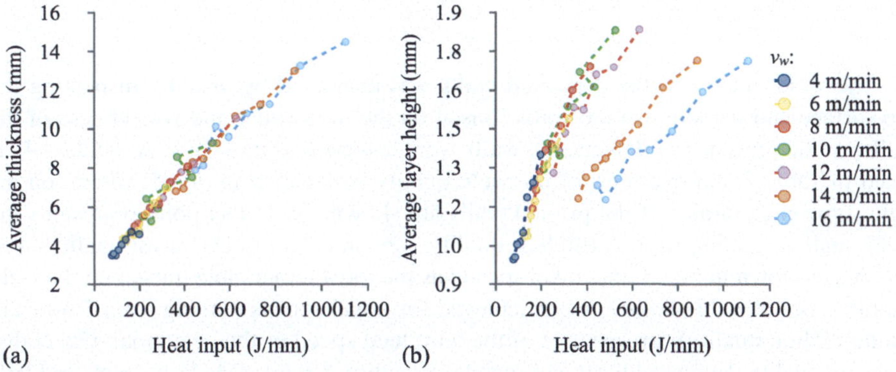

Fig. 3. Variation of (**a**) the average thickness and (**b**) the average layer height with the heat input

Generally, as expected, the higher the heat input, the greater the average thickness and average layer height. As shown in Fig. 3b, in the case of $v_w = 14$ m/min and $v_w = 16$ m/min, the variations of the average layer height exhibit a pronounced shift to the right. This shift is attributed to the pre-programmed synergic settings of the power source employed herein, which relate the wire feed speeds to the nominal power (equal to $I_{nom} \times V_{nom}$). As shown in Fig. 4, in the region $12 < v_w \leq 15$ m/min, the slope of the synergic relationship is higher than the slope within the $v_w < 12$ m/min region. This resulted in a disproportionate increase in the nominal heat input (which is proportional to the nominal power) in the case of $v_w = 14$ m/min and $v_w = 16$ m/min since it was not followed by a corresponding increase in the volume of the deposited material (the increase in the wire feed speed remained proportional between the printing strategies).

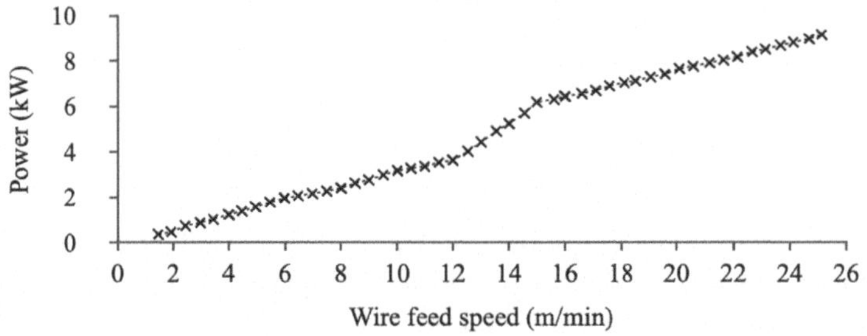

Fig. 4. Synergic relationship of the power source relating the wire feed speed with the nominal power

The weld quality of the deposited walls was evaluated by visually inspecting the top surface and the sides of the walls. Based on the uniformity and smoothness of the walls, or the presence of defects, the walls were categorised into Class A to Class 4, as listed in Table 2. An overview of the weld quality is depicted in Fig. 5, where, on the right, typical examples of the printed walls are shown. Each data point represents the weld quality, ranging from A (highest quality—green color) to D (lowest quality—red color). As shown in Fig. 5, the travel speed has the most remarkable impact on the weld quality—better weld qualities were achieved for travel speeds ranging from 5 mm/s to 7 mm/s (High quality), irrespective of the wire feed speed or the heat input. Generally, the weld quality decreased for travel speeds exceeding 8 mm/s (Medium level) and falls to even lower levels when the travel speed exceeded 10 mm/s (Low quality).

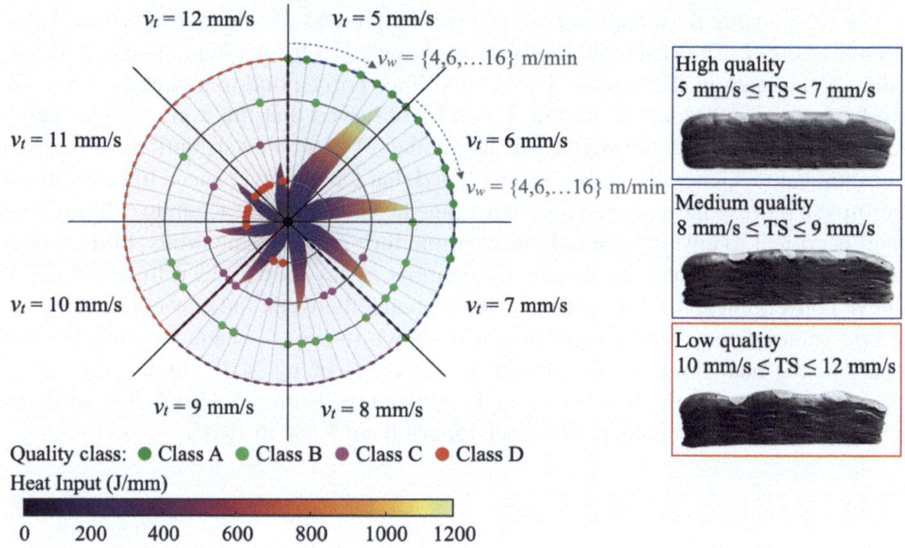

Fig. 5. Weld quality mapping based on the travel speed, the wire feed speed and the heat input

3.2 Sensitivity of Average Layer Height

To determine the sensitivity of the average layer height to the number of printed layers, taller walls were printed using fifteen of the strategies that were previously classified as Class A or Class B welds. Again, the chosen strategies covered a wide range of heat inputs. The overall wall heights were measured at 10-layer increments, based on the strategy with the smallest average layer height (i.e. Strategy 4). The average layer height was considered to have converged when the measurements deviated by less than 1% from the previous measurement. The walls of this investigation are shown in Fig. 6a. The measurements were taken at the third points, as illustrated in Fig. 6b.

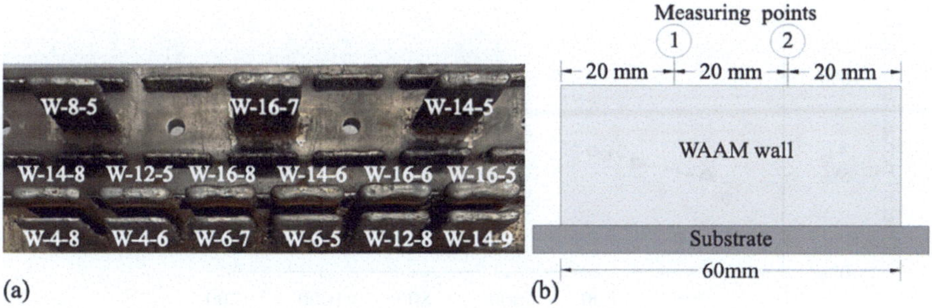

Fig. 6. (**a**) Taller WAAM wall specimens; (**b**) location of measurement points

The relationship between the average layer height and the number of printed layers is shown in Fig. 7. The heat input levels are represented by the marker colours, as defined in the colour scale in the legend. The dashed lines correspond to strategies where the wire feed speed exceeded 12 m/min. It can be observed that the average layer height decreases gradually with the increase in the number of printed layers until it converges to a constant value. Generally, it can be concluded that tens of layers need to be printed to determine a representative value for the average layer height, the accurate calibration of which is critical to ensure minimal intervention during large scale prints. Furthermore, the higher the heat input, the greater the number of layers that need to be printed to achieve convergence. The overall percentage change between the converged value and the first value of the average layer height is shown in Fig. 8, where its variation with respect to the heat input is also shown. In the cases of $v_w \leq 12$ m/min, the overall percentage change in the average layer height ranges from 1.8% to 6.2%, while for $v_w > 12$ m/min the percentage difference ranges from 5.6% to 13.4%.

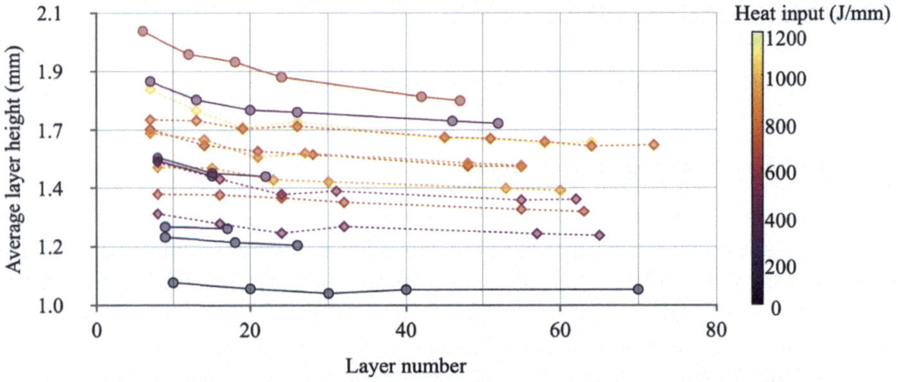

Fig. 7. Variation of average layer height with the number of printed layers

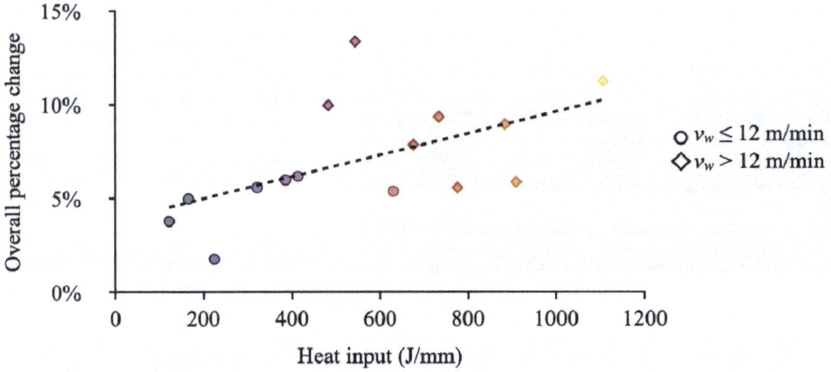

Fig. 8. Variation of the overall percentage change between the converged value and the first value of the average layer height with respect to the heat input

4 Conclusions

A comprehensive experimental programme has been conducted to investigate the manner in which the key process parameters in wire arc additive manufacturing (WAAM), namely the wire feed speed and the travel speed, affect the geometric characteristics and weld quality of thin-walled WAAM steel elements. The study comprised fifty-six combinations of wire feed speeds and travel speeds that were employed for the production of short WAAM walls, covering heat input values ranging from 81 J/mm to 1032 J/mm. The geometric measurements demonstrated a strong positive correlation between the heat input and the geometric characteristics of the WAAM walls, namely their thickness and average layer height. It is thus shown that the heat input is a key controlling parameter with regards to the printed geometry of thin-walled WAAM elements.

To assess the weld quality of the fifty-six combinations adopted herein, the WAAM walls were categorised in Class A (highest quality) to Class D (lowest quality) by means of visual inspection that considered criteria such as the surface roughness and the presence of defects. It is thus shown that, as far as the weld quality is concerned, the travel speed is the key controlling parameter. Class A and Class B cases were most frequent for travel speeds ranging between 5 mm/s and 7 mm/s, irrespective of the adopted wire feed speed. At travel speeds exceeding 8 mm/s, the weld quality decreased, indicating that the productivity of the WAAM process can be limited by the travel speed.

To examine the sensitivity of the average layer height to the number of printed layers, taller WAAM walls were also printed, based on several Class A and Class B cases, as determined from the initial short wall investigation. The analysis of the results demonstrated that the average layer height decreases gradually with the number of printed layers, with higher heat inputs requiring a higher number of printed layers before convergence is achieved.

Acknowledgements. The research was supported by the Hellenic Foundation for Research and Innovation (H.F.R.I.) under the "2nd Call for H.F.R.I. Research Projects to support Faculty Members and Researchers," AMOSS project: "Additively Manufactured Optimized 3D Printed Steel Structures," (Project Number: 02779).

References

1. Shah, A., Aliyev, R., Zeidler, H., Krinke, S.: A review of the recent developments and challenges in wire arc additive manufacturing (WAAM) process. J. Manuf. Mater. Process. **7**(3), 97 (2023)
2. Karayel, E., Bozkurt, Y.: Additive manufacturing method and different welding applications. J. Mater. Res. Technol. **9**(5), 11424–11438 (2020)
3. Delgado, C.D., Clayton, P., O'Brien, W.J., Seepersad, C., Juenger, M., Ferron, R., Salamone, S.: Applications of additive manufacturing in the construction industry—a forward-looking review. Autom. Constr. **89**, 110–119 (2018)
4. Evans, S.I., Wang, J., Qin, J., He, Y., Shepherd, P., Ding, J.: A review of WAAM for steel construction—manufacturing, material and geometric properties, design, and future directions. Structure. **44**, 1506–1522 (2022)

5. Singh, S.R., Khanna, P.: Wire arc additive manufacturing (WAAM): a new process to shape engineering materials. Mater. Today Proc. **44**, 118–128 (2021)
6. Xu, X., Ding, J., Ganguly, S., Diao, C., Williams, S.: Oxide accumulation effects on wire + arc layer-by-layer additive manufacture process. J. Mater. Process. Technol. **252**, 739–750 (2018)
7. Dhinakaran, V., Stalin, B., Ravichandran, M., Balasubramanian, M., Anand, C.C., Pritima, D.: Wire Arc additive manufacturing perspectives and recent developments. IOP Conf. Ser. Mater. Sci. Eng. **988**(1), 012102 (2020)
8. Tawfik, M.M., Nemat-Alla, M.M., Dewidar, M.M.: Enhancing the properties of aluminum alloys fabricated using wire + arc additive manufacturing technique—a review. J. Mater. Res. Technol. **13**, 754–768 (2021)
9. Raut, L.P., Taiwade, R.V.: Wire arc additive manufacturing: a comprehensive review and research directions. J. Mater. Eng. Perform. **30**(7), 4768–4791 (2021)
10. Rosli, N.A., Alkahari, M.R., Abdollah, M.F.B., Maidin, S., Ramli, F.R., Herawan, S.G.: Review on effect of heat input for wire arc additive manufacturing process. J. Mater. Res. Technol. **11**, 2127–2145 (2021)
11. Mohd, M.M.S., Raja, S., Yusof, F., Muhamad, M.R., Manurung, Y.H., Adenan, M.S., Hussein, N.I.S., Ren, J.: Integrated approach to wire arc additive manufacturing (WAAM) optimization: harnessing the synergy of process parameters and deposition strategies. J. Mater. Res. Technol. **30**, 2478–2499 (2024)
12. Ma, Y., Cuiuri, D., Shen, C., Li, H., Pan, Z.: Effect of interpass temperature on in-situ alloying and additive manufacturing of titanium aluminides using gas tungsten arc welding. Addit. Manuf. **8**, 71–77 (2015)
13. Gokhale, N.P., Kala, P., Sharma, V.: Thin-walled metal deposition with GTAW welding-based additive manufacturing process. J. Braz. Soc. Mech. Sci. Eng. **41**(12), 41–569 (2019)
14. Evans, S.I., Hadjipantelis, N., Wang, J.: Stub column tests on wire arc additively manufactured equal-leg angle sections. Eng. Struct. **317**, 118591 (2024)
15. Huang, C., Kyvelou, P., Zhang, R., Ben, B.T., Gardner, L.: Mechanical testing and microstructural analysis of wire arc additively manufactured steels. Mater. Des. **216**, 110544 (2022)
16. Kyvelou, P., Slack, H., Daskalaki, M.D., Wadee, M.A., Britton, T.B., Buchanan, C., Gardner, L.: Mechanical and microstructural testing of wire and arc additively manufactured sheet material. Mater. Des. **192**, 108675 (2020)
17. Hadjipantelis, N., Weber, B., Buchanan, C., Gardner, L.: Description of anisotropic material response of wire and arc additively manufactured thin-walled stainless steel elements. Thin-Walled Struct. **171**, 108634 (2022)
18. Yildiz, A.S., Davut, K., Koc, B., Yilmaz, O.: Wire arc additive manufacturing of high-strength low alloy steels: study of process parameters and their influence on the bead geometry and mechanical characteristics. Int. J. Adv. Manuf. Technol. **108**(11–12), 3391–3404 (2020)
19. Bharat, K.C.H., Anandakrishnan, V.: Experimental investigations on the effect of wire arc additive manufacturing process parameters on the layer geometry of Inconel 825. Mater. Today Proc. **21**, 622–627 (2020)
20. EN 1011-1:2009 Welding: Recommendations for welding of metallic materials. In: General Guidance for Arc Welding. European Committee for Standardization, Brussels (2009)

Bridging Traditional Sculpture Methods with 3D Scanning/Printing

G.-Fivos Sargentis(✉), Matina Kougia, and Nikos D. Lagaros

School of Civil Engineering, National Technical University of Athens, Heroon Polytechneiou 9, 157 80 Athens, Greece
fivos@itia.ntua.gr

Abstract. This project explores the dynamic intersection of traditional sculpture techniques and state-of-the-art 3D scanning and 3D printing technology. Two intricate clay sculptures were crafted by G.-Fivos Sargentis, serving as the initial embodiment of the creative vision. Following this traditional sculpting phase, each model underwent a transformative process, marking a synthesis of classical artistry and modern innovation. To bridge the analogue and digital realms, the clay sculptures were 3D scanned using mobile apps for smartphones, capturing the fine details and nuances of the original handcrafted forms. This digital replication paved the way for further digital sculpture processing and the utilization of 3D printing technology, as the scanned models were recreated in tangible, three-dimensional form. The printed sculptures became a canvas for the artist's continued exploration, fostering a unique dialogue between the physical and the virtual. The final models were printed in various materials, including plastic, resin (for casting) and wax (for casting). Some were then cast in bronze using the lost-wax casting method.

Keywords: sculpture · artwork · 3D printing · 3D scanning · human progress

1 Introduction

In the ever-evolving landscape of art, the convergence of traditional craftsmanship and state of the art technology has become a dynamic arena for creative exploration [1–3]. This project delves into this intersection, unravelling the symbiotic relationship between traditional sculpture techniques and state-of-the-art 3D scanning and printing technologies. Initiated in the summer of 2023, the endeavor was led by Dr. Engineer and sculptor G.-Fivos Sargentis, who crafted two intricate clay sculptures as the cornerstone of this innovative exploration.

The project transcends the boundaries of conventional artistry, seamlessly integrating classical sculpting methodologies with the transformative potential of modern technology. Following the initial sculpting phase, where the artist brought life to the clay, each model underwent a revolutionary metamorphosis, exemplifying the synthesis of time-honoured artistry and contemporary advancements.

N. D. Lagaros et al. (Eds.): ADDOPTML 2024, LNCE 646, pp. 225–236, 2025.
https://doi.org/10.1007/978-3-031-92029-5_21

Central to this creativity is the pivotal role played by 3D scanning and 3D printing technology.

Specialized mobile applications were employed to capture meticulously the essence of the handcrafted sculptures, preserving their fine details and nuances in a digital realm. This digital replication then became the gateway to a new frontier, where 3D printing technology took center stage. The scanned models were printed into tangible, 3D sculptures in wax, serving as a canvas for the artist's ongoing exploration and dialogue between the physical and the virtual.

In a blend of tradition and innovation, the final phase saw the bronze casting of these models using the lost-wax casting method. This transformation from digital to physical, from clay to wax and finally to bronze, epitomizes the project's commitment to pushing the boundaries of artistic expression while honoring the enduring legacy of traditional craftsmanship.

As we delve deeper into the intricacies of this endeavor, we present a harmonious marriage of classical artistry and modern technology, where the tangible and intangible converge to redefine the possibilities within the realm of contemporary sculpture.

This project not only showcases the transformative power of technology in the realm of sculpture but also underscores the importance of preserving and adapting traditional artistic techniques in the face of rapid technological advancements. The resulting collection is a testament to the seamless integration of classical artistry [4, 5] and digital innovation, creating a harmonious synthesis that pushes the boundaries of artistic expression in the ever-evolving landscape of contemporary art [6–10].

2 The Creation of the Clay Models

In the summer of 2023, the artist (G.-Fivos Sargentis) created two sculptures. The first was inspired by the relief of the symbol of National Technical University of Athens (NTUA) named Προμηθεύς Πυρφόρος (English translation: Prometheus the torchbearer)

(a) (b)

Fig. 1. (**a**) The symbol of NTUA Προμηθεύς Πυρφόρος (English translation: Prometheus the torchbearer); (**b**) Gia.

designed by Nikos Chatzikyriakos-Gikas (Fig. 1a) and the second was crafted in the memory of his friend's aunt, named Gia (Fig. 1b). The two sculptures were created in the region of Northern Euboea.

Inspired by the anaglyph of Nikos Chatzikyriakos-Gikas, the artist created the bust of Prometheus [11]. The phases of the creation of two sculptures in clay are depicted in Figs. 2 and 3.

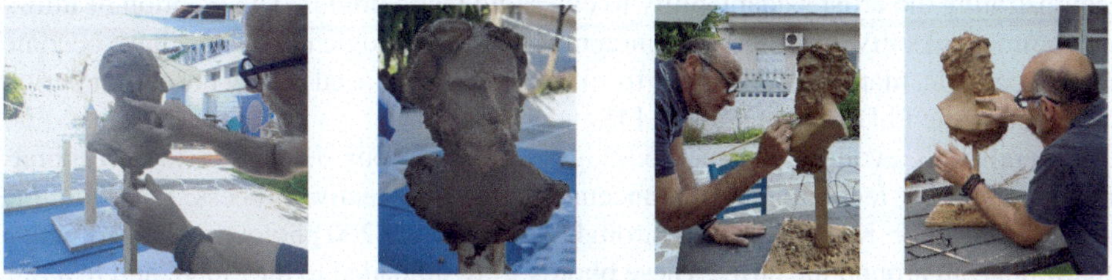

Fig. 2. Shaping of the clay on the armature for the bust of Prometheus.

Fig. 3. The complete model of the bust of Gia.

3 3D Scanning of the Clay Models

Completing the prototypes in clay in North Euboea, the artist found that the necessary specialized materials (rubber and specific plaster for art-use) required for creating molds were not available there and the prototypes could be lost as clay is not a stable material. Therefore, exploring the use of scanning the artworks was considered as an alternative However, 3D scanners [12, 13] were also unavailable.

The historical context provided underscores the evolution of 3D scanning techniques over the years. In 2005, G.-Fivos Sargentis explored photogrammetric methods for 3D scanning as part of his PhD thesis entitled the "Use and Technical Aspects of Materials in Sculpture" [14]. However, despite the initial exploration of these methods, their practical implementation was limited by the lack of sufficient computational power and commercial software at the time. The absence of these critical components limited the widespread use of photogrammetry for 3D modeling sculptures, showcasing the technological constraints of the time [15]. This historical backdrop emphasises the significant advancements that have occurred since then, as technological capabilities and software accessibility have improved, enabling more efficient and widespread utilisation of photogrammetric methods in contemporary artistic endeavors. Recently, by capturing around 200 photos from various perspectives of the sculptures, we embarked on the challenging task of composing a 3D model using commercial software. This hands-on approach signifies a transition from theoretical exploration to practical implementation, demonstrating the artist's adaptability to emerging technologies. The mention of utilizing commercial software reflects the integration of technological tools that have become more accessible, marking a departure from the limitations faced during the earlier period described in the PhD thesis in 2005 [14].

Overall, this evolution in methodology showcases our dedication to incorporating contemporary technological advancements into the creative process. We found the mobile application KIRI ENGINE, through which around 200 photos were taken with a smartphone all around the busts. These photos were uploaded to the cloud, and in a few hours, the application had composed the 3D model, which is also available online [16]. Figure 4 depicts the view of 3D model of Prometheus [17, 18].

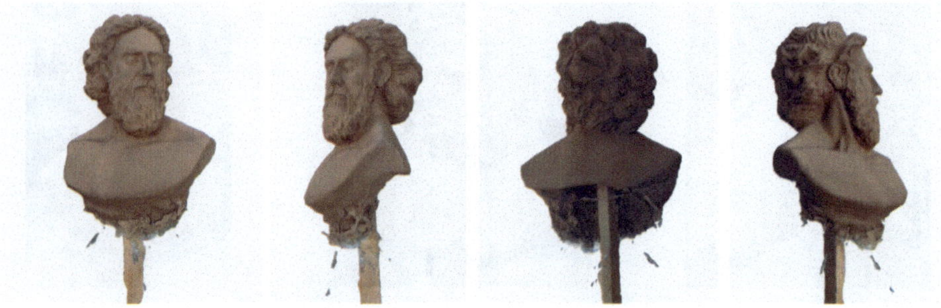

Fig. 4. The 3D model of Prometheus by KIRI Engine

4 Digital Processing of the Artwork

Nikos Chatzikyriakos-Gikas had created Προμηθεύς Πυρφόρος (English translation: Prometheus the Torchbearer), the symbol of NTUA but he had also created the emblems of the schools. Figure 5a is the embroidered gold emblem of the School of Civil Engineering, which is preserved in the Historical Rectory of the University and in Fig. 5b is the linear representation of the emblem.

Fig. 5. (a) The emblem of the School of Civil Engineering; (b) Linear representation of the emblem

Following the expression of Nikos Chatzikyriakos-Gikas, which is considered to be an abstract mix of Ancient Greek and Byzantine writing, G.-Fivos Sargentis composed the letters Προμηθεύς Πυρφόρος (English translation: Prometheus the Torch-bearer) (Fig. 17) trying to follow the artistic intention of Nikos Chatzikyriakos-Gikas. In Fig. 6, we decrypt the steps of artistic composition for the name Προμηθεύς (English translation: Prometheus) and in Fig. 7, we decrypt the steps of the artistic composition for the characterisation Πυρφόρος (English translation: the Torchbearer).

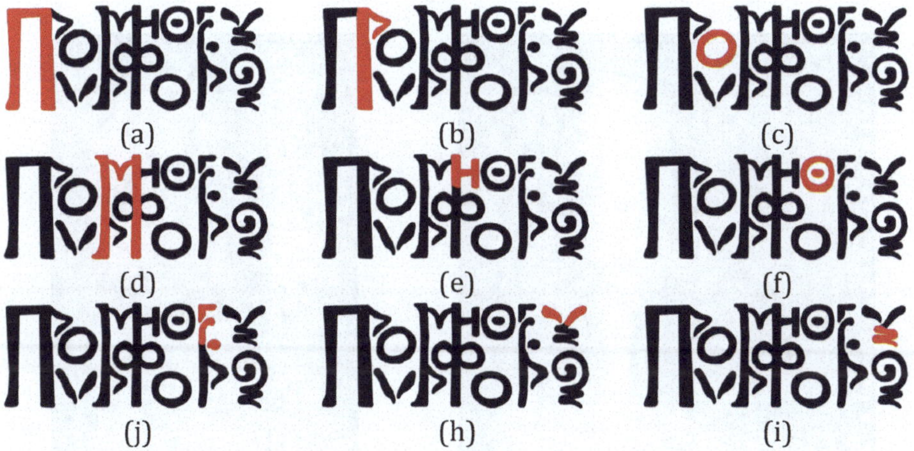

Fig. 6. Composition of the name Προμηθεύς (English translation: Prometheus) (a) Π; (b) P; (c) O; (d) M; (e) H; (f) Θ; (g) E; (h) Y; (i) Σ.

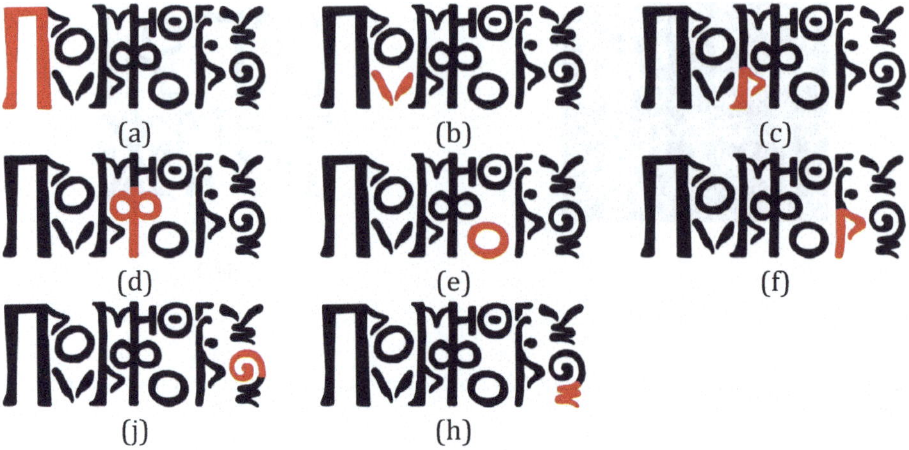

Fig. 7. Composition for the characterization Πυρφόρος (English translation Torchbearer). (a) Π; (b) Y; (c) P; (d) Φ; (e) O; (f) P; (g) O; (h) Σ.

With appropriate manipulation using software Blender, thickness was added to the black areas of Figs. 16b and 17. This process resulted in the creation of three-dimensional models for these artworks to be used as labels for the bust of Prometheus.

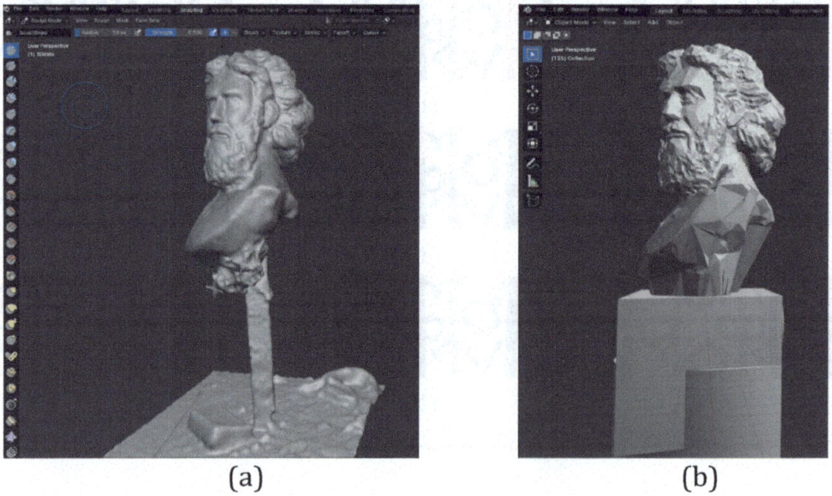

Fig. 8. (a) Scanned 3D model imported in Blender; (b) The digital processing for composition of texture

Further digital sculpture processing was converted into Blender. We imported the scanned artwork in.obj file in Blender (Fig. 8a), and with available tools for modelling we composed an abstract texture of the sculpture (Fig. 8b). The digital processing in Blender, allowed us to compose the base of the 3D model (Fig. 9) (available presentation in youtube: https://youtu.be/xW3mc9lcHlE).

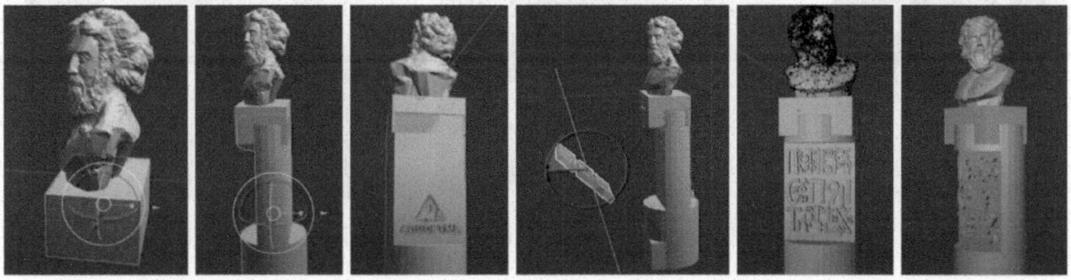

Fig. 9. Steps of the synthesis of the base of Prometheus with Blender

5 3D Printing in Wax and Synthesis of the Base and Labels

This section outlines the phase of the artistic process, focusing on the transition from digital models to tangible sculptures [3, 19]. The decision to print the models in wax for casting is notable, as it allows for the faithful reproduction of intricate details [20, 21]. The choice of various sizes suggests a nuanced approach to presentation and a consideration of spatial dynamics within the artistic context.

The accurate reproduction of object details in the wax, highlights the efficacy of the 3D printing process in preserving the artist's vision. The concluding step of assembling bases and labels in the artist's sculpture studio underscores the hands-on nature of the artistic process, combining traditional craftsmanship with modern technology. Overall, the passage provides insights into the detailed and multi-faceted nature of the artistic process, covering digital design, 3D printing, and the final physical assembly in the artist's workshop.

The importance of 3D printing in artwork lies in its ability to transform the traditional artistic process and offer unique advantages as the capability to produce the same artwork at different scales. This customization provides a level of flexibility that is challenging to achieve with traditional methods, offering artists the opportunity to explore diverse aesthetic expressions.

In addition, 3D printing facilitates the creation of replicas with consistent quality. This is particularly valuable in making art more accessible to a wider audience. Reproducing artworks at various scales ensures that individuals with different spatial constraints or preferences can appreciate and own a version of the piece, promoting inclusivity and democratization of art. This iterative process allows for fine-tuning the design and understanding how the artwork's impact may vary at different scales.

Figure 10 depicts the evolution of the composition of the three-dimensional model and the manipulation of the details of Prometheus with its base and the label.

Fig. 10. Synthesis of 3D printing model of Prometheus in sculpture workshop

6 Lost-wax Casting Process and Metal Processing for Sculpture Finishes

The lost-wax casting process was carried out at the Manos Vardaxis foundry.

The wax models were placed in what is called a casting tree. Around the wax models, wax channels were positioned, which, when melted in the second stage, along with the wax of the pattern, will form the inputs for the metal casting (Fig. 11a, b). After the mold was covered with casting plaster, it was placed in an oven, baked for 2 days and the wax was removed. Metal was then cast into the mold. The sculptures, after casting, are shown in the Fig. 11c.

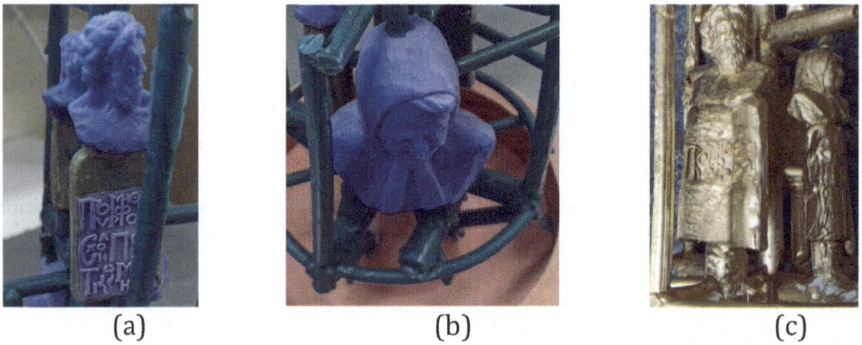

Fig. 11. (a, b) The casting trees; (c) The casted models in bronze

The process of shaping the metal is one more battle with the material that follows specific steps. It starts with cutting the casting channels, local grinding with specific tools, and further grinding to smooth out the created edges. One more objective is to level the base to prevent the sculpture from tilting, which is not a simple task even with the appropriate tools (Fig. 12).

Fig. 12. Shaping the metal (a) the beginning of the process; (b) cutting the casting channels

Following the completion of shaping and finalizing the volumes of the artwork, the patina is crafted. A specialized substance is applied to the metal, inducing swift oxidation on its surface, resulting in a uniform darkening (Fig. 33). Careful and meticulous lighting adjustments are necessary to achieve the desired effect. Specific areas are delicately rubbed with a wire sponge designed for the process. After the metal has dried, further processing takes place to accentuate the desired parts for its ultimate shine. Lastly, a solution of natural wax of pine resin in turpentine is carefully brushed onto the sculpture, acting as a protective varnish.

The final shine of the artwork is achieved after the wax solution in turpentine has dried, with systematic rubbing of the sculpture using a soft cloth (Fig. 34).

7 The Completed Sculpture Projects

The completed sculpture projects are presented in Figs. 13, 14 and 15.

Fig. 13. The bust of Prometheus. 3D printing in wax and cast in bronze

Fig. 14. The bust of Prometheus. 3D printing in plastic

Fig. 15. The bust of Gia. 3D printing in wax and cast in bronze

8 Conclusions

The works which were presented in this case study, Prometheus and Gia, would have been lost if there had been a commitment only to traditional mold-making practices, as the region where they were created lacked the necessary tools and materials at that time.

Using the 3D scanning method, the artworks were preserved in digital models and could be printed and reproduced through 3D printing. This process is estimated not only to have preserved artistic creation but also, to have added value, provided freedom of expression, adaptation to various sizes and ultimately capitalised on artistic creation.

In essence, 3D printing in artwork not only introduces novel possibilities for artistic expression but also addresses practical considerations of accessibility, efficiency and technological innovation. The ability to replicate artworks at various scales marks a transformative shift in the art landscape, enhancing both the creative process and the audience's engagement with art.

Acknowledgments. This research was supported by the ADDOPTML project: "ADDitively Manufactured OPTimized Structures by means of Machine Learning" (No: 101007595) belonging to the Marie Skłodowska-Curie Actions (MSCA) Research and Innovation Staff Exchange (RISE) H2020-MSCA-RISE-2020.

References

1. Bourella, D.L., Beaman, J.J., Leub, M.C., Rosenc, D.W.: History of additive manufacturing and the 2009 roadmap for additive manufacturing: looking back and looking ahead. In: Proceedings of the RapidTech 2009: US-TURKEY Workshop on Rapid Technologies, Istanbul, pp. 1–8 (2009)
2. Horvath, J.: A brief history of 3D printing. In: Mastering 3D Printing. Apress, Berkeley, CA (2014)
3. Yu, J.: The application of 3D printing technology in sculpture. In: MacIntyre, J., Zhao, J., Ma, X. (eds.) Advances in Intelligent Systems and Computing, Proceedings of the 2020 International Conference on Machine Learning and Big Data Analytics for IoT Security and Privacy, SPIOT 2020, vol. 1283, p. 1283. Springer, Cham (2021)
4. Mattusch, C.: The berlin foundry cup: the casting of Greek Bronze statuary in the early fifth century BC. Am. J. Archaeol. **84**, 435–444 (1980)
5. Hasaki, E.: Craft apprenticeship in ancient Greece: reaching beyond the masters. In: Wendrich, W. (ed.) Archaeology and Apperenticeship, Body Knowledge, Identity and Communities of Practice. The University of Arizona Press, Tucson, AZ (2012)
6. Sargentis, G.-F., Frangedaki, E., Chiotinis, M., Koutsoyiannis, D., Camarinopoulos, S., Camarinopoulos, A., Lagaros, N.D.: 3D scanning/printing: a technological stride in sculpture. Technologies. **10**, 9 (2022). https://doi.org/10.3390/technologies10010009
7. Sargentis, G.-F., Hadjibiros, K., Papagiannakis, I., Papagiannakis, E.: Plastiras Lake: influence of the relief on the revelation of the water presence. In: 9th International Conference on Environmental Science and Technology. Department of Environmental Studies, University of the Aegean, Rhodes Island (2005)
8. Sargentis, G.-F., Dimitriadis, P., Koutsoyiannis, D.: Aesthetical issues of Leonardo da vinci's and Pablo picasso's paintings with stochastic evaluation. Heritage. **3**, 283–305 (2020). https://doi.org/10.3390/heritage3020017
9. Sargentis, G.-F., Dimitriadis, P., Iliopoulou, T., Koutsoyiannis, D.: A stochastic view of varying styles in art paintings. Heritage. **4**, 333–348 (2021). https://doi.org/10.3390/heritage4010021
10. Sargentis, G.-F., Ioannidis, R., Chiotinis, M., Dimitriadis, P., Koutsoyiannis, D.: Aesthetical issues with stochastic evaluation. In: Belhi, A., Bouras, A., Al-Ali, A.K., Sadka, A.H. (eds.) Data Analytics for Cultural Heritage. Springer, Cham (2021). https://doi.org/10.1007/978-3-030-66777-1_8
11. Chatzimpiros, P., Lagos, G., Sargentis, G.-F.: Man and city. In: Myth and vision, 10th International Conference on Environmental Science and Technology. Department of Environmental Studies, University of the Aegean, Kos Island (2007)
12. Heritage, L.G., Large, A.R.G.: Principles of 3D laser scanning. In: Heritage, L.G., Large, A.R.G. (eds.) Laser Scanning for the Environmental Sciences. Wiley-Blackwell, Oxford (2009)

13. Levoy, M., Pulli, K., Curless, B., Rusinkiewicz, S., Koller, D., Pereira, L., Ginzton, M., Anderson, S., Davis, J., Ginsberg, J., et al.: The Digital Michelangelo Project: 3D Scanning of Large Statues. University of Stanford, Stanford, CA (2000)
14. Sargentis, G.-F.: Use and Technical Aspects of Materials in Sculpture. Ph.D. Thesis, School of Architecture, National Technical University of Athens, Athens, Greece (2005)
15. Williams, L.D.: Additive manufacturing or 3d scanning and printing. In: Manufacturing Engineering Handbook. McGraw-Hill, New York, NY (2015)
16. The bust of Prometheus in kiriengine: https://www.kiriengine.app/share/ShareModel?code=NT4JIH&serialize=301660d7c96940b8b792f189c69a99ce
17. Luhmann, T.: A historical review on panorama photogrammetry. Int. Arch. Photogramm. Remote. Sens. Spat. Inf. Sci. **34** (2008) https://www.researchgate.net/publication/228766550_A_historical_review_on_panorama_photogrammetry
18. Kiriengine. https://www.kiriengine.app/
19. Hughes, B., Wilson, G.: 3D/additive printing manufacturing: A brief history and purchasing guide. Technol. Eng. Teach. **75**, 18 (2015)
20. Mattusch, C.: The Casting of Greek Bronzes: Variation and Repetition. In: Proceeding of the Symposium of Small Bronze Sculpture from the Ancient World, March 1989 at the Getty Museum. The J. Paul Getty Museum, Los Angeles, CA (1990)
21. Koroneos, A., Sargentis, G.-F.: Casting of the Bust of Professor A. Prokopiou, Research Report. Laboratory of Building Materials, National Technical University of Athens, Athens (2005)

Fast-Track Modeling of the Landscape for Hydraulic Studies, Using Drones and Photogrammetry in Field Research

G.-Fivos Sargentis(✉), Evangelia I. Frangedaki, Theano Iliopoulou, Panayiotis Dimitriadis, Stefanos Voulgaris, and Nikos D. Lagaros

School of Civil Engineering, National Technical University of Athens, Heroon Polytechneiou 9, 157 80 Athens, Greece
fivos@itia.ntua.gr

Abstract. Hydraulic studies involve assumptions and uncertainties due to the complexity of physical systems. Variations in the soil and its properties, which affect the flow of water, contribute to these uncertainties. Techniques such as soil modeling by field research are used to constrain this parameter. Digital terrain models play a critical role in representing the surface, recording elevations and slope, which determines both the path and speed of water. An effective method to minimize uncertainty is to take successive images of the landscape by unmanned aerial vehicles, which with appropriate photogrammetric processing, can produce the 3D terrain model in detail, with the materials and texture of the soil. These models offer a more accurate understanding of the area compared than field research and digital terrain models. By incorporating critical parameters such as terrain and channel roughness, engineers can calibrate the simulation of water movement on the ground. Consequently, the combination of digital soil models, water flow parameters and advanced simulation techniques optimizes the accuracy of hydraulic models.

Keywords: flood risk analysis · urbanization · technological progress · Unmanned Aerial Vehicle (UAV) · photogrammetry

1 Introduction

The role of hydraulic studies is to understand and predict the movement of water in the landscape [1]. This depends on the characteristics of the ground such as altitude, slope and roughness, which significantly influence the behavior of water [2]. As these parameters contain inherent uncertainties due to their complexity, hydraulic studies consequently face great challenges. Traditional field methods often fall short of providing accurate and comprehensive terrain representations, as they can be limited by the resolution of data collection and are subject to various human and environmental factors that introduce uncertainty and unpredictability [3–5].

In recent years, there has been a notable shift towards the incorporation of digital tools in hydrologic and hydraulic research, driven by technological advances in remote

N. D. Lagaros et al. (Eds.): ADDOPTML 2024, LNCE 646, pp. 237–244, 2025.
https://doi.org/10.1007/978-3-031-92029-5_22

sensing and computational power [6, 7]. Traditional methods [8], such as manual survey measurements or basic aerial photography, often fail to capture fine details necessary for accurate modeling of water flow, especially in areas with complex topographies. To address these issues, various methods are employed, most often including digital terrain modeling (DTM) as well as the incorporation of water flow parameters with specialized software [1, 9–11].

Field research is a critical step of the methodology which is necessary to describe the state of flow and channel roughness giving necessary inputs to hydraulic modeling [12]. Since field research from the ground has a limited view, an advantage method can give the creation of the anaglyph by Unmanned Aerial Vehicles (UAV) [13]. UAVs equipped with cameras have enabled researchers to create detailed DTMs with photogrammetric processing with appropriate software. These models can capture topographic variations that directly affect water flow and are essential for developing more reliable hydraulic simulations [13, 14].

UAVs offer flexibility, lower costs, and the ability to access areas that are difficult or dangerous for human surveyors. Photogrammetry, the science of making measurements from photographs, can further enhance this process by allowing the creation of three-dimensional models from successive images taken from multiple angles. This combination has proven effective for mapping complex terrain, by reducing uncertainties related to the physical properties of the landscape.

This research explores the use of advanced digital modeling techniques, particularly the use of UAVs and photogrammetry, to produce high-resolution, three-dimensional terrain models that better support hydraulic simulations and engineering analyses critical to societal resilience [15–18].

2 Methodology

Since the movement of water is highly dependent on the topography of the terrain and friction due to roughness, accurate mapping of the area to input data into flood study software is extremely important for estimating various hydraulic properties, such as the Manning's roughness coefficient for riverbeds [19]. To date, field inspections have proven to be crucial in flood studies providing detailed information and resolving ambiguities that are not visible in the available satellite imagery or the digital terrain model. Although this information is conveyed through reports and relevant photographs, it may contain ambiguities for the researcher creating the model in the office using a computational fluid dynamics software, who has not taken field measurements or even not seen the landscape firsthand.

To minimize this ambiguity, the methodology used in this study involves a multi-step process to produce accurate, high-resolution DTMs using UAVs and photogrammetry. First, a commercially available camera-equipped UAV is used to capture multiple overlapping images of the target area from different angles and heights. The images are then processed by photogrammetric software, which constructs a three-dimensional terrain model by aligning the images based on common features detected across multiple photos.

The next step is to calibrate the model using specific terrain parameters such as slope, roughness, and elevation, which are critical for hydraulic simulations. Additional data on water flow parameters, including channel roughness, are incorporated to further refine the model. These calibrated DTMs are used as input to simulation software that simulates the movement of water across the landscape, helping to predict flow paths, potential flood areas, flood risk and hazard, and the effect of terrain features on water distribution.

3 Case Study: Application in a River Channel

To illustrate the effectiveness of this method, a case study was carried out in a river channel in the Attica region (Fig. 1a), where traditional field methods had previously been used to assess hydraulic characteristics (Fig. 1b). The selected area had a varied landscape with both hydraulic structures and natural riverbeds, which significantly influence water flow patterns during heavy rainfall events.

Fig. 1. (a) Attica region. Landmark 38° 2′31.29″N; 23°49′37.80″E (b) The view from the ground

Using UAVs, the research team captured images over the study area, covering a critical part of the landscape that could not be adequately measured using traditional methods (Fig. 2).

Successive overlapping images were taken to create the 3D model. The images were imported into the web-based photogrammetry software Kiriengine [20] and a high-resolution 3D file was created that accurately reflected the topography of the area, including subtle changes in elevation that are crucial for predicting water accumulation zones. This file could be exported in OBJ (*.obj) or STL (*.stl) file formats. It is important to note that the channel roughness can be estimated from this model, which is critical for estimating the riverbed and floodplain Manning coefficients.

The 3D file was then used as an input to hydraulic modeling software such as HEC-RAS [21], which simulated potential flood events under different rainfall conditions.

Fig. 2. Characteristic images by UAV. Coordinates: 38° 2′31.29″N; 23°49′37.80″E

The results showed potential flood paths and identified areas at high risk of water accumulation, information that was invaluable to local authorities in planning effective flood mitigation strategies [9].

Figure 3 shows the resulting 3D model generated by a combination of photogrammetry and UAV-based field inspections is shown. This method captures highly detailed spatial data, allowing an accurate visual representation of key topographic features such as texture and roughness, as well as the morphology of the riverbed and the channel slopes. These fine details are essential for incorporating accurate data into a hydraulic model, as they allow for a better understanding and representation of the physical conditions that affect water flow, sediment transport, and overall channel hydraulics. With these detailed insights, the hydraulic model can be calibrated to more closely reflect real-world conditions, improving the model's reliability in simulating water dynamics and predicting outcomes under different scenarios.

Fig. 3. Characteristic snapshots from 3D model [22]. Coordinates: 38° 2′31.29"N; 23°49′37.80″E

On the other hand, Fig. 4 shows a high quality DTM in the same area as Fig. 3, based on a 2 m × 2 m grid, which, despite its high resolution, lacks surface roughness information. The lack of roughness data limits our ability to capture the variability in surface texture that significantly affects hydraulic calculations. Without this information, the DTM cannot fully support the requirements of the hydraulic model, as roughness values are critical in estimating flow resistance and other hydraulic parameters. Therefore, while the DTM provides accurate elevation data, it does not provide the necessary information on the local land cover, which hinders the ability to make accurate parameter estimate for hydraulic modeling purposes. This lack of texture and roughness data can lead to oversimplified modelling and potentially inaccurate predictions when simulating natural river dynamics. Specifically in this case, the necessary land cover information is either retrieved from open land-cover databases [23], which are often of limited local accuracy, or obtained through field visits, which are not always feasible. In contrast, a UAV-generated DTM model captures both elevation and surface condition information and can be easily updated when necessary (e.g., after a disaster).

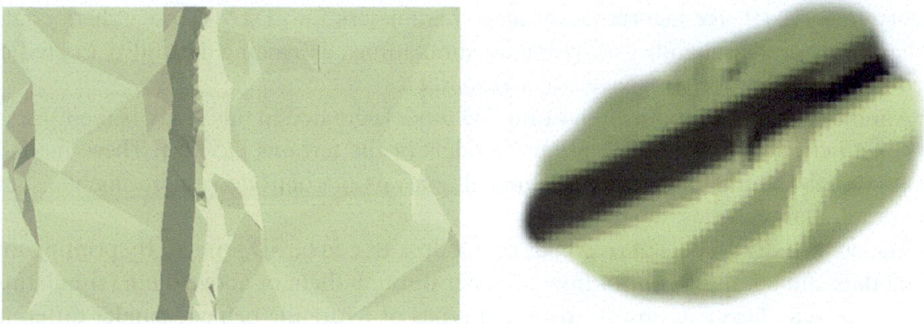

Fig. 4. The view of the same region of Fig. 3 from the 3D model in ArchGIS. Coordinates: 38° 2′31.29″N; 23°49′37.80″E

4 Challenges and Limitations

While the integration of UAVs and photogrammetry in hydraulic studies offers significant benefits, it is not without its challenges. One major limitation is the dependence of UAV flights on weather conditions; strong winds, rain, or fog can affect image quality or prevent data collection altogether. Additionally, photogrammetry software requires significant computing power, especially when processing large datasets, which can be a limiting factor for researchers with limited resources.

Moreover, the accuracy of photogrammetric models depends on careful calibration and validation against real-world measurements. Soil features, vegetation, or water surfaces can distort the photogrammetry process. To mitigate these uncertainties, data collected by ground-based observation (field research) or remote sensing techniques such as LiDAR can be integrated into this process to verify the accuracy of photogrammetric models.

5 Future Directions

As technology evolves, the use of UAVs in hydraulic studies is likely to increase. The evolution of drones such as longer flight times, image analysis, autonomous navigation techniques and various sensors integrated into drones over time give possibilities that are difficult to imagine at this stage. In addition, advances in photogrammetry software, driven by artificial intelligence and machine learning, are expected to streamline the data processing phase and improve model accuracy. The integration of multi-source data from different sensors and satellite imagery promise to produce even more detailed DTMs, enabling researchers to achieve unprecedented accuracy in hydraulic modeling.

Future studies could also focus on refining UAV methods to capture specific terrain features that are known to influence water flow, such as soil moisture and vegetation density [e.g., [24]]. By incorporating these parameters into DTMs, researchers could further improve the reliability of hydraulic simulations, enhancing the ability to predict flood risk and effectively manage water resources.

Another promising application of the 3D models created in this study is their use in the construction of physical 3D printed models of the terrains [25–27]. These models can serve as valuable tools for experimental analysis in controlled environments, such as hydraulic laboratories.

The created 3D models of riverbeds or floodplains can be 3D printed. By printing the 3D models and adding a control flow of water through them or above them (simulating the rain), researchers can observe different types of hydraulic behavior under different conditions. This approach allows the capture of flow patterns, turbidity, sediment transport and erosion in ways that may be difficult to model with computational methods. In addition, physical models can verify computational ones and provide an alternative approach that optimizes the understanding and accuracy of hydraulic studies and can be used in educational process.

6 Conclusion

This paper highlights how the integration of UAVs technology and photogrammetry can optimize hydraulic studies. More specifically, hydraulic studies, which contain many uncertainties, could mitigate these uncertainties concerning water flow and soil roughness, from accurate 3D models, which can capture all the details of the soil of the terrain (morphology, roughness, permeability) and can be created with a commercial UAV and an on-line photogrammetric application.

This approach offers many advantages such as: faster data collection, especially in large or hard-to-reach areas and greater accuracy and monitoring of landscape changes, as digital terrain models can be updated more frequently. By minimizing the uncertainties of ground representation, this method can improve the reliability of hydraulic models, allowing engineers to make more accurate estimates in conjunction with computer models.

Acknowledgments. This research was supported by the ADDOPTML project: "ADDitively Manufactured OPTimized Structures by means of Machine Learning" (No: 101007595) belonging to the Marie Skłodowska-Curie Actions (MSCA) Research and Innovation Staff Exchange (RISE) H2020-MSCA-RISE-2020.

References

1. Sigourou, S., et al.: An Advanced Methodology for Field Visits Towards Efficient Flood Management on Building Block Level. EGU General Assembly 2023, Vienna (2023. 24–28 Apr 2023, EGU23-16168). https://doi.org/10.5194/egusphere-egu23-16168
2. Wang, N., et al.: How can changes in the human-flood distance mitigate flood fatalities and displacements? Geophys. Res. Lett. **50**, e2023GL105064 (2023). https://doi.org/10.1029/2023GL105064
3. Sargentis, G.-F., Hadjibiros, K., Papagiannakis, I., Papagiannakis, E.: Plastiras lake: influence of the relief on the revelation of the water presence. In: 9th International Conference on Environmental Science and Technology. Department of Environmental Studies, University of the Aegean, Rhodes Island (2005). https://doi.org/10.5281/zenodo.3767050
4. Sargentis, G.-F., Dimitriadis, P., Iliopoulou, T., Koutsoyiannis, D.: A stochastic view of varying styles in art paintings. Heritage. **4**, 333–348 (2021). https://doi.org/10.3390/heritage4010021
5. Sargentis, G.-F.: The Aesthetic Element in Water, Hydraulics and Dams. Master's Thesis, Department of Civil Engineer, NTUA, Athens, Greece (1998). https://doi.org/10.5281/zenodo.3770806
6. Kougia, M., et al.: Modern vs Traditional Mapping Methods for Flood Risk Estimation: A Case Study for the River Pikrodafni, Athens, Greece. EGU General Assembly 2024, Vienna, Austria., 14–19 Apr 2024, EGU24-6459 (2024). https://doi.org/10.5194/egusphere-egu24-6459
7. Campbell-Kelly, M., Aspray, W.F., Yost, J.R., Tinn, H., Díaz, G.C.: Computer: A History of the Information Machine. Routledge, London (2023)
8. Presidential Decree No. 696/1974, ΦΕΚ Α'301/8-10-1974. https://www.et.gr/api/DownloadFeksApi/?fek_pdf=19740100301
9. Tsouni, A., et al.: Multi-parameter Flood Risk Assessment Towards Efficient Flood Management in Highly Dense Urban River Basins in the Region of Attica, Greece. EGU General Assembly 2023, Vienna, Austria., 24–28 Apr 2023, EGU23-12624 (2023). https://doi.org/10.5194/egusphere-egu23-12624
10. Sargentis, G.-F., Sigourou, S.: Multiparametric flood risk assessment. Management planning in high spatial resolution in Attica. KTIRIO. **5**(2024), 22–23 (2024)
11. Iliopoulou, T., Dimitriadis, P., Koutsoyiannis, D.: Pluvial flood risk assessment in urban areas: a case study for the archaeological site of the Roman Agora, Athens. Heritage. **6**(11), 7230–7243 (2023)
12. Moraiti, K., et al.: Documenting the changing floodplain of Nileas basin in north Euboea (Greece) before and after Storms Daniel and Elias. Rural. Reg. Dev. **2**(3), 10013 (2024). https://doi.org/10.35534/rrd.2024.10013
13. Sargentis, G.-F., Moraiti, K., Benekos, I., Ioannidis, R., Mamassis, N.: Fast-track documentation of the alterations on the landscape, before and after a natural hazard—case study: north Euboea Greece before and after Storms Daniel and Elias. Rural. Reg. Dev. **2**, 10016 (2024). https://doi.org/10.70322/rrd.2024.10016
14. Moraiti, K., Sigourou, S., Kougia, M., Sargentis, G.-F., Koutsoyiannis, D.: On-site inspection in monitoring of water flow for the calibration of hydraulic models and the statistical analysis of the hydraulic output. In: Case Study: The river Nileas in North Euboea Greece, before and after the storm Elias in September 2023. EGU General Assembly 2024, Vienna, Austria., 14–19 Apr 2024, EGU24-5186 (2024). https://doi.org/10.5194/egusphere-egu24-5186

15. Storm Elias: Dramatic situation in Euboea—Images of complete destruction. https://www.cnn.gr/ellada/story/384064/kakokairia-elias-dramatiki-i-katastasi-stin-eyvoia-eikones-apolytis-katastrofis. Accessed 14 May 2024
16. Sargentis, G.-F., Koutsoyiannis, D.: The function of money in water–energy–food and land nexus. Land. **12**, 669 (2023). https://doi.org/10.3390/land12030669
17. Markantonis, D., et al.: Stochastic evaluation of the investment risk by the scale of water infrastructures—Case study: The Municipality of West Mani (Greece). World. **4**, 1–20 (2023). https://doi.org/10.3390/world4010001
18. Sargentis, G.F.: Issues of Prosperity: Stochastic Evaluation of Data Related to Environment, Infrastructures, Economy and Society. PhD thesis. National Technical University of Athens (2022)
19. Dimitriadis, P., et al.: Comparative evaluation of 1D and quasi-2D hydraulic models based on benchmark and real-world applications for uncertainty assessment in flood mapping. J. Hydrol. **534**, 478–492 (2016). https://doi.org/10.1016/j.jhydrol.2016.01.020
20. Kiriengine. https://www.kiriengine.app/
21. HEC-RAS 5.0: Reference Manual. (2016) http://www.hec.usace.army.mil/software/hec-ras/documentation/HEC-RAS%205.0%20Reference%20Manual.pdf. Accessed 10 Sept 2017
22. 3D model by kiriengine. https://www.kiriengine.app/share/ShareModel?code=NT4JIH&serialize=a8855d73a6f4402ea7a2571b26b7faf3
23. CORINE Land Cover 2018 (vector/raster 100 m), Europe, 6-yearly. https://doi.org/10.2909/960998c1-1870-4e82-8051-6485205ebbac
24. Alexandris, S., et al.: Integrating drone technology into an innovative agrometeorological methodology for the precise and real-time estimation of crop water requirements. Hydrology. **8**(3), 131 (2021)
25. Sargentis, G.-F., Frangedaki, E., Iliopoulou, T., Dimitriadis, P., Lagaros, N.D.: The Use of 3D Printed Models of the Landscape in Hydraulic Studies, ADDOPTML2024, OPTARCH2024, and OPT-ii2024, Irbid - Jordan, Jordan, 1st-4th Oktober 2024. https://www.itia.ntua.gr/el/docinfo/2494/
26. Sargentis, G.-F., et al.: 3D scanning/printing: a technological stride in sculpture. Technologies. **10**, 9 (2022). https://doi.org/10.3390/technologies10010009
27. Sargentis, G.-F.: Use and Technical Aspects of Materials in Sculpture. Ph.D. Thesis, School of Architecture, National Technical University of Athens, Athens, Greece (2005)

Advanced Numerical Simulation of a 3D-Printed Spinal Brace

Iason Rossetos[1(✉)], Charis J. Gantes[1], George Kazakis[1], Stephanos Voulgaris[1], Nikos D. Lagaros[1], Konstantinos Soultanis[2], and Dimitrios Galanis[2]

[1] School of Civil Engineering, National Technical University of Athens, Athens, Greece
iasonas.ross@gmail.com

[2] School of Medicine, National and Kapodistrian University of Athens, Athens, Greece

Abstract. The use of spinal braces is the most common solution to address adolescent idiopathic scoliosis by limiting the progression of spinal inclination. From a mechanics point of view, spinal braces are shell-type structures, relying on the fundamental engineering principles of three-point bending and inversion forces. A research project is ongoing to develop a procedure comprising laser scanning, advanced numerical simulation, topology optimization and additive manufacturing, to design and 3D print lightweight, personalized braces. In the present paper recent advances in the numerical simulation procedure are reported.

Finite element simulation of the brace and its interaction with the patient's body is employed, to evaluate the developing deformations and stresses during the brace's use. For that purpose, the brace is modeled with shell finite elements and the body with a mirror surface of appropriate geometry, considering deformability and detachment by means of contact elements with an appropriate pressure – overclosure relation, employing realistic values of human body stiffness at its different parts. The mechanical properties of the brace material employed in the simulation are adopted from experimental tensile tests on 3D-printed Polylactic Acid (PLA) specimens.

From the numerical simulation, von Mises stresses and displacements of the spinal brace are obtained and are used to interpret its structural behavior. To understand how the patient's body is affected by the use of the brace, contact pressures are presented, reflecting the interaction between the brace and the body.

Keywords: Spinal Brace Simulation · Finite Element Analysis · 3D Printing · Scoliosis · Numerical Simulation

1 Introduction

Idiopathic scoliosis (IS) is a complex spinal condition characterized by a three-dimensional deformity affecting several individuals, predominantly teenagers and adolescents, as described by Chan et al. [1]. A non-surgical approach to managing this condition involves the use of spinal braces designed to control the progression of the spinal inclination, described by Negrini et al. [2], Kuroki [3] and Ali et al. [4]. Currently,

N. D. Lagaros et al. (Eds.): ADDOPTML 2024, LNCE 646, pp. 245–254, 2025.
https://doi.org/10.1007/978-3-031-92029-5_23

the most commonly used braces in clinical settings are the Boston Brace and the Milwaukee Brace. However, these braces are often described as bulky and uncomfortable for the wearers, as reported by Zarea et al. [5], Olafsson et al. [6] and Khan et al. [7]. The objective of this study is to present the application of finite element simulation in the development process of spinal braces, as part of a methodology to achieve 3D printed, lightweight, personalized braces for comfortable use by patients. Similar studies have been reported in the literature like Aubin et al. [8], Ronca et al. [9], Redaelli et al. [10] and Rossetos et al. [11], focusing on simulating the structural behavior of a spinal brace with the finite element method.

In this study, Abaqus CAE v. 2021 software [12] is used, and CAD scans of a real spinal brace (generated with Meshmixer v. 3.5.474 by Autodesk [13]) are applied for numerical simulation. Laser scanning technology facilitates the capture of precise anatomical data without invasive procedures, enabling customization of product designs for medical purposes, as demonstrated by Kim et al. [14] and Taneva et al. [15]. The main purpose of this study is to simulate a 3d printed spinal brace utilizing the mechanical properties of PLA (Polylactic Acid), obtained from laboratory testing as presented in research [11].

A main aspect of simulating a spinal brace is the loads and boundary conditions. A main load scenario is based on the concept of three-point bending and inversion forces, aiming to restrict the progression of spinal inclination. The present study focuses and develops the loading scenario described in previous research by the same authors [11, 16], utilizing contact elements in order to simulate the interaction between body and brace. This study builds upon the authors' previous data and methodology [11, 16], with specific modifications. Notably, it incorporates more accurate data regarding the stiffness of human skin and aims to capture the nonlinear behavior of the interaction between the human body and the brace. Most of the research in the literature [17, 18] adopt simplified approaches to boundary conditions, often employing pinned or fixed supports at specific parts of the brace. However, these approaches neglect the deformability of the body, and tend to overestimate the pressures exerted on the body from contact with the brace. The results include von Mises stresses and displacements of the brace, as well as contact pressures, which explain how the human body is affected by the spinal brace.

2 Laser Scanning and Model Preparation

The process of laser scanning the human torso employs advanced three-dimensional imaging technologies to create a detailed digital model of the upper body, from the chest to the spine. This procedure involves several key steps: (a) Pre-scan preparation, where patients are dressed in fitted, non-reflective clothing. (b) Calibration of the scanning equipment according to the manufacturer's guidelines to ensure precise measurements. (c) Positioning the patient within a designated scanning zone. (d) Execution of the laser scan, in which laser beams are projected onto the patient's torso to capture distance measurements from multiple points on the surface. This generates a 'point cloud' through data collection from various angles, typically across multiple scanning sessions. (e) Integration of the scanning data to form a cohesive and accurate 3D model by aligning the individual point clouds. (f) Conversion of the point clouds into a mesh, composed of interconnected triangular elements (Fig. 1).

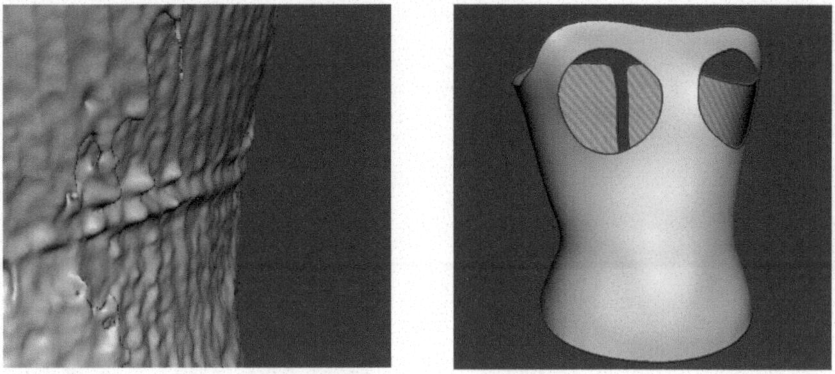

(a) (b)

Fig. 1. (a) Scanning imperfections, (b) Modified 3D Model

After generating the torso model, the next phase involves preparing it for spinal brace design using the CAD software Meshmixer by Autodesk. A major challenge during this stage is managing sudden geometric inconsistencies within the scan, which can obstruct the brace modeling process. These imperfections, particularly near the brace edges, are smoothed to create a more uniform surface. Finally, the model's thickness is removed to accommodate shell finite elements in the modeling process, and the model is exported as an STL file for further use.

3 Geometry and Numerical Model

Following the scanning procedure, the STL file is imported into the Abaqus CAE software environment. The shape and representative dimensions of the examined spinal brace are illustrated in Fig. 2.

A numerical model of the spinal brace utilizing shell finite elements has been developed, with a cross-sectional thickness of 4 mm. Due to the brace's complex geometry, characterized by numerous curves and asymmetry, discrete surfaces were defined to improve clarity and precision in the application of loads and boundary conditions. For this purpose, the brace was partitioned along planes parallel to the three principal axes, as shown in Fig. 3a. A free meshing technique using triangular elements, depicted in Fig. 3b, was applied to accommodate the brace's curved surface. This results in an unstructured mesh, which is inherently adaptable to diverse geometric configurations. The average surface area of each triangular element is approximately 10.3 mm^2.

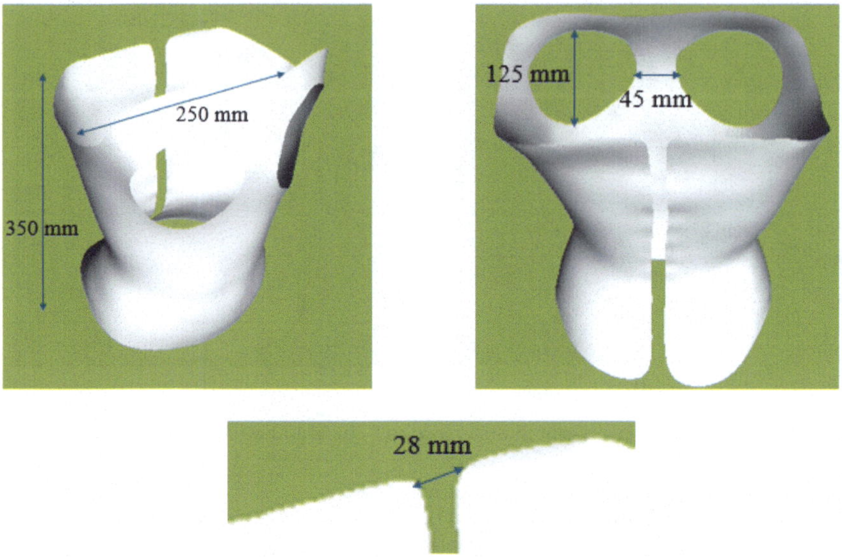

Fig. 2. Dimensions of the brace

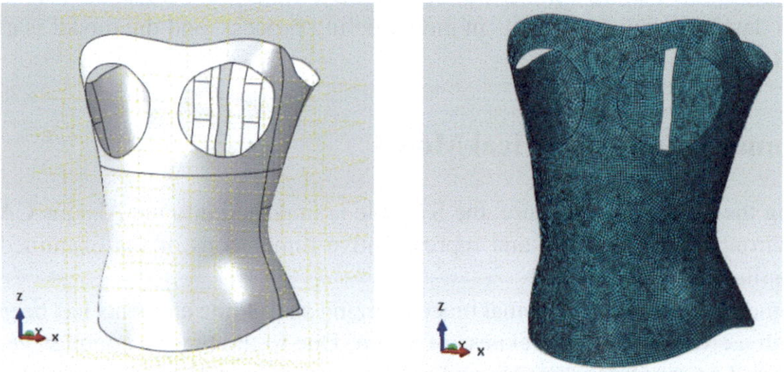

(a) (b)

Fig. 3. (a) Regions defined by parallel planes, (b) Free meshing technique

4 Material

This research focuses on describing the structural behavior of a 3d printed spinal brace. The mechanical properties of the material are obtained by previous research [11], where PLA (Polylactic Acid) characteristics were determined through laboratory tensile testing of 3d printed specimens. The material properties are presented in Table 1.

The tensile strength of PLA in directions 1 and 2 was found to be approximately 49 MPa and 35 MPa, respectively.

Table 1. Material properties of the 3D printed PLA used in the simulation

Material	E_1	E_2	v_{12}	G_{12}	G_{13}	G_{23}
	MPa	MPa		MPa	MPa	MPa
PLA	3045	2898	0.33	987	1012	987

5 Loads and Boundary Conditions

5.1 Loads

The brace is tightened through the activation of three straps located at the back, as presented in Fig. 4. In the numerical model, this was simulated by applying concentrated forces of 10 N, at the center of each strap area. Alternatively, imposed displacements could be applied to the model for similar simulation purposes.

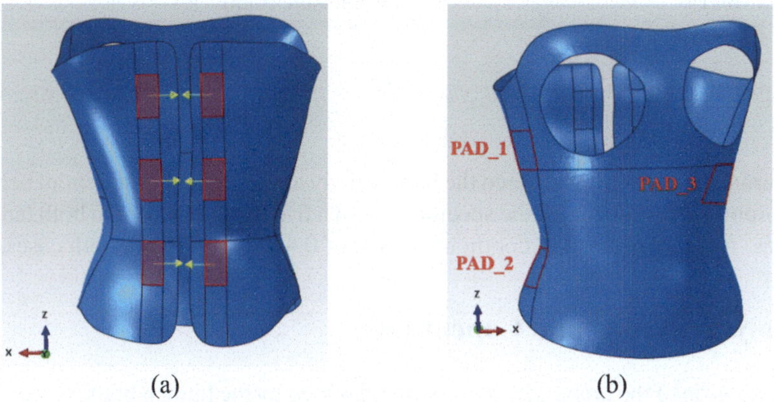

Fig. 4. (a) Strap surfaces and locations of concentrated forces applied in the model. (b) Supporting pad surfaces used to simulate contact with the body

5.2 Boundary Conditions

Accurately defining boundary conditions is essential for realistic spinal brace simulations. While most existing studies rely on simplified models, such as pinned or fixed supports at designated points, these methods fail to account for the body's natural flexibility. As a result, they often lead to an overestimation of the pressure distribution between the brace and the body. In this study, a homothetic part, which presents the human body, is created, as presented in Fig. 5. The body part and the brace part interact by transferring pressures only when they are in contact. The interaction relationship between the two parts is presented in Table 2, utilizing data referring to stiffness of the human skin [19–23].

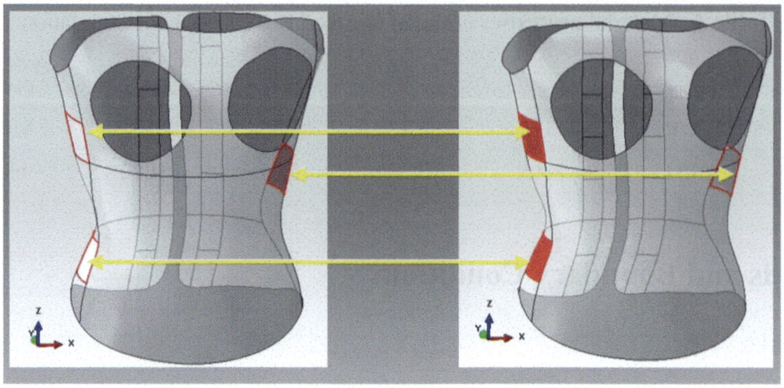

Fig. 5. Pad areas interacting by contact properties. Brace part (left side) and body part (right side)

Table 2. Pressure-overclosure relationship

Pressure (MPa)	Overclosure (mm)
0	0
0.02	0.7

To simulate the friction between the human body and the brace, "tangential behavior" was incorporated, allowing for the specification of a friction coefficient in both tangential directions. A standard friction coefficient value of 0.4 was applied in both cases.

6 Analysis Results and Conclusions

The deformation of the brace, when pressure is applied by the human body, is represented by the ascending portion of the relationship in Table 2. In contrast, when the brace deforms in such a way that it loses contact with the body, no pressure is exerted, resulting in zero contact force. This behavior introduces nonlinearity, necessitating a nonlinear analysis. To solve the nonlinear equilibrium equations, the full Newton's method was selected due to its superior convergence rate compared to other methods, particularly for the types of nonlinear problems commonly addressed using ABAQUS CAE software. The analysis was set with a maximum of 100 increments, an initial increment size of 0.1, and minimum and maximum increment values of 0.1.

6.1 Von Mises Stresses and Displacements

At the end of the analysis, results are extracted that describe the developing stresses (Fig. 6) as well as the displacements (Fig. 7) in the three axes over the entire surface of the brace. The results are shown in the following figures.

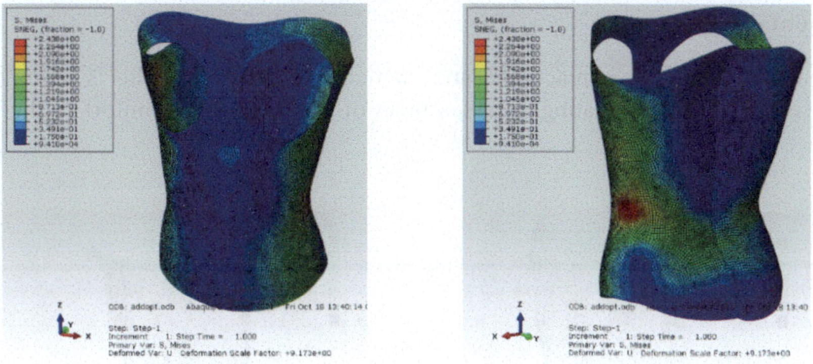

(a) (b)

Fig. 6. Von Mises stresses results (a) On the front side of the brace (b) On the rear side of the brace

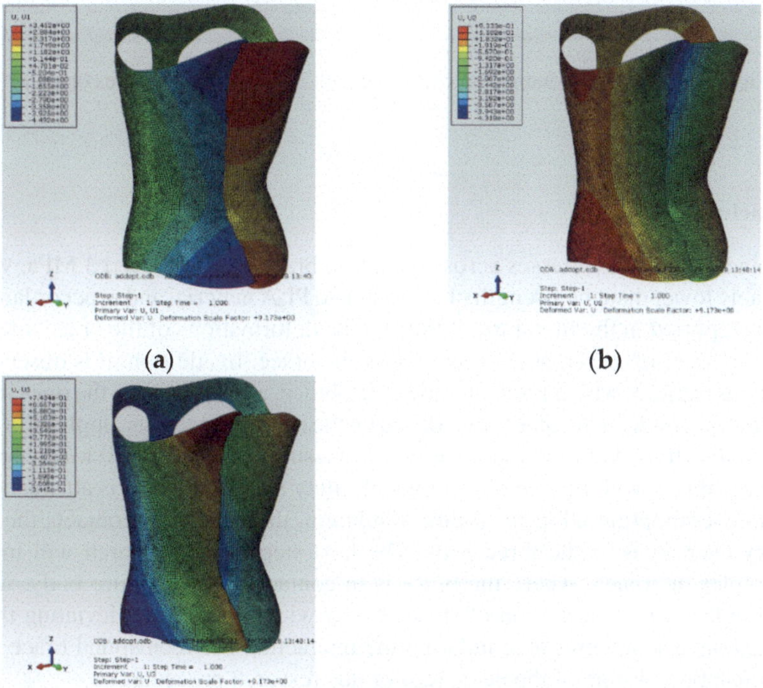

(a) (b)

(c)

Fig. 7 Displacements of the brace (a) in X-axis (b) in Y-axis (c) in Z-axis

6.2 Contact Pressures

The results pertaining to contact pressures are delineated in Fig. 8. This is an important response parameter, facilitating an assessment of the safety and comfort of the brace design.

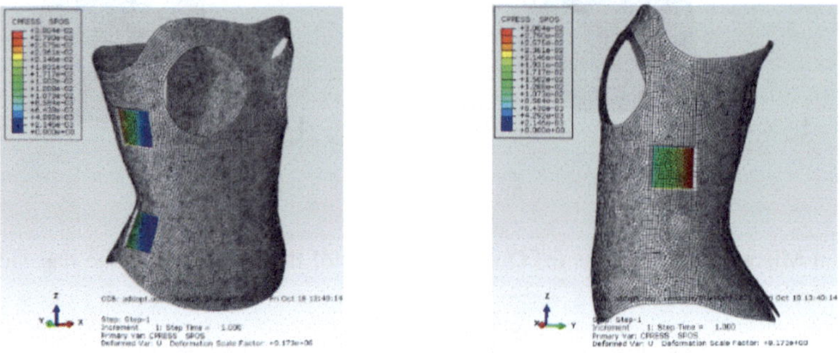

(a) (b)

Fig. 8. (a) Contact pressures on pad 1 and pad 2 (b) contact pressures on pad 3

6.3 Conclusions

The obtained von Mises stresses across the entire brace do not exceed 3 MPa, which is considerably lower than the strength of the assumed PLA material and in accordance with the results reported in the literature. Referring the deformations of the brace, maximum displacements are observed at the rear face of the brace. In addition, it is observed that the pressures at pad 3, which is on one side of the brace, are higher than the ones in pads 2 and 3, which are both on the other side. Based on the numerical results, applying a tension of 10 N to the straps led to the generation of pressures in order of 30 kPa, suggesting a local incapability with human physiological safety standards. This is attributed to the unfavorable assumption adopted in this simulation, that the brace contacts the human body only over areas of the three pads. The next step of our research will involve a more complex approach, where the brace is in contact with the entire body, which is expected to lead to a distribution of stresses over wider area, thus alleviating the local stress maxima. The attempt of simulating the interaction between spinal brace and the entire human body is one of the next steps of this research procedure.

Acknowledgements. This research has been financed by the European Union – Next Generation EU in the context of the OrThOP3Dics project: "Topology optimization of 3D printed patient-specific spinal braces" (No.: TAEDK-06191) belonging to the National Recovery and Resilience Plan, Greece 2.0 Project.

References

1. Chan, W.Y., et al.: Mechanical and clinical evaluation of a shape memory alloy and conventional struts in a flexible scoliotic brace. Ann. Biomed. Eng. **46**, 1194–1205 (2018)
2. Negrini, S., Aulisa, A., Cerny, P., de Mauroy, J., McAviney, J., Mills, A., et al.: The classification of scoliosis braces developed by SOSORT with SRS, ISPO, and POSNA and approved by ESPRM. Eur. Spine J. **31**, 980–989 (2022)
3. Kuroki, H.: Brace treatment for adolescent idiopathic scoliosis. J. Clin. Med. **7**, 136 (2018)
4. Ali, A., Fontanari, V., Fontana, M., Schmolz, W.: Spinal deformities and advancement in corrective orthoses. Bioengineering. **8**, 2 (2020)
5. Zarea, M., Aminian, G., Khosravi, M., Baghaei, R.: Effect of Milwaukee brace on quality of life in adolescents with idiopathic scoliosis. J. Clin. Physiother. Res. **5**, e13 (2020)
6. Olafsson, Y., Saraste, H., Söderlund, V., Hoffsten, M.: Boston brace in the treatment of idiopathic scoliosis. J. Pediatr. Orthop. **15**, 524 (1995)
7. Khan, M.J., Srinivasan, V.M., Jea, A.H.: The history of bracing for scoliosis. Clin. Pediatr. **55**, 320–325 (2015)
8. Aubin, C., et al.: Improved brace design combining CAD/CAM and finite element simulation for the conservative treatment of adolescent idiopathic scoliosis (AIS): preliminary results of a randomized control trial. In: Proceedings of the 10th Meeting of the International Research Society of Spinal Deformities (IRSSD 2014 Sapporo), Sapporo, Japan, 29 June–2 July 2014 (2014)
9. Ronca, A., et al.: Comparative study for material selection in 3D printing of scoliosis back brace. Materials. **15**, 5724 (2022)
10. Redaelli, D.F., et al.: 3D printing orthopedic scoliosis braces: a test comparing FDM with thermoforming. Int. J. Adv. Manuf. Technol. **111**, 1707–1720 (2020)
11. Rossetos, I., et al.: Numerical modeling and nonlinear finite element analysis of conventional and 3D-printed spinal braces. Appl. Sci. **14**, 1735 (2024)
12. Simulia Abaqus CAE. https://www.3ds.com/products/simulia/abaqus/cae. Accessed 24 Jan 2024
13. Meshmixer by Autodesk. https://meshmixer.com. Accessed 24 Jan 2024
14. Kim, J., Heo, G., Lagravère, M.O.: Accuracy of laser-scanned models compared to plaster models and cone-beam computed tomography. Angle Orthod. **84**(3), 443–450 (2013). https://doi.org/10.2319/051213-365.1
15. Taneva, E., Kusnoto, B., Evans, C.A.: 3D scanning, imaging, and printing in orthodontics. In: Issues in Contemporary Orthodontics. IntechOpen, London (2015) https://www.intechopen.com/chapters/48165. Accessed 24 Jan 2024
16. Rossetos, I., Voulgaris, S., Kazakis, G., Gantes, C.J., Lagaros, N.D.: Numerical investigation of 3D-printed spinal braces. In: Proceedings of the IASS 2024 Symposium Redefining the Art of Structural Design, August 26–30, 2024, Zurich Switzerland
17. Grycuk, S., Mrozak, P.: Scoliosis brace finite element model and preliminary experimental testing using electronic speckle pattern interferometry. Appl. Sci. **12**, 3876 (2022)
18. Mafi, M., Niyaval, A.M.K., Khanmohammadi, S.: Brace simulation for patients with scoliosis: biomechanical comparison with three different polymeric materials. BKK Med. J. **15**, 161 (2019)
19. Cua, A.-B., Wilhelm, K.-P., Maibach, H.-I.: Elastic properties of human skin: relation to age, sex, and anatomical region. Arch. Dermatol. Res. **282**, 283–288 (1989)
20. Griffin, M.-F., Leung, B.-C., Premakumar, Y., Szarko, M., Butler, P.-E.: Comparison of the mechanical properties of different skin sites for auricular and nasal reconstruction. J. Otolaryngol. Head Neck Surg. **46**, 33 (2017)

21. Graham, H.-K., McConnell, J.-C., Limbert, G., Sherratt, M.-J.: How stiff is skin. Exp. Dermatol. **28**(suppl. 1), 4–9 (2018)
22. Wei, J.-C.-J., Edwards, G.-A., Martin, D.-J., Huang, H., Crichton, M.-L., Kendall, M.-F.: Allometric scaling of skin thickness, elasticity, viscoelasticity to mass for micro-medical device translation: from mice, rats, rabbits, pigs to humans. Sci. Rep. **7**, 15885 (2017)
23. Zhang, M., Mak, A.-F.-T.: In vivo friction properties of human skin. Prosthet. Orthot. Int. **23**, 135–141 (1999)

Numerical Simulation and Nonlinear Finite Element Analysis of 3D-Printed Joints for Deployable Structures

Georgios D. Pantazis[1(✉)], Charis J. Gantes[1], and Nikos D. Lagaros[2]

[1] Institute of Steel Structures, National Technical University of Athens, 9 Iroon Polytechneiou Str., Zografou Campus, 15772 Athens, Greece
gpantazisdom@yahoo.gr, chgantes@central.ntua.gr

[2] Laboratory of Structural Analysis and Antiseismic Research, National Technical University of Athens, 9 Iroon Polytechneiou Str., Zografou Campus, 15772 Athens, Greece
nlagaros@central.ntua.gr

Abstract. Objective of the present study is the simulation of an experimental process for the design of 3D-printed joints of a scissor-based deployable shelter for emergency response [1]. The test specimen consists of a 3D-printed joint made of Polylactic Acid (PLA) and four aluminum bars with rectangular hollow sections hinged to the joint using steel bolts. At their other ends the aluminum bars are hinged to a rigid steel base, and the load is applied by a tension rod system to the center of the joint.

Geometry and material nonlinear analyses (GMNA) were carried out in Abaqus CAE software Version 2021 [2], where all members of the specimen were simulated by 3D solid finite elements, with the aim of predicting the evolution of the experimental process but also determining the deviations between numerical and experimental findings. Nonlinear material laws were adopted for the joints, the bars and the bolts, representing PLA [3], aluminum, and steel, respectively. In addition, all possible contacts of the members were simulated through contact elements.

The analysis results demonstrated a high concentration of stresses and material yielding around the bearings of the bars and the joint, already at relatively low load levels, while high stresses develop also inside the joint at higher loads. Significant margins for optimization of the joint topology were established, as inactive areas of the joint were observed for the applied loads.

Keywords: deployable structures · 3D printing · finite element analysis · nonlinear analysis · contact nonlinearity · topology optimization

1 Introduction

The main advantage of deployable structures is their ability to be transported from one place to another, their quick erection and dismantling, and the large ratio of their volume in the open configuration to the one in the closed configuration. On the other hand, their main disadvantages are the complexity of their structural configuration and connections, and the high cost of their analysis and structural design [4].

N. D. Lagaros et al. (Eds.): ADDOPTML 2024, LNCE 646, pp. 255–266, 2025.
https://doi.org/10.1007/978-3-031-92029-5_24

For such structures it is important to minimize their weight as much as possible, to facilitate transportation and erection. Studying a scissor deployable shelter for emergency use [5], it was found that the steel joints between the scissor elements weigh about 45% of the total weight of the aluminum scissor elements [6], which is a particularly high percentage. Thus, there is a need for topology optimization [10, 11] of the joints as well as adopting different and lighter materials for the reduction of the weight ratio.

The aim of the present study is to simulate an experimental process for the design of 3D-printed joints. A 1:2.5-scale test specimen was designed, consisting of a 3D-printed solid joint made of Polylactic Acid (PLA) and four aluminum bars with rectangular hollow sections hinged to the joint using steel bolts. At their other ends the aluminum bars are hinged to a rigid steel base, and the load is applied by a tension rod system to the center of the joint.

2 Numerical Simulation

The numerical simulation of the experiment was carried out by solid finite elements in the Abaqus CAE software Version 2021 [2]. Non-linear geometry and material analyses (GMNA) were performed to determine the internal stresses of the joint as well as the other members of the specimen.

2.1 Specimen

The simulated specimen consists of a PLA 3D-printed joint, four aluminum bars and four steel bolts (Fig. 1).

Fig. 1. Specimen members: 3D-printed joint, Aluminum bar, Steel bolt

The joint was designed with the smallest possible dimensions so that the four aluminum bars could be connected to it through the bolts. The shape of the joint is symmetrical to its center point, with its external dimensions being 64 × 64 mm and its thickness 7 mm. The size of the holes, both in the joint and the bars, is equal to the diameter of the bolts, namely 3 mm. In the center of the joint there is a 6 mm diameter hole for applying the load through a tension rod system.

The bars are of hollow rectangular aluminum section (EN-AW 6063-T5 [7]) with dimensions of 30 × 15 mm, 1.5 mm thickness. The bolts used are M3 and their material is grade 8.8 structural steel.

To determine the load to be applied to the specimen, a deployable canopy comprised of scissor elements [6] was studied, to identify the distribution of internal forces in the four beams. From the entire structure, the joint where the maximum tensile forces develop in the beams connected to it, was selected. In this position it was observed that the tensile forces in one direction are much larger than in the other direction, while all other internal forces were much smaller in both directions. For this reason, the point where the bars are hinged to the rigid steel base in one direction (X) is higher than the hinges in the other direction (Y), almost at the same level with the loading point of the joint, so that higher tensile forces are developed in the bars in the Y direction (Fig. 2).

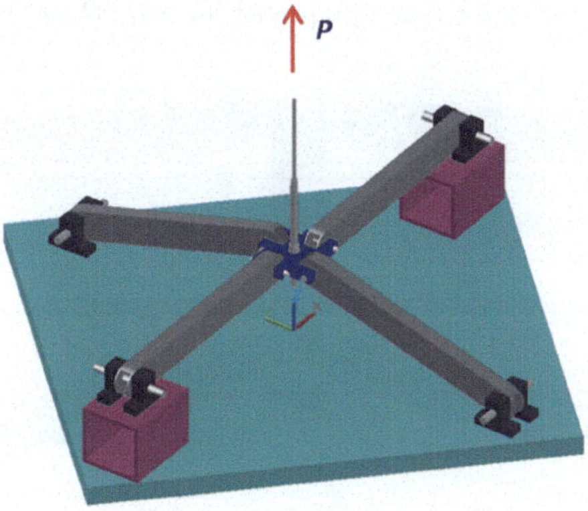

Fig. 2. Experimental setup

2.2 Materials

Joint Material. PLA (Polylactic Acid), commonly used in 3D-printing, was chosen as the material for the joint, as it is a popular versatile, biodegradable, and eco-friendly thermoplastic material, made of renewable resources such as corn starch and sugarcane. For the simulation of that material, the literature was searched for a material law describing the mechanical properties of PLA. Rossetos et al. [3], performed tensile tests on dog-bone specimens printed on a three-dimensional printer in two printing directions. The test results showed that the behavior of the specimens is different depending on the printing direction of the specimen, so the printed product has an orthotropic behavior. The modulus of elasticity is on average similar in the two cases, approximately equal to 3.000 MPa. The ultimate strength is different for the two directions, about 49 MPa for 0° and 35 MPa for 90° printing. In the present work, the PLA material was considered as bilinear with zero strain hardening modulus, modulus of elasticity equal to 3.000 MPa, and yield strength equal to 35 MPa. The Poisson's ratio was assumed equal to 0.33 [7].

Aluminum. Bars' Material. Aluminum alloy EN-AW 6063-T5 was chosen as the material for the bars, which according to EN1999 [7] has a characteristic value of the yield strength $f_o = 130$ N/mm^2. Aluminum was simulated as bilinear with zero strain hardening modulus, with a yield strength equal to 130 MPa. The modulus of elasticity was considered equal to 70.000 MPa while the Poisson's ratio equal to 0.33 [8].

Bolt's Material. The material of the bolts is grade 8.8 structural steel. It was simulated as bilinear with zero strain hardening, with a yield strength of 640 MPa. The modulus of elasticity was considered as 210.000 MPa and the Poisson's ratio equal to 0.30 [9].

2.3 Boundary Conditions, Contacts, Load

Boundary Conditions. For the support of the specimen, hinges were considered at the ends of the bars so that all degrees of freedom are fixed, except of the rotations about the axis that is perpendicular to the side surfaces of the bars. The hinges were applied to the surfaces of the holes of the supports, and specifically on the side of the holes that is in contact with the bolts, due to the tension of the bars (Fig. 3).

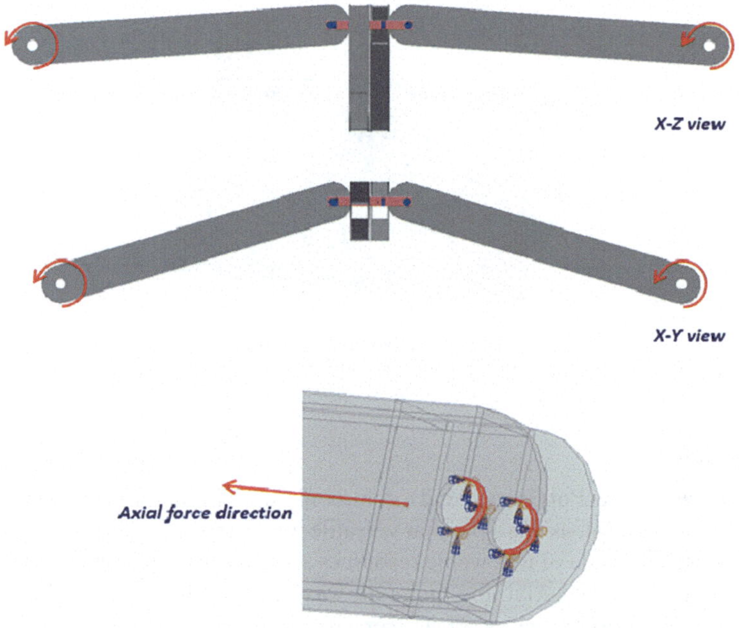

Fig. 3. Boundary conditions

Contact Elements. All possible contacts between the specimen members were simulated through contact elements, as below:

- Bolt heads/nuts – Joint external faces
- Bolt shanks – Joint bolt holes
- Bars' external faces – Joint internal faces
- Bars' bolt holes– Bolt shanks

Load. The load was simulated as a pressure exerted at the bottom of the joint, around the loading hole, with an upward (+Z) direction. The loading surface corresponds to the interface surface of the joint with the tension rod system (Fig. 4).

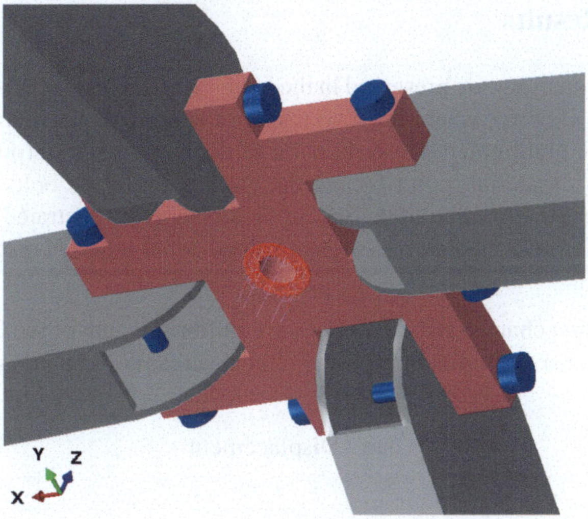

Fig. 4. Loading surface and direction

2.4 Analysis Method

As the contacts of the specimen members were simulated, a non-linear approach to the problem was necessary. First, geometry nonlinear analyses (GNA) were performed with an elastic consideration of materials, and then geometry and material nonlinear analyses (GMNA) were performed to determine the influence of the material nonlinearity. In these analyses, the materials were modeled as bilinear without strain hardening, as mentioned in Sect. 2.2.

To achieve convergence of the nonlinear solution algorithm and proper actuation of the contact elements, a sufficiently dense finite element mesh had to be defined. After testing, the grid was defined as rectangular, with an edge size of about 2 mm and an aspect ratio of about 1(Fig. 5).

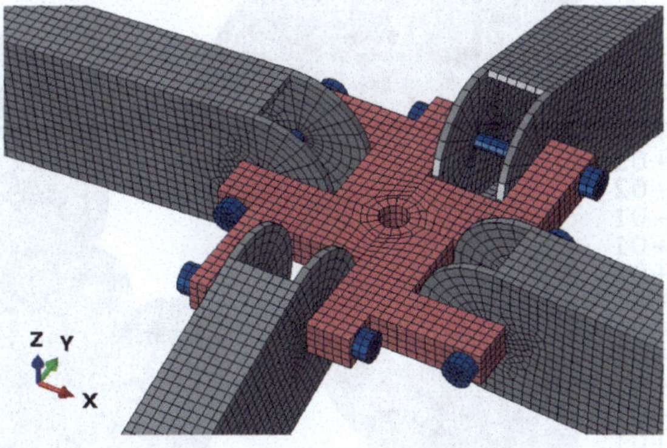

Fig. 5. Meshed members

3 Analysis Results

The results of the analyses are presented in the load – vertical displacements of the center of the joint curve (Fig. 6), where the load resultant is used. Three characteristic points of the diagram are highlighted: the yield point of the bars' material (point A), the yield point of the joint's material (point B) and the yield point of the bolts' material (point C), where the stress distribution of the specimen members is illustrated in the following and discussed. Moreover, the deformations for load equal to 775N and 1500N are also presented.

Point A. At the first characteristic point of the equilibrium path of the GMNA analysis, the yielding of the bars' aluminum is observed in the areas corresponding to the surfaces

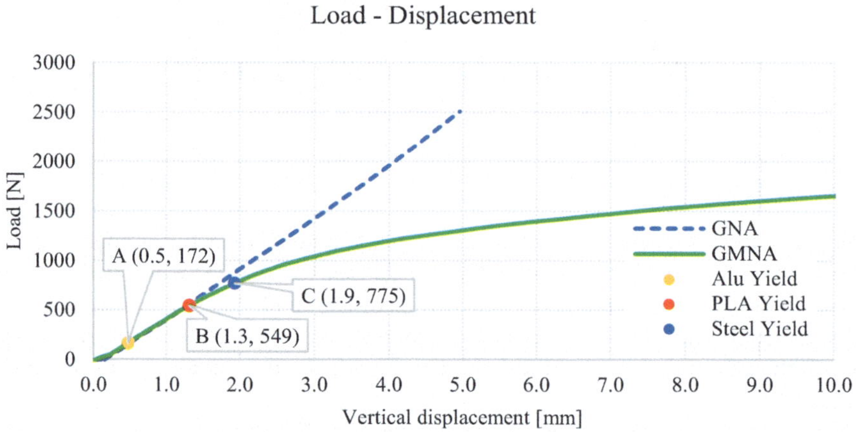

Fig. 6. Load – displacement curve

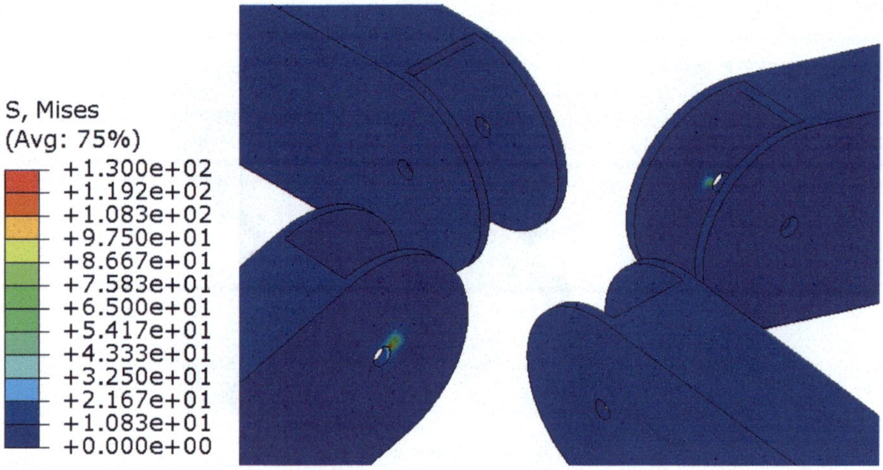

Fig. 7. Von Mises stresses in the bars (max σ_{vM} = 130 MPa)

of the holes in connection with the joint (Fig. 7). The load corresponding to this point is equal to 172 N and the vertical displacement is equal to 0.50 mm (Fig. 6). At the same point of the curve, the maximum von Mises stress in the joint is found at lower values than the yield point of the PLA (15 MPa < $f_{y,PLA}$ = 35 MPa) (Fig. 8), while the maximum stress in the bolts is only 95 MPa.

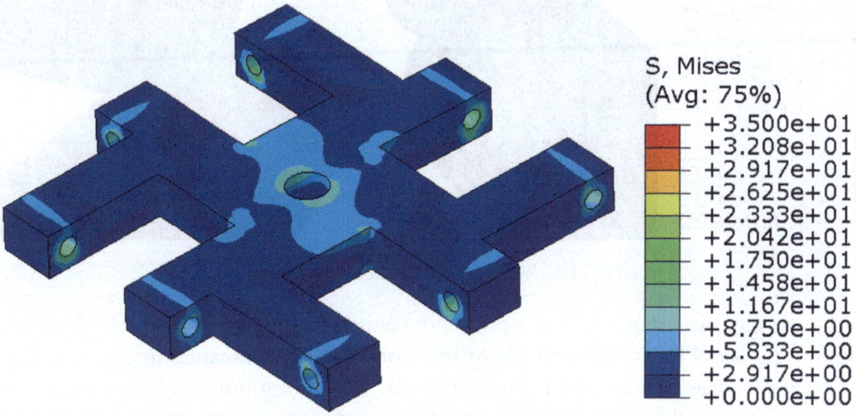

Fig. 8. Von Mises stresses in the joint (max σ_{vM} = 15 MPa)

Point B. The second characteristic point of the equilibrium path corresponds to the initiation of yielding of the joint's material. At this point the force is equal to 549N while the vertical displacement is 1.3 mm (Fig. 6.). The maximum von Mises stress in the joint is equal to 35 MPa, located at the surfaces of the holes in connection with the bolts, but also at the center of the joint, around the loading hole (Fig. 9). At the same point the stresses around the surfaces of the holes in the aluminum bars are higher (Fig. 10), while the maximum stress of the bolts is equal to 420 MPa, quite close to the steel's yield stress.

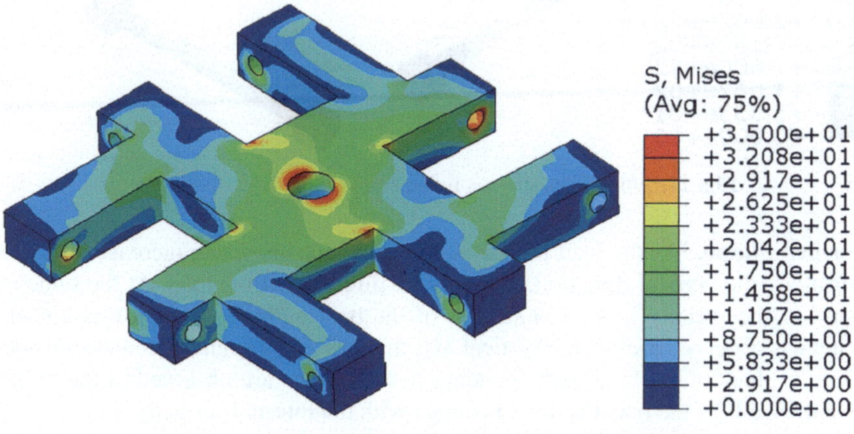

Fig. 9. Von Mises stresses in the joint (max σ_{vM} = 35 MPa)

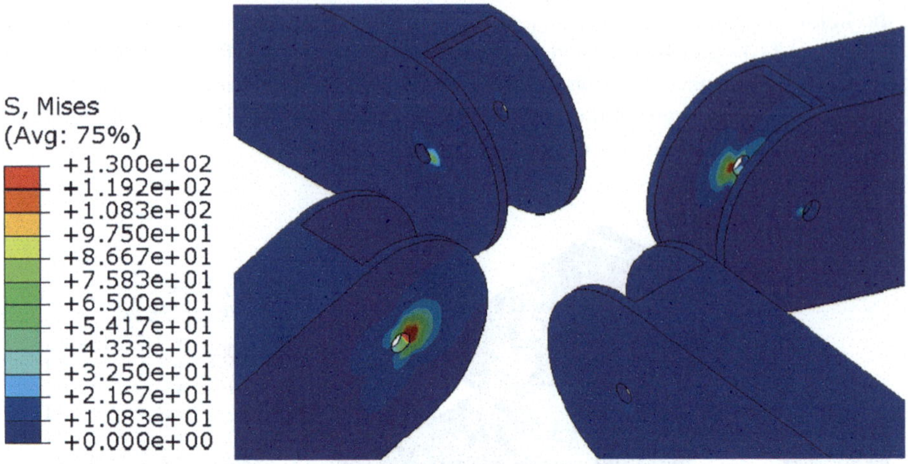

Fig. 10. Von Mises stresses in the bars

Point C. At the third characteristic point of the equilibrium path of the GMNA analysis, yielding of the steel bolts is observed. At this point the force is equal to 775N while the vertical displacement is equal to 1.9 mm (Fig. 6). The maximum von Mises stress in the bolts is observed at the contact regions with the surfaces of the bars' holes, where the stresses are higher (Fig. 11). Also at the joint, higher stresses are observed, both on the surfaces of the holes and over the internal regions around the loading hole (Fig. 12).

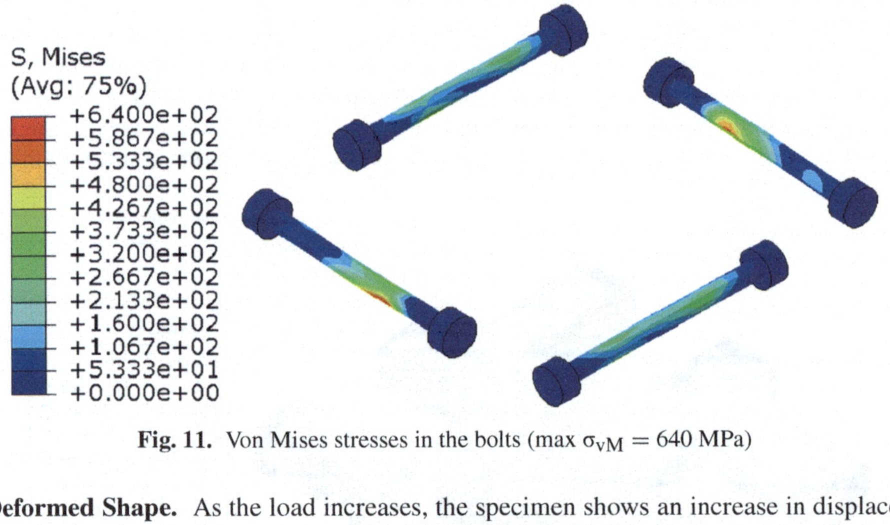

Fig. 11. Von Mises stresses in the bolts (max $\sigma_{vM} = 640$ MPa)

Deformed Shape. As the load increases, the specimen shows an increase in displacements along the vertical direction, which is the direction of the load. At the same time, due to the eccentricity in the connection of the bars with the joint, a rotation of the specimen is observed around the vertical axis that passes through the center of the joint. For large values of the load, large bending deformations are observed in the bolts but also in the walls of the bars that are in contact with the internal surfaces of the joint. The

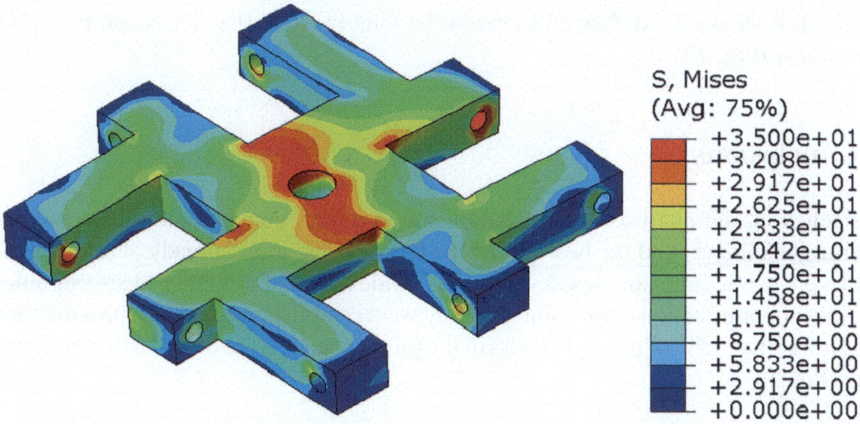

Fig. 12. Von Mises stresses in the joint

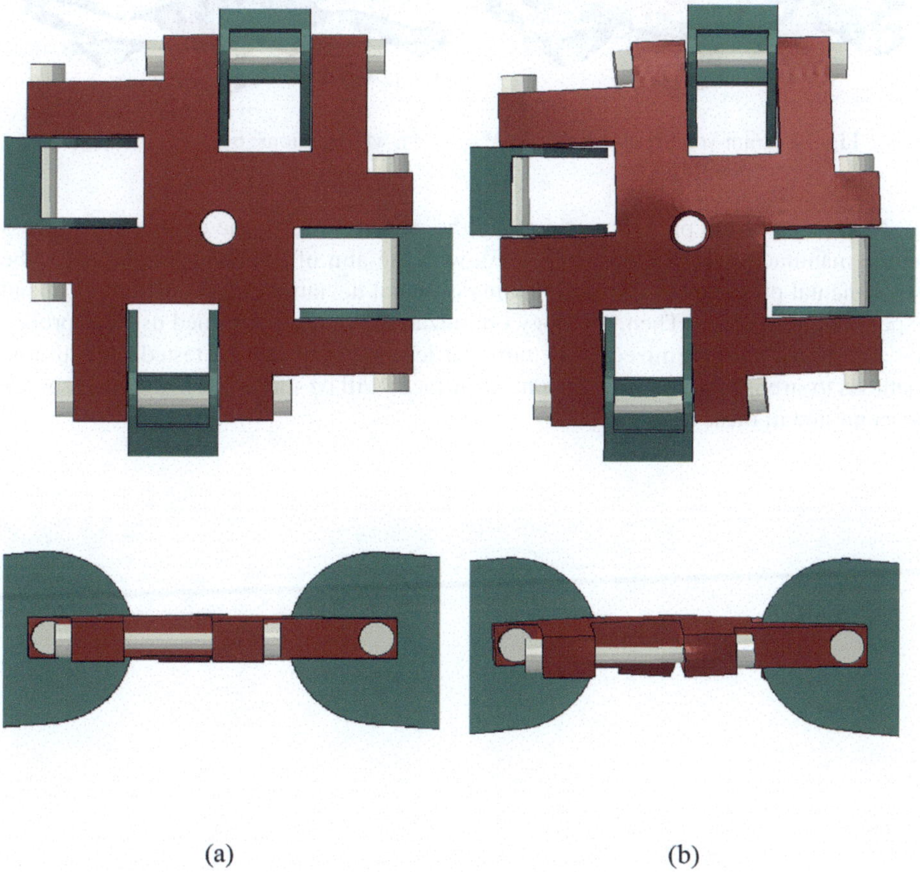

Fig. 13. Deformed shape for P = 775N (a) and P = 1500N (b)

figure below shows the deformed shape of the joint for load P = 775N and P = 1500N respectively (Fig. 13).

4 Conclusions

In conclusion, the analysis results demonstrated a high concentration of stresses and material yielding around the bearings of the bars and the joint, already at relatively low load levels, while high stresses develop also inside the joint at higher loads. Significant margins for optimization of the joint topology were established, as areas with a maximum stress lower than 15% of the yield stress of the joint material were observed for the applied loads (Fig. 14).

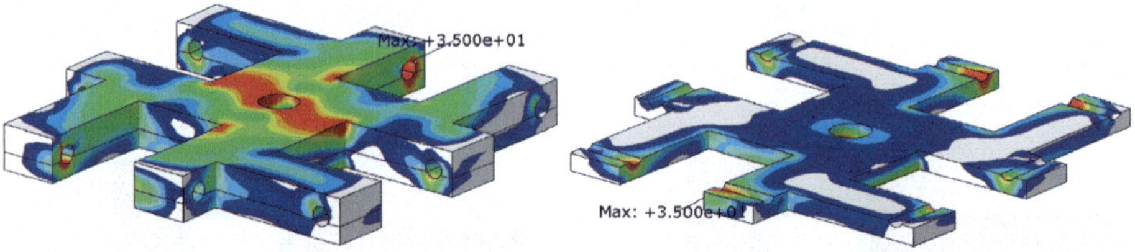

Fig. 14. Joint von Mises stresses for P = 775N, white regions: $\sigma_{vM} < 0.15_{fy,PLA}$

Next steps of the present research are the performance of the experiment process on the manufactured specimens (Fig. 15), with the aim of predicting evolution of the experimental process, but also determining eventual deviations between numerical and experimental findings. Then, topology optimization will be performed using appropriate algorithms and optimized joints utilizing less material will be tested. In addition, methods to strengthen the bar hole contact surfaces will be sought, as higher stresses are concentrated in these areas.

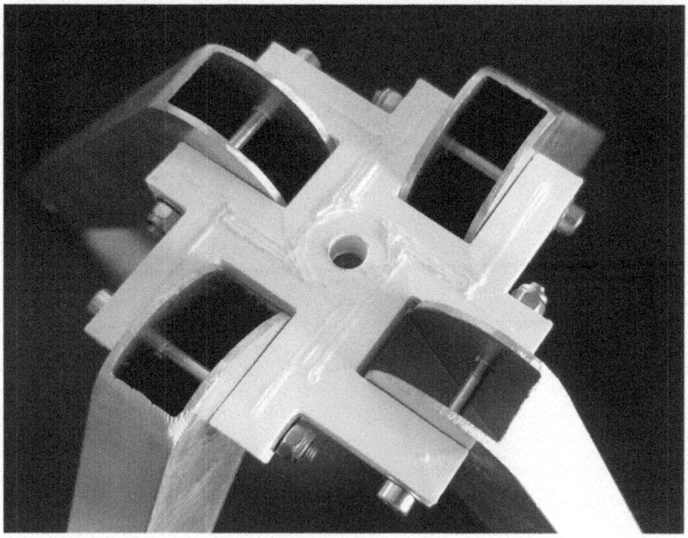

Fig. 15. Physical model of the discussed specimen

Acknowledgements. This research was supported by the ADDOPTML project: "ADDitively Manufactured OPTimized Structures by means of Machine Learning" (No: 101007595) belonging to the Marie Skłodowska-Curie Actions (MSCA) Research and Innovation Staff Exchange (RISE) H2020-MSCA-RISE-2020. Their support is highly acknowledged.

References

1. Gantes, C.J.: Deployable Structures – Analysis and Design. WIT Press, Southampton (2001)
2. Simulia Abaqus CAE. https://www.3ds.com/products/simulia/abaqus/cae. Accessed 30 Nov 2024
3. Rossetos, I., et al.: Numerical modeling and nonlinear finite element analysis of conventional and 3D-printed spinal braces. Appl. Sci. **14**(5), 1735 (2024)
4. De Temmerman, N.: Design and analysis of deployable bar structures for mobile architectural applications. PhD Thesis, Vrije Universiteit Brussel, Brussel (2007)
5. Koumar, A.: Deployable scissor structures for disaster relief: multi – criteria optimization and design adaptability. PhD Thesis, Vrije Universiteit Brussel, Brussel (2016)
6. Pantazis, G.D.: Design of Deployable Shelters for Emergency Response. Diploma Thesis, National Technical University of Athens, Greece (2022) (in Greek)
7. European Committee for Standardization.: Eurocode 9: Design of aluminium structures. (2007)
8. Ferreira, R.,T.,L., Amatte, I.,C., Dutra, T.,A., Bürger, D.: Experimental characterization and micrography of 3D printed PLA and PLA reinforced with short carbon fibers. Compos. Part B: Eng. **124**, 88–100 (2017)
9. European Committee for Standardization: Eurocode 3: Design of steel structures (2005)

10. Pérez-Valcárcel, J., López-César, I.R., Muñoz-Vidal, M., Freire-Tellado, M.J., Suárez-Riestra, F.: Deployable tubular bar structures with laterally confined flattened ends. Structures **55**, 112–122 (2023)
11. Seifi, H., Javan, A.R., Xu, S., Zhao, Y., Xie, Y.M.: Design optimization and additive manufacturing of nodes in gridshell structures. Eng. Struct. **160**, 161–170 (2018)

Measuring and Evaluating Layer Height to Width Ratio in 3DCP Towards Higher Geometric Conformity

Odysseas Kontovourkis[1](✉), Christos Georgiou[1], Alexis Andreou[2], Vasilis Andreou[3], and Savvas Dimitriou[1]

[1] Department of Architecture, University of Cyprus, Nicosia, Cyprus
kontovourkis.odysseas@ucy.ac.cy
[2] Independent Researcher, Nicosia, Cyprus
[3] Biosynthesis Tech Research Ltd, Nicosia, Cyprus

Abstract. Geometric conformity is an essential aspect of 3D Concrete Printing (3DCP) due to its importance in the accurate production of design outcomes. This work aims to discuss the potential for higher geometric accuracy by investigating the height-to-width ratio of overall and individual layer filaments in 3D printed results through a series of robotic 3DCP sample prototypes using different material mixtures. The results are examined using manual, semi-automated, and automated algorithmic procedures to analyze a series of cross-section shapes derived from 3D printed samples. This enables the extraction of measurable data that includes, among others, the height and width of overall cross-sections as well as each individual printed filament. The collection of data allows for comparison between different parameters influencing 3DCP, including material mixtures, robotic speed, and adjusted height/width of layer filaments. The results show that printing parameters such as material mixtures and robotic speed play an important role during 3DCP and can be applied to control the process in future design and 3DCP production steps.

Keywords: 3D Concrete Printing · Processing Parameters · Geometric Accuracy · 3D Scanning · Cross-section Images

1 Introduction

3D Concrete Printing (3DCP) is an innovative and promising alternative technology in the concrete construction industry, based on the layer-by-layer deposition of concrete materials. Known advantages over conventional construction techniques include the reduction of material waste and construction time, as well as increased flexibility in constructing complex structures. Although there is a rapid increase in examples that implement 3DCP technology in both the industrial and academic sectors, many issues remain to be investigated before this technology can be widely used in the construction industry.

N. D. Lagaros et al. (Eds.): ADDOPTML 2024, LNCE 646, pp. 267–279, 2025.
https://doi.org/10.1007/978-3-031-92029-5_25

Achieving precise geometric conformity, i.e. the accuracy with which the printed structure matches the intended design, is a vital issue that plays a major role in the success of 3DCP [1]. This accuracy is affected by several factors, such as the material composition and the printing parameters that lead to different height-to-width ratio of the printed layers. Measuring, predicting and controlling 3D printed height-to-width ratio enhance quality and improves geometric conformity. This ratio is considered as an important factor towards an effective control of 3DCP.

Various examples can be found in the literature that conduct testing methods to measure and evaluate the performance of 3DCP results [2], which directly or indirectly incorporate techniques for measuring the layer height and width of printed results by applying manual or automated techniques. In the work by [3] a method to test the shape stability was described that incorporated a camera and a ruler for measuring the deformation of layers before and after deposition of the next layer. The work by [4] introduced a variable-size square nozzle to examine its potential application for manufacturing architectural ornaments. To check the geometric conformity between filament's cross-section and nozzle size ruler-based measurements were conducted. Also, in the work by [5] an extrusion-based 3D concrete printing with variable width size was proposed and a series of test printing samples were performed and measured (width of filaments) using manual ruler.

In other examples, cameras and scanning techniques for image analysis of 3DCP results were introduced to extract measurable data and evaluate printing performance. In the work by [6] a method for assessing the quality of 3DCP multilayered structures was proposed using 3D scanner, point cloud processing and parameters definition. In the work by [7] a methodology for inline monitoring of 3DCP results consisting of image acquisition, interlayer line and layer segmentation, as well as local layer characterization and anomaly detection with the use of a camera was introduced. Also, in another work by [8] a real time vision-based control using vision sensor (camera) and image-processing techniques was introduced for layer-width setting in concrete 3D printing applications. Furthermore, in the work by [9] a real-time height measurement and control system for 3DCP was suggested and in the work by [10] a 2D laser profile sensing system was suggested that extract cross-sectional information of concrete filament including width, thickness, and area. Finally, apart from measuring the width and height of cross-sections using scanner and image processing techniques, research works suggested predictive models to simulate filament geometry outcomes based on printing inputs [11] and introduced machine vision to detect defects and deviations but also predicted the motion speed of robot to maintain predetermined layer width [12]. Although a wide range of manual and automated techniques have been found in the literature, comparative analysis and critical reflection have been rarely observed.

The objective of this study is to develop and present a range of manual to automated methods for measuring the height-to-width ratio of printed layers and hence to enhance and control the geometric conformity of 3D printed results. Also, to compare methods and highlight the advantages and limitations of each method in terms of accuracy, efficiency but also applicability in real case scenarios. Finally, the research findings aspire to provide valuable information towards optimization of 3D printing parameters to achieve control of improved geometric accuracy during 3DCP and provide valuable insight for future research development. To achieve this, three case studies are presented, focusing on measuring and evaluating the dimensions of cross-sections in predetermined points of 3D printed results that are produced through planar and non-planar 3DCP.

2 Methodology

To check the geometric conformity of 3D printed results, the study proposes a methodology that combines manual, semi-automated and automated techniques for measuring and evaluating geometric data from 3D printed samples that have emerged from three (3) main case studies: two (2) non-planar 3DCP cases and one (1) planar 3DCP case. Each case study utilizes specific techniques of measuring and evaluating the height-to-width ratio of 3D printed results.

2.1 Case Study A: Non-planar 3DCP

The recording and analyzing of the 3DCP samples in Case Study A is based on a manual technique using conventional means.

Experimental Setup. The study involves an experimental setup that follows three main research steps, which are analytically presented in [13]. The shape development, the toolpath planning and the robotic 3DCP execution. The study also focuses on examining the compositions of cement-based materials, the various surface curvatures, and the maximum possible 3D printing angles in non-planar surfaces. The shape development is based on parametric modeling of non-planar surfaces for 3DCP. Then, three (3) non-planar surfaces ('I', 'S', and 'U' shapes) out of forty (40) surfaces, as samples for testing, are selected.

Subsequently, the toolpath preparation through parametric design is conducted (Fig. 1), followed by the 3DCP execution using a robotic 3DCP mechanism (Fig. 2). A total of nineteen (19) trials using a hand-made formwork with a polystyrene base in non-planar shapes are 3D printed, and physical prototypes on a 1:2 scale are produced in the laboratory. This procedure attempts to achieve the optimal material mixture and in parallel to evaluate the printing process in relation to construction-scale applications.

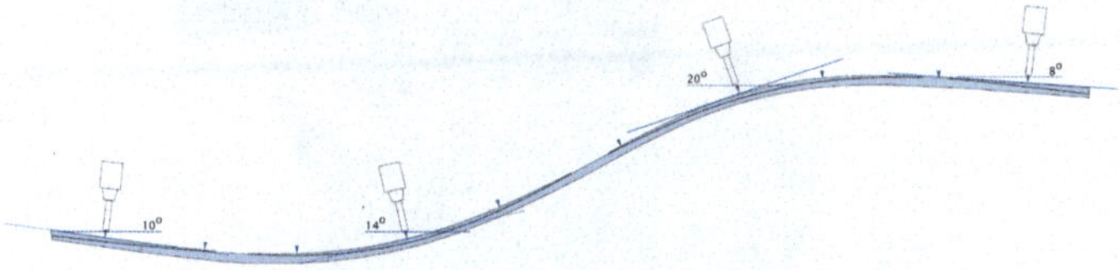

Fig. 1. Non-planar travelling of extruder during 3D printing in case of 'S' shape.

Data Collection and Analysis. The different trials aim at examining the 3D printed results in relation to their optimal material mixtures and printing speed, but also in relation to height and thickness info based on manual means of investigation and specifically

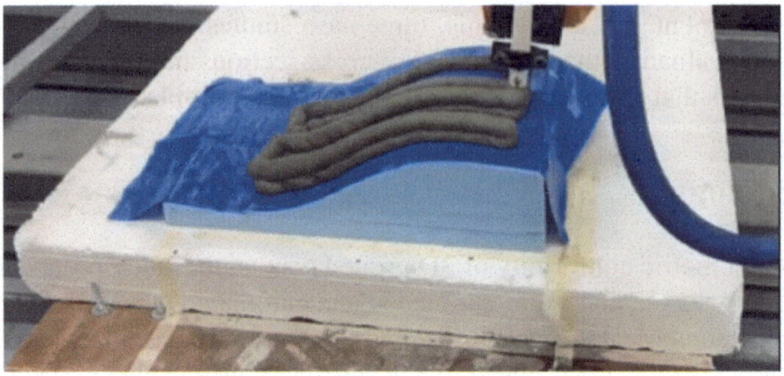

Fig. 2. Non-planar robotic 3DCP execution of 'S' shape.

by using conventional scale meter and ruler. Finally, the trials are examined in relation to physical to digital height and width discrepancies of printed filaments that lead to correlation and comparison between physical and digital results.

2.2 Case Study B: Non-planar 3DCP

Case Study B investigation is based on a semi-automated technique using digital means for recording and analyzing.

Experimental Setup. This study aims, through the design of a specific workflow that is analytically presented in [14], to suggest a 3DCP technique that is influenced by traditional architecture of cross-vaults in Cyprus. The suggested technique discussed in this paper involves the design and production of cross-vault structural elements using non-planar 3DCP. The toolpath planning implements an algorithm that generates non-planar slicing based on static analysis and specifically it follows curves perpendicular to the force-flow lines to direct the movement of 3DCP extruder (Fig. 3).

Fig. 3. Non-planar 3D printing of a structural component of cross-vault.

Data Collection and Analysis. The non-planar 3D printed trials are examined in relation to their optimal material mixtures and printing speed. Also, info related to layers' height and thickness is extracted based on a semi-automated technique that involves cutting of selected cross-sections and scanning using a 3D Structure Light Scanner. Then, the generated Point Cloud is imported in 3D modeling software Rhino for manual outline of cross-section geometry and measurement of overall cross-section and individual filament dimensions (Fig. 4). Finally, a comparison between physical and digital layers' height and thickness is presented and discussed.

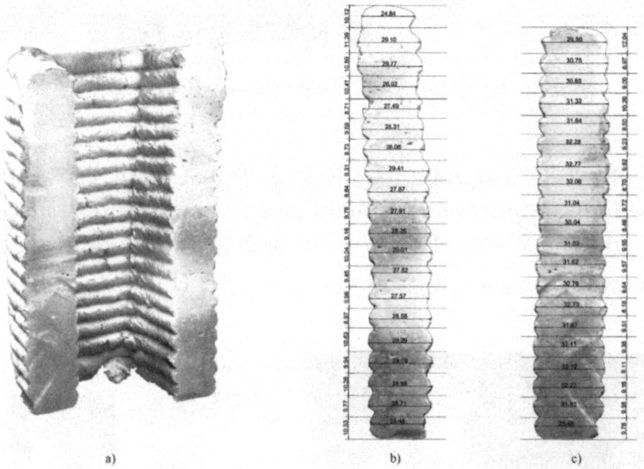

Fig. 4. Point Cloud of the segmented scanned physical model. (a) 3D scanned model, (b) and (c) Selected cross-sections 1 and 2.

2.3 Case Study C: Planar 3DCP

Case Study C investigation is based on an automated technique using digital means for recording and analyzing.

Experimental Setup. In this study, planar 3D printed samples are assessed in terms of their geometric accuracy and conformity. The investigation includes 3D parametric modeling of samples and robotic execution to ensure precision based on planar 3DCP. The study also examines the composition of cement-based materials mixtures (Fig. 5).

Data Collection and Analysis. The examination of 3D printed samples focuses on identifying optimal material mixtures and 3DCP processing parameters in relation to height and width measurements (overall and individual layers) of physical prototypes and their comparison with digital counterparts. The process of collecting and analyzing the data is based on an automated algorithmic approach that incorporates the recording of dimensions and statistical analysis.

Specifically, the approach utilizes an automated algorithm for sectioning, automated measuring and analyzing the ratio of height-to-width of the printing filaments. The step involves sectioning of 3D printed samples at specific predefined positions using a circular

Fig. 5. Planar 3DCP execution of samples.

saw for both perpendicular and diagonal cross-sections. Then, cross-section images are imported in Rhino/Grasshopper environment and by using an in-house automated algorithm, the total height and width of sections as well as the average width and maximum height for each layer are measured for further analysis (Fig. 6).

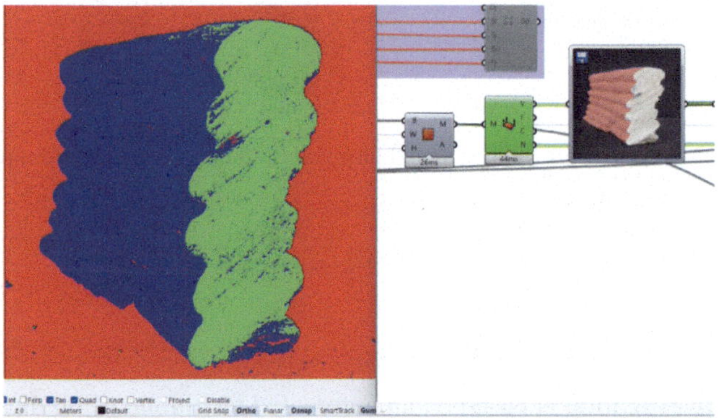

Fig. 6. Generation of cross section outline using an in-house automated algorithm.

3 Results

3.1 Case Study A—Data Collection and Analysis of Results

Data Collection. Data extracted from the 3D printing trials of Case Study A is collected and further analyzed. Data recording includes cement-based material mixture ratios applied in the different cases and the measurements of height and thickness of each filament. In addition, the 3DCP processing parameters applied in Case Study A are examined and evaluated in accordance with the results of height and thickness. The best layer height is found to be 10 mm in the first layer and 12 mm in the initial layers along with printing thickness of 20 mm.

Correlation Between Physical and Digital Results. The study compares physical against digital 3DCP results in the three (3) shapes under investigation. The following figures demonstrate results derived from case 'S', that show good correlation between physical and digital outcomes. The differences in heights indicate variations ranging from 1 mm to 6 mm in the case of the 4-layer printed trial (Fig. 7).

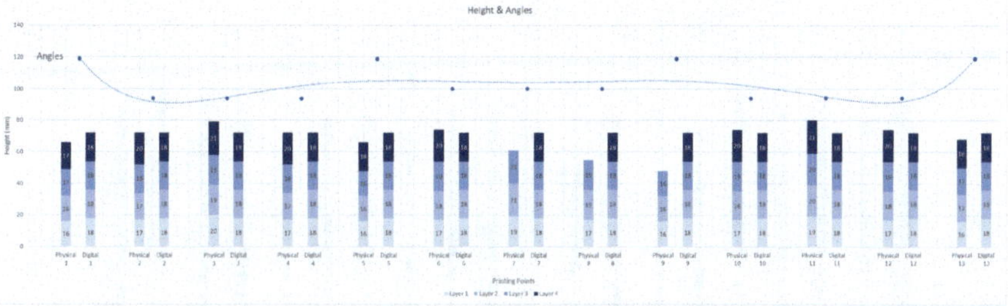

Fig. 7. 3D printed heights (physical vs digital) in case of 'S' shape (layer by layer) for a total of 4 printed layers.

The comparison between digital and physical thicknesses reveals significant differences. In all shapes under investigation, the printed thickness exceeds the digital size. In case of 'S' shape the thickness of physical outcomes is measured between 26 mm to 30 mm, compared to 20 mm thickness in digital samples (Fig. 8). The increased thickness is primarily due to material deformation caused by the self-weight of the printed parts and the extrusion pressure during the 3D printing process. To achieve better results, the use of thinner nozzles and acceleration chemicals would be necessary.

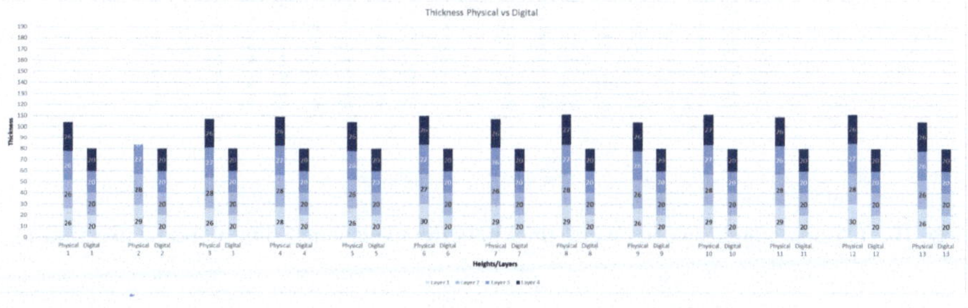

Fig. 8. 3D printed thickness (physical vs digital) in case of 'S' shape (layer by layer) for a total of 4 printed layers.

The Case Study A findings highlight that data collection and analysis of results is complex due to non-planar 3D printing. To achieve geometric conformity and best correlation between physical and digital results in non-planar 3DCP, careful consideration regarding material mixtures and 3D printing parameters is necessary.

3.2 Case Study B—Data Collection and Analysis of Results

Data Collection. The 3D printed trials of Case Study B are collected and further analyzed. The material mixture for the experimental trial consists of 12,000 g of cement, 5250 g of water, 12,400 g of sand (<2 mm), and 50 mL of superplasticizer. The measurements of height and thickness of the physical model and their comparison with digital information are demonstrated in Figs. 9 and 10.

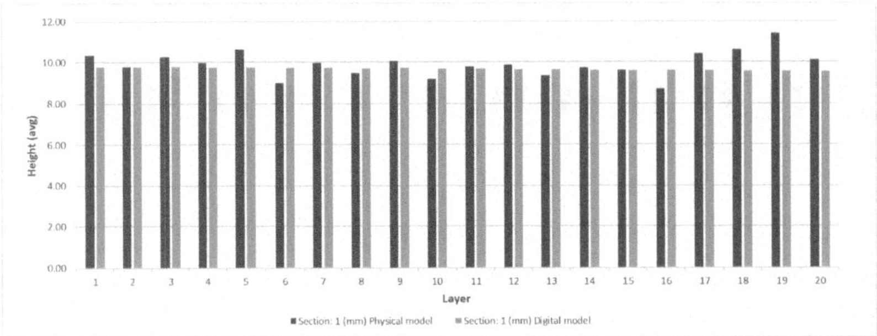

Fig. 9. Comparison between layers' height in physical and digital model (Section 1).

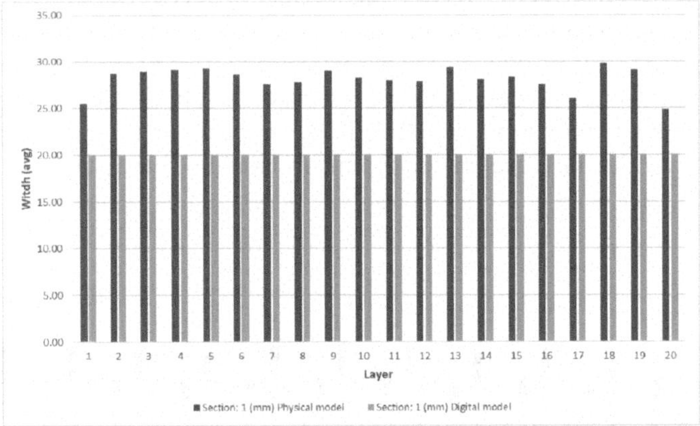

Fig. 10. Comparison between layers' thickness in physical and digital model (Section 1).

Correlation Between Physical and Digital Results. The comparison between the digital and physical layers' height and thickness reveals obvious differences. Initially, an increase in the maximum average height of each layer is observed. This is a consequence of increasing the slope of the printing toolpath, which causes less deformation due to the self-weight of the structure. In some cases, smaller heights in the physical model are observed due to variable heights and vertical nozzle angle. Furthermore, an increase in physical thickness compared to the digital thickness is noted, likely due to the lower robotic printing speed and nozzle dimensions. These observations indicate the need for further refinement of material mixtures and 3DCP processing parameters.

3.3 Case Study C—Data Collection and Analysis of Results

Data Collection. In Case Study C, a series of cross-section images for each sample are imported into the automated algorithm to extract data related to the height and width measurements of overall sections and individual layers. The data provide insights into the geometric accuracy and performance of the printed samples, highlighting key measurements that contribute to the overall analysis.

Correlation Between Physical and Digital Results. The layer height analysis shows that deviations typically fall within the bounds of +20% and −20%. The initial layer consistently shows a height less than the intended design, aligning with expectations due to concrete shrinkage during setting and the weight of subsequent layers (Fig. 11).

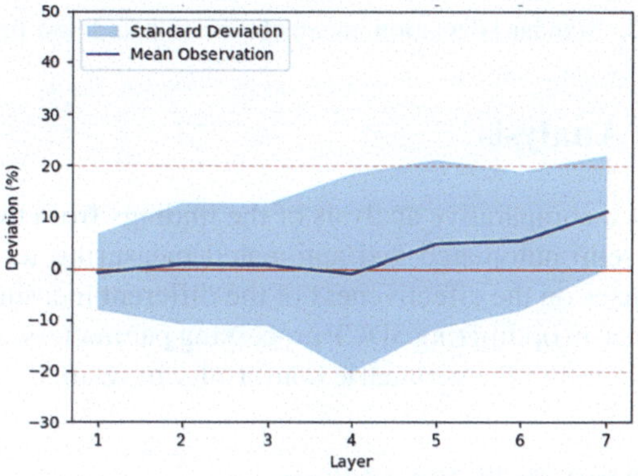

Fig. 11. Layer height analysis of the selected 3D printed sample.

The layer height versus layer width analysis reveals a clear negative correlation, indicating that width decreases as height increases. A best fit line was applied, yielding the equation $y = -1.45x + 54.58$, meaning that for every unit increase in height, the width reduces by approximately 1.45 units. This observation occurs at a robot speed of 48 mm/sec (Fig. 12).

The Case Study C findings show that the automated algorithm successfully extracts critical measurements from the cross-section images, highlighting deviations and performance metrics that provide detailed insights and contribute to the overall analysis of the geometric accuracy of the printed samples.

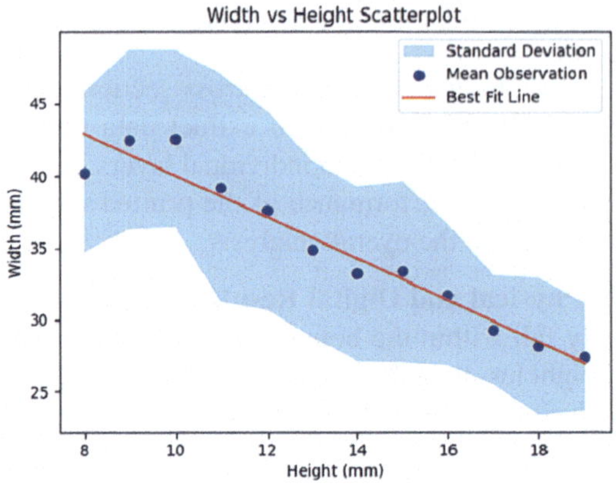

Fig. 12. Standard deviation, mean observation and best fit line.

4 Comparative Analysis

This section provides a comparative analysis of the findings from the three case studies that utilize manual, semi-automated and automated measuring techniques. The comparative analysis focuses on the effectiveness of the different measurement and analysis methods, their influence in optimizing 3DCP processing parameters, and their differences and similarities in relation to the geometric conformity of results.

4.1 Methods of Measurement and Analysis

Regarding the effectiveness of the three different methods of measuring and analyzing data collected from 3DCP, the results reveal the following observations. The manual method of measuring height and width dimensions can be time-consuming and less accurate due to the active human involvement in the process. Despite these challenges, the manual method might be helpful in cases where measuring results, due to the ambiguity of the physical prototypes, are difficult to be extract using semi or fully automated methods, especially in the case of complex non-planar 3DCP.

The measurement and analysis of results based on the automated approaches reveals that the applied automated algorithm can provide precise and consistent measurements of height and width. This technique can decrease human error during measuring and in parallel can increase efficiency of data collection and analysis, especially when a large amount of data is analyzed. Through the detailed and precise extraction of height-to-width ratio, more accurate conclusions regarding geometric conformity both in planar and non-planar printing can be provided.

A disadvantage might be the inability of the automated algorithm to record measurements of sections that present failures or errors in printing. This can be resolved by combining manual, semi-automated, and automated methods under a hybrid operating framework, which can help towards efficient recording of printed results.

4.2 3DCP Processing Parameters

The influence of measurement and analysis methods of heigh-to-width ratio in optimizing 3DCP processing parameters, especially in adjusting the material composition and robotic printing parameters through a series of printing trials, is important both in the cases of non-planar and planar 3D printing.

The calculation of height-to-width ratio allows investigation into optimal material mixtures and control of geometric conformity results towards higher precision. For instance, the control of robotic printing speed is valuable for controlling height and thickness for the initial and subsequent layers, allowing repeatability of results.

However, the implementation of manual methods in data collection, analysis and extraction of consistent results is a challenge, especially in the case of non-planar 3D printing due to the complexity and variability of height-to-width ratio. On the other hand, the accurate extraction of measurable data can benefit from the implementation of semi and fully automated algorithms.

4.3 Geometric Conformity

Differences and similarities of methods in relation to the geometric conformity of results are as follows. In non-planar 3D printed samples, significant variations in height and width measurements, and hence lower geometric conformity, are observed when manual and semi-automated methods are applied. This discrepancy between physical and digital prototypes results not only from the methods adopted but also from material falling during non-planar printing, where successive layers do not have the consistency of planar printing, leading to failures.

In general, the consistency of non-planar printing was lower compared to planar printing, indicating that non-planar surfaces pose greater challenges for achieving geometric conformity. On the contrary, planar printing shows more consistent results with fewer deviations from the digital models. This is not only a result of the automated height and width measurement method used but also of the 3DCP technique adopted, which is based on planar layers.

5 Conclusion

The study demonstrated the importance of measuring and analyzing the geometric conformity of 3DCP outcomes by investigating the height-to-width ratio of 3D printed samples using manual, semi-automated, and automated methods. This was done by presenting three case studies that involved non-planar and planar 3DCP. The comparative analysis provided valuable insights into the advantages and disadvantages of each measurement method, their influence on optimizing 3DCP processing parameters, and their impact on the geometric conformity of results.

The comparison of different methods of measurement demonstrated the importance of sectioning in measuring and evaluating the deviations between digital and physical 3D printed results. Although the extraction and analysis of results based on manual techniques were less accurate than automated methods, they demonstrated significant findings. Moreover, the application of semi-automated and automated methods showed that a higher level of accuracy can be achieved, providing accurate and reliable data for evaluating geometric conformity.

All case studies demonstrated that 3DCP processing parameters, such as material mixtures and robotic printing speed, play an important role and can significantly determine the height-to-width ratio. This allows for greater control and hence superior geometric conformity between the digital and physical printed samples. By optimizing the processing parameters, the printed samples can better and more accurately correspond to the desired geometries. Additionally, optimized parameters can be applied to control the 3DCP process in future design and production steps.

Future studies might benefit from this investigation to further explore and refine automated methods and printing processing parameters. These improvements can be applied to complex design tasks to enhance geometric conformity and the quality of printed results, as well as to increase efficiency, scalability, and applicability.

References

1. Buswell, R., et al.: Geometric quality assurance for 3D concrete printing and hybrid construction manufacturing using a standardised test part for benchmarking capability. Cem. Concr. Res. **156**, 106773 (2022)
2. Kaliyavaradhan, S.K., Ambily, P.S., Prem, P.R., Ghodke, S.B.: Test methods for 3D printable concrete. Autom. Constr. **142**, 104529 (2022)
3. Kazemian, A., Yuan, X., Cochran, E., Khoshnevis, B.: Cementitious materials for construction-scale 3D printing: laboratory testing of fresh printing mixture. Constr. Build. Mater. **145**, 639–647 (2017)
4. Xu, J., Ding, L., Cai, L., Zhang, L., Luo, H., Qin, W.: Volume-forming 3D concrete printing using a variable-size square nozzle. Autom. Constr. **104**, 95–106 (2019)
5. Yuan, P.F., Zhan, Q., Wu, H., Beh, H.S., Zhang, L.: Real-time toolpath planning and extrusion control (RTPEC) method for variable-width 3D concrete printing. J. Build. Eng. **46**, 103716 (2022)
6. Chen, Y., Zhang, Y., Liu, Z., Zhang, W., Pang, B., Zhang, Y., Wang, D.: Quantitative surface quality evaluation for 3D-printed concrete with coarse aggregate through 3D scanning. Case Stud. Constr. Mater. **20**, e03077 (2024)
7. Rill-García, R., et al.: Inline monitoring of 3D concrete printing using computer vision. Addit. Manuf. **60**, 103175 (2022)
8. Barjuei, E.S., Courteille, E., Rangeard, D., Marie, F., Perrot, A.: Real-time vision-based control of industrial manipulators for layer-width setting in concrete 3D printing applications. Adv. Ind. Manuf. Eng. **5**, 100094 (2022)
9. Wolfs, R.J.M., Bos, F.P., van Strien, E.C.F., Salet, T.A.M.: A real-time height measurement and feedback system for 3D concrete printing. In: Hordijk, D., Luković, M. (eds.) High Tech Concrete: Where Technology and Engineering Meet. Springer, Cham (2018)
10. Jhun, J., Lee, D.H., Ur Rehman, A., Kang, S., Kim, J.H.: Development of a real-time geometric quality monitoring system for extruded filaments of 3D concrete printing construction. IEEE Access. **12**, 69981–69999 (2024)
11. Alhussain, A., Duarte, J.P., Brown, N.C.: Developing a data-driven filament shape prediction model for 3D concrete printing. Front. Built Environ. **10**, 1363370 (2024)
12. Silva, J.M., et al.: Real-time precision in 3D concrete printing: controlling layer morphology via machine vision and learning algorithms. Inventions. **9**(4), 80 (2024)

13. Georgiou, C., Kontovourkis, O.: Robotic 3D printing on inclined surfaces using adaptive formwork principles for prefabricated curve-like structures. In: Block, P., Boller, G., DeWolf, C., Pauli, J., Kaufmann, W. (eds.) Proceedings of the IASS 2024 Symposium, Redefining the Art of Structural Design. ETH, Zurich (2024)
14. Dimitriou, S., Kontovourkis, O.: Force-flow based 3D printing of structural elements inspired by the traditional architecture of cross-vaults. In: Herrera, P.C., Gomez, P., Estevez, A.T., Torreblanca-Diaz, D.A. (eds.) Proceedings of the SIGraDi 2024 XXVI 11 International Conference of the Ibero-American Society of Digital Graphics, pp. 1529–1540. Barcelona, UIC (2024)

Driving Behavior Optimization Based on Acceleration Metrics via Genetic Algorithm

Spyros Damikoukas, Giorgos Kazakis, Nikos Ath. Kallioras(✉), and Nikos D. Lagaros

National Technical University of Athens, Athens, Greece
nkallioras@yahoo.com

Abstract. In this work, a holistic approach to optimizing driving behavior by exploiting acceleration-based metrics is proposed. By employing a Genetic Algorithm (GA), a search for smooth and safe acceleration profiles, particularly minimal "jerk," which is the time derivative of acceleration, that also satisfy practical constraints such as maximum acceleration limits, traveling time, and specific target distance is executed. Contrasting conventional gradient-based optimizers, GA presents the ability to effectively explore a vast and potentially non-smooth solution area, thereby overcoming the risk of converging to local minima or infeasible solutions. A detailed formulation of the optimization problem, reflecting the constraints and objectives incorporated in a GA formulation is presented. Optimization results demonstrate that GA-based solutions identify smooth acceleration profiles, maximize passenger safety, and satisfy key performance metrics, such as comfort, and constraints.

Keywords: driving behavior · acceleration optimization · genetic algorithm · jerk minimization · vehicle dynamics · travel time constraints

1 Introduction

The optimization of driving behavior has attracted increasing attention in the context of intelligent transportation systems, vehicle safety, and emerging automated driving technologies. Among the many ways to characterize driving behavior, acceleration and deceleration stand out as the most critical metrics influencing passenger comfort, safety, vehicle stability and fuel efficiency. Rapid, high-amplitude accelerations and decelerations tend to degrade vehicle handling, increase mechanical stress, and cause discomfort. Equally important is the rate of change of acceleration (i.e., jerk), which strongly impacts perceived ride quality and can contribute to heightened collision risks when abrupt maneuvers are executed.

Traditional optimization approaches in automotive control often rely on gradient-based or model-predictive methods, which assume relatively smooth objective functions and constraints. While efficient, they may struggle with complex, discrete, or non-smooth constraints inherent in real-world driving (e.g., traffic signal constraints, speed bumps, or abrupt velocity changes). Consequently, there is growing interest in more robust meta-heuristic algorithms—particularly Genetic Algorithms (GAs)—that can handle these complexities.

N. D. Lagaros et al. (Eds.): ADDOPTML 2024, LNCE 646, pp. 280–290, 2025.
https://doi.org/10.1007/978-3-031-92029-5_26

In this work, the use of Genetic Algorithm for driving behavior optimization with a focus on minimizing jerk while observing realistic acceleration bounds, travel time and distance constraints is proposed. The underlying problem formulation carefully accounts for key aspects such as initial conditions, final conditions, and maximum permissible accelerations. A description of the implementation details using MATLAB, drawing direct connections between the theoretical problem formulation and a practical GA-based code is also discussed.

2 Literature Review

2.1 Acceleration as a Key Metric via Jerk Minimization

Acceleration and deceleration metrics have been identified as key variables influencing safety and comfort in vehicle control in relevant up to date research work [1]. It is commonly stated that acceleration pikes are often responsible for wheel slip, large stopping distances and reduced vehicle stability, especially in the cases of adverse road conditions. Additionally, passenger comfort is dependent to both magnitude and frequency of acceleration differentiations. As a result, jerk has been proposed as a metric for acceleration differentiation in relevant literature [2].

Jerk is defined as the derivative of acceleration with respect to time. While acceleration itself influences forces on a vehicle, jerk impacts how rapidly these forces are changed. Swift changes often shock a human driver or passenger, leading to disorientation or discomfort [3]. In semi-autonomous or fully autonomous vehicles, jerk minimization is commonly recognized as critical for maintaining user trust towards the technology. This is particularly relevant to ride-hailing services, where comfort is one of the major quality metrics [4].

In the recent past, work has been presented in classifying driving behavior with respect to sensor recorded acceleration data [5, 6]. For this study, a dataset collected from user cell phones, comprising acceleration metrics such as longitudinal and lateral acceleration was used in combination to a dataset provided in the work of Cojocaru et al. [6]. The data was gathered from a diverse range of drivers over various road conditions, capturing a wide spectrum of driving behaviors. Preprocessing involving removal of outliers and noise, normalization and feature engineering was performed in the data set. This dataset is used in the current work for driving behavior optimization with respect to safety and comfort.

2.2 Genetic Algorithm Applications

Genetic Algorithms (GAs) where initially inspired by the Darwinian evolution processes. In detail, they focus on maintaining a population of candidate solutions, characterized as individuals and refining their quality in an iterative manner through selection, crossover, and mutation processes [7]. As an optimization solver, GA has been successfully applied in several problems ranging from trajectory planning [8] to powertrain optimization [9]. GAs' population-based search naturally explores multiple areas of the solution space, effectively reducing the probability of entrapment in local minima that usually trap single-point algorithms such as gradient descent or Sequential Quadratic Programming (SQP).

In the problem of driver behavior optimization in terms of safety, the solution search space can be described as extremely non-convex and nonlinear. In many real-world driving scenarios, acceleration profiles may need to incorporate abrupt transitions—for instance, in response to traffic lights or yield signs. GAs remain robust even when faced with such discontinuities [10], supporting the decision to be selected for dealing with the problem at hand.

3 Methodology Overview

3.1 Genetic Algorithms for Driving Profile Optimization

The proposed methodology relies on representing an acceleration profile over a specified time frame as a chromosome in GA. Each gene corresponds to the acceleration at a discrete time step. By creating a population of such profiles, the GA can explore various ways to accelerate or decelerate while respecting:

1. Physical limits (maximum acceleration, etc.).
2. Travel time and distance constraints.
3. Safety and comfort measures (minimizing jerk).

3.2 Integration with Vehicle Dynamics

To connect these profiles with the vehicle kinetic data, velocity is computed by integrating acceleration over time and subsequently computing distance by integrating velocity. This approach enables the enforcement of a target distance to be covered within a fixed travel period. The GA-based solution must satisfy all of these constraints. In the different case where the identified solution fails to satisfy the constraints, it is penalized. The optimization process focuses on reducing the sum of squared jerk in the trip designed. In Sect. 4, we show the exact mathematical model that matches our MATLAB GA implementation.

4 Problem Formulation

In this section, the driving behavior optimization problem with an explicit objective function and corresponding constraints, directly aligned with the previously generated GA code is presented in detail. The scope of the proposed formulation is to:

1. Minimize jerk (via the sum of its squared values).
2. Impose and respect acceleration bounds.
3. Respect initial and final acceleration conditions.
4. Satisfy a travel distance constraint within a specified time T.

4.1 Decision Variables

Let us discretize the total travel time T into N intervals, each of length: $\Delta t = T/N$. We denote by a_i the acceleration value at the ith time step, for $i = 1,2,...,N$.

Consequently, our set of decision variables is:

$$a = [\, a\_1, a\_2, \ldots, a\_N \,]$$

4.2 Objective Function: Sum of Squared Jerk

We define jerk j_i at time step i as:

$$j\_i = (a\_(i+1) - a\_i)/\Delta t$$

Because sudden differentiations in acceleration negatively affect safety and can undermine comfort, the goal is to minimize the sum of squared jerk:

$$J(a) = \sum_{i=1}^{N-1} \left[\frac{a\_(i+1) - a\_i}{\Delta t} \right]\hat{}2$$

Thus, the optimization objective function is:

$$\text{Minimize} \quad (a) \quad J(a) = \sum_{i=1}^{N-1} \left[\frac{a\_(i+1) - a\_i}{\Delta t} \right]\hat{}2$$

4.3 Constraints

4.3.1 Acceleration Bounds

Each i must lie within the maximum allowable acceleration $\pm a_{max}$:

$$-a\_max <= a\_i <= a\_max$$

for $i = 1,...,N$. These bounds represent the physical limits of the vehicle's powertrain and safety imposed thresholds.

4.3.2 Initial and Final Acceleration Conditions

We also impose conditions on the starting and ending acceleration values:

$$\alpha_1 = \alpha_{\text{start}}, \alpha_{\text{N}} = \alpha_{\text{end}}$$

In many driving scenarios, these instances are set to zero—i.e., the vehicle begins at resting and ends at resting in terms of acceleration. However, one could, as well, choose non-zero values if the vehicle is already accelerating when the scenario starts (rolling start scenario).

4.3.3 Travel Distance Constraint

Let v_i be the velocity at the i_{th} time step, starting from an initial velocity v_{start}. We compute velocity by discrete integration of α_i:

$$v_i + 1 = v_i + a_i \ \Delta t, \text{ where } I = 1, \ldots, N - 1.$$

with $v_I = v_{start}$.

We then compute the total distance traveled d over the time horizon T by integrating velocity (using, for example, the trapezoidal rule):

$$d = \sum_{i=1}^{N-1} \frac{v_i + v_{i+1}}{2} \Delta t$$

The constraint is that this distance must match a target distance D:

$$d \ = D$$

This ensures that the desired distance will be covered and solutions that tend to minimize jerk by reducing traveling distance will be avoided.

4.3.4 Travel Time and Discretization

The total time T is fixed and divided into N equal intervals. Consequently, each step is of duration $\Delta t = \frac{T}{N}$ The GA uses this discretization to evaluate the objective and constraints at each time step.

4.3.5 Optional Jerk Limits

In some cases, we may also wish to enforce:

$$|j_i| \leq j_{max}, \quad \textit{for all } i.$$

When included, these upper and lower bounds on jerk become additional inequality constraints. Some scenarios may require strict jerk limits for regulatory compliance or extreme passenger comfort. These constraints can be handled within the GA by penalizing individuals (candidate solutions) that violate

$$|j_i| \geq j_{max}$$

4.4 Connection to the Code Implementation

The code snippet discussed previously aligns directly with this formulation:

1. **Objective Function:**
   The objective function used in the proposed formulation of GA calculates the sum of squared

$$\sum_{i=1}^{N-1} \left[\frac{a_{i+1} - a_i}{\Delta t} \right] \wedge 2$$

2. **Nonlinear Constraints:**

- Fix α_1 and α_N to meet initial and final acceleration constraints.
- Integrate acceleration to get velocity, then integrate velocity to confirm the distance traveled equals *D*.
- Optionally, enforce jerk bounds if desired.

The genetic algorithm then iterates over possible acceleration profiles to find one that minimizes this sum of squared jerks while satisfying all constraints.

5 Genetic Algorithm Implementation

5.1 Chromosome Representation

In the GA, each candidate solution, i.e., each chromosome, consists of *N* real-valued genes representing the acceleration α_i at each discrete time step. Formally, a chromosome is:

$$[\alpha_1, \alpha_2, \ldots, \alpha_N]$$

A population of such chromosomes is maintained.

5.2 Fitness Function and Penalization

Due to the fact that most GA libraries are designed to maximize a fitness function, this function in the current work is defined as:

$$\text{Fitness(a)} = -\text{J(a)}$$

so that maximizing Fitness corresponds to minimizing the sum of squared jerk $J(a)$.

Penalization can be used to handle constraints. When a candidate solution violates constraints (e.g., distance is not exactly equal to *D*, or jerk bounds are exceeded), the fitness can be reduced proportionally to the magnitude of violation. Another common approach is to embed constraints in a "nonlinear constraint" function that discards or repairs infeasible solutions.

5.3 Selection, Crossover, and Mutation

1. Selection: Methods like tournament selection or roulette wheel selection can be employed to choose which individuals reproduce.
2. Crossover: Offspring are formed by combining genes from two parent solutions. For real-valued problems, one might use single-point, two-point, or BLX-α crossover.
3. Mutation: A small percentage of genes are randomly perturbed (e.g., by adding a small random Gaussian value) to maintain population diversity and avoid premature convergence.

5.4 Handling Constraints

- Acceleration Bounds: Definition of lower ($-\alpha_{\text{max}}$) and upper ($+\alpha_{\text{max}}$) bounds for each gene.
- Distance Constraint: In MATLAB, a function is formulated that calculates the distance traveled for a given acceleration profile and enforces $\text{d} = \text{D}$ via an equality constraint.
- Initial/Final Acceleration: We either fix α_1 and α_{N} outright or include them in the constraint function to ensure they equal α_{start} and α_{end}.
- Jerk Limits: a penalization is used penalize for solutions that exceed j_{max}.

5.5 Algorithmic Flow

1. **Initialization**: Generate a population of random acceleration profiles within $[-\alpha_{\text{max}}, +\alpha_{\text{max}}]$.
2. **Evaluation**: For each individual, calculate jerk, sum of squared jerk, velocity, distance traveled, and constraint violations.
3. **Selection**: Choose parents based on their fitness (lower jerk sum and fewer constraint violations yield higher fitness).
4. **Crossover**: Form new offspring by mixing parents' genes.
5. **Mutation**: Randomly perturb some offspring accelerations to maintain diversity.
6. **Constraint Repair/ Penalization**: Fix or penalize any profiles that violate bounds or distance constraints.
7. **Iteration**: Replace the old population with the new one and repeat until a termination criterion is met (e.g., max generations or minimal improvement).

6 Simulation Setup

6.1 Parameter Choices

In the proposed formulation, the parameters' values were set as follows:

- $T = 100$ s,
- $N = 1000$ time steps,
- $\Delta t = 0.1$ s,
- $\alpha_{\text{max}} = 2.5$ m/s^2,
- $D = 1000$ m
- $a_{start} = 0$ m/s^2,
- $a_{end} = 0$ m/s,
- $v_{start} = 0$ m/s.

These values reflect a scenario where the vehicle must travel 1000 m in 100 s, starting and ending with zero acceleration.

6.2 GA Settings

With respect to the algorithmic parameters, the selected values were:

- Population size: 100,
- Max generations: 300,
- Crossover fraction: 0.8,
- Mutation rate: 0.01,
- Selection method: Tournament selection,
- Crossover: Blend crossover (BLX-α\alphaα) or two-point crossover.

7 Results and Discussion

7.1 Sample Acceleration Profile

Figure 1a shows a typical acceleration profile returned by the GA. The acceleration ramps up from 0, stays within ± 2.5 m/s^2, and ends near 0 as required by the final acceleration condition. Overall, the profile is smooth, reflecting minimal jerk.

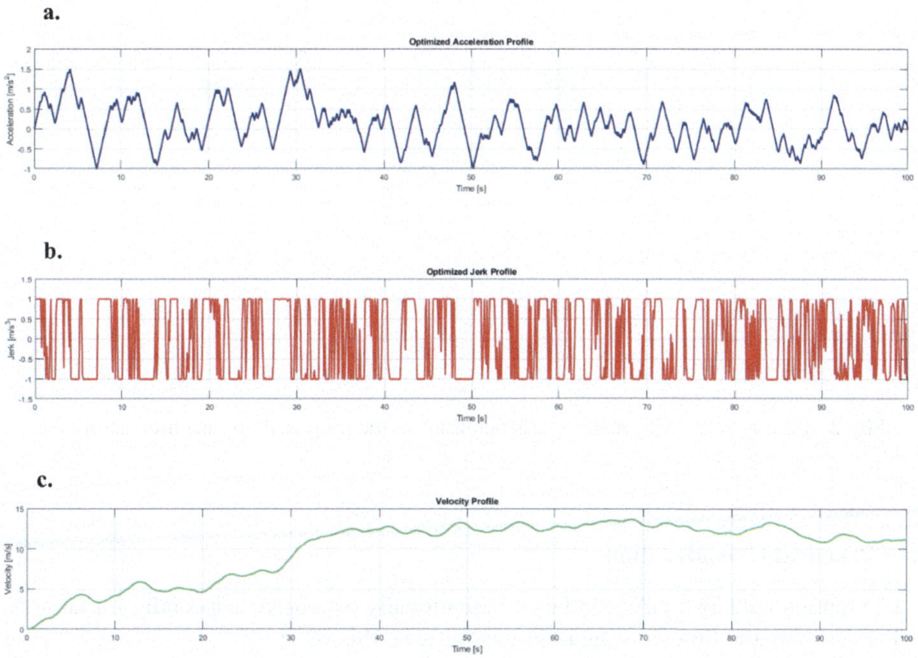

Fig. 1. .

7.2 Jerk and Velocity Profiles

Figure 1b presents the jerk over time. Spikes in jerk are penalized through the objective function, so we observe that the GA solution exhibits small fluctuations in jerk. The corresponding velocity profile (Fig. 1c) shows a gradual increase from 0 m/s to a peak velocity, then a gentle decrease to satisfy the distance constraint by $t = 100$ s.

7.3 Convergence Characteristics

The GA typically converges to near-optimal profiles within 100–150 generations, although it was allowed to continue up to 200 or 300 generations for thoroughness. Convergence can be tracked via the best fitness value in each generation as presented in Fig. 2. It can be witnessed that occasional "jump improvements" where due to crossover or mutation targeting significantly better solutions.

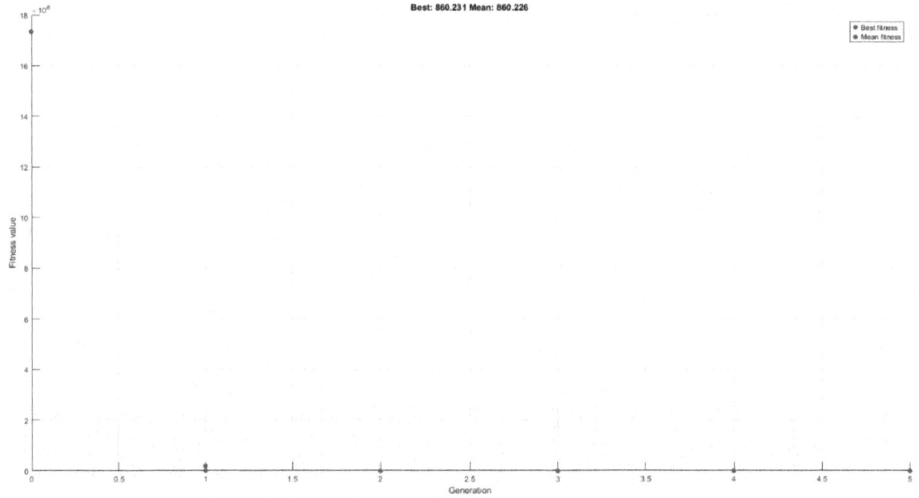

Fig. 2. Fitness value with respect to generations in the proposed optimization solution.

7.4 Practical Considerations

1. Computational Cost: GAs can be computationally expensive, especially for large N. Parallelizing the fitness evaluations can mitigate this issue.
2. Real-time Constraints: Implementing a full GA in real-time driving systems remains challenging; offline or receding-horizon strategies may be more practical.
3. Incorporating Traffic Rules: Real roads have speed limits, signals, and curvature constraints. The same approach can be extended by adding these constraints into the GA's evaluation function.
4. Multi-objective Extensions: Combining jerk minimization with fuel-efficiency or emission constraints is a promising direction for future research.

8 Conclusions and Future Work

In this study, a Genetic Algorithm (GA) framework was proposed and formulated for solving the problem of optimizing driving behavior based solely on acceleration data, subject to constraints relevant to maximum acceleration, travel time and travel distance. By formulating a discrete-time problem with explicit velocity integration, it was guaranteed that solutions produced by the optimization algorithm would satisfy the requirements mimicking a real case scenario. The performed simulations validated that GA-based solutions can propose feasible acceleration profiles via optimization, providing safer and more comfortable trips.

It is worth pointing out that GAs have proven to be competent in handling the non-smooth, nonlinear constraints that often arise in real-world optimization problems. Additionally, by imposing bounds and distance constraints, physically realizable solutions are imposed. Moreover, the same approach can be extended to different driving situations, from urban stop-and-go to highway cruising.

Future research ideas include:

1. Extend the problem from single to multi-objective in order to incorporate fuel efficiency, ecological driving, battery usage (in electric or hybrid vehicles), or emission constraints.
2. Develop an online version of the optimization tool where acceleration decisions are updated at each step based on sensor or traffic information.
3. Incorporation of preceding vehicle dynamics for scenarios like adaptive cruise control which is commonly used by current vehicles.
4. Investigate how these smooth acceleration profiles can be adapted to driver preferences or training.

Overall, it is presented that this GA-based approach for acceleration profile optimization holds promise for next-generation advanced driver-assistance systems, potentially enhancing both passenger comfort and overall traffic safety.

Acknowledgments. The present research was carried out within the research project "OptiMo-Optimising Driver Behaviour for Safe, Green and Energy Efficient Mobility", which is carried out within the framework of the National Recovery and Resilience Plan Greece 2.0, funded by the European Union – NextGenerationEU (Implementation body: HFRI).

References

1. Kiencke, U., Nielsen, L.: Automotive Control Systems. Springer, Berlin, Heidelberg (2005)
2. Rajamani, R.: Vehicle Dynamics and Control. Springer, New York, NY (2012)
3. Mitschke, M., Wallentowitz, H.: Dynamik der Kraftfahrzeuge. Springer, Berlin (2004)
4. Taheri, S., Rakheja, S.: A generalized vehicle ride model for human comfort analysis. Veh. Syst. Dyn. **43**(12), 905–921 (2005)
5. I. Cojocaru, P.S. Popescu, Building a Driving Behaviour Dataset, Proceedings of RoCHI 2022
6. Cojocaru, I., Popescu, P.S., Mihaescu, C.: Driver Behaviour Analysis based on Deep Learning Algorithms, Proceedings of RoCHI (2022). https://doi.org/10.37789/rochi.2022.1.1.18

7. Holland, J.H.: Adaptation in Natural and Artificial Systems. University of Michigan Press, Ann Arbor, MI (1975)
8. Coello, C.A.C., Van Veldhuizen, D.A., Lamont, G.B.: Evolutionary Algorithms for Solving Multi-Objective Problems. Kluwer Academic, Boston, MA (2002)
9. Guzzella, L., Sciarretta, A.: Vehicle Propulsion Systems: Introduction to Modeling and Optimization. Springer, Berlin (2013)
10. Goldberg, D.E.: Genetic Algorithms in Search, Optimization, and Machine Learning. Addison-Wesley, Boston (1989)
11. Deb, K.: Multi-Objective Optimization using Evolutionary Algorithms. Wiley, Hoboken, NJ (2001)
12. Bouscayrol, A., Lemaire-Semail, B., Trigui, R.: Hybrid Vehicles: From Components to System-Level Techniques. IEEE Press, Piscataway, NJ (2013)
13. Li, S., Li, K., Wang, J.: Economy-oriented vehicle adaptive cruise control with coordinating multiple objectives function. Control. Eng. Pract. **18**(9), 1007–1014 (2010)
14. Mitchell, M.: An Introduction to Genetic Algorithms. MIT Press, Cambridge, MA (1998)

Embryo Surface Metrics as a Growth Proxy for Blastocyst Prediction

Pantelis Tsakalis[1](✉), Nikos Kallioras[1], Nikos D. Lagaros[1], Dimitra Christopikou[2], and Stephen Davies[3]

[1] National Technical University of Athens, 15772 Athens, Greece
tsakalis.pantelis@gmail.com
[2] Assisting Nature, Thessaloniki, Greece
[3] Embryogenesis, Athens, Greece

Abstract. Prior research has demonstrated that morphological characteristics and cell division timings can indicate the quality and developmental potential of an embryo. However, conventional evaluation techniques miss minute quantitative changes that can be identified by sophisticated computer vision analysis. Time-lapse data and embryo classifications from 2170 embryos grown for at least 110 h at a single IVF centre were examined in this retrospective cohort study. An extra 326 embryos were set aside for blind testing so that the model's performance could be compared to that of an embryologist.

The metrics were extracted by training an Attention U-Net model to segment the embryos in time-lapse images, enabling a robust analysis of the morphological patterns. The final machine learning mode was trained using extracted data from embryo surface metrics time series to perform a binary classification task, in order to discriminate between embryos that reached blastocyst stage and those who did not. The machine learning model was able to predict if an embryo will reach the blastocyst stage with an AUC of 0.85 [95% CI 0.88–0.93] in a blind test. The model was able to outperform the embryologist's prediction in accuracy by 12% (0.791 vs 0.696) and in F1 Score by 8.5% (0.811 vs 0.742).

Keywords: Computer Vision · Time Series · Feature Extraction · Machine Learning

1 Introduction

Despite advances in the area of assisted human reproduction such as personalized ovarian stimulation regimes, intracytoplasmic sperm injection (ICSI) pre-implantation genetic testing, extended embryo culture, and cryopreservation, approximately one-third of all IVF cycles result in a sustained clinical pregnancy [1]. A major challenge in the implementation of the IVF procedure is posed by the inability to understand the entire set of features that are most predictive for a healthy livebirth with reduction in the number of failed IVF attempts, miscarriages and reduced time to pregnancy. The application of deep neural networks has already been introduced as means of creating more consistent

N. D. Lagaros et al. (Eds.): ADDOPTML 2024, LNCE 646, pp. 291–299, 2025.
https://doi.org/10.1007/978-3-031-92029-5_27

and objective embryo evaluation tools during IVF treatment. This paper investigates a pipeline using an Attention U-Net-based model for segmenting embryo images, extracting relevant features and predicting their developmental outcomes, addressing existing gaps in IVF embryo evaluation.

2 Data Collection

Morphokinetic data from fresh embryos fertilised and cultivated in a time-lapse incubator between January 2011 and December 2020 were retrospectively analysed in this work. The dataset contained embryos from oocyte donation, vitrified oocyte warming, and fresh oocyte retrieval. Patients from a range of demographic backgrounds and medical histories were included in the sample of embryos that were biopsy-performed for pre-implantation genetic testing. In order to accurately forecast the formation of blastocysts, the inclusion criteria made sure that embryos had enough time to acquire morphokinetic data. To prevent model bias, embryos with faulty or missing time-lapse data were not included.

3 Data Processing

3.1 Binary Mask Extraction

The first step is to robustly identify and accurately locate the embryo throughout all the developmental stages. The time lapse approximately 600 images per embryo.

After extensive experimentation with typical segmentation methods (Binary threshold, Watershed algorithms etc) the focus was shifted to deep learning segmentation models. And particularly a variant of the U-Net architecture (Fig. 1) the Attention U-Net. Approximately 800 embryo images were segmented manually using segmentation tools.

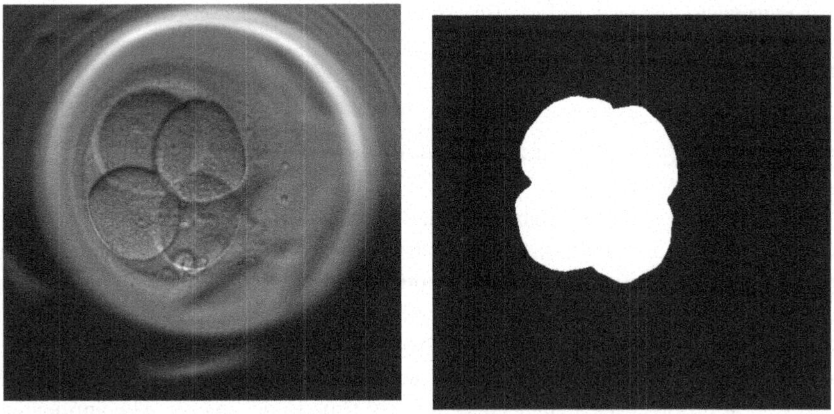

Fig. 1. Raw image(left) and extracted binary mask(right) at 48 h since insemination.

The Attention U-Net model was trained successfully on the segmented images and produced very good results (Fig. 2). The model was able to capture with great accuracy the outline of the embryo in the complete time lapse. The Attention U-Net output is a binary mask (Fig. 2) which is what will be used in the extraction of the surface related metrics. A binary mask is a representation where pixels are classified as either part of the embryo or the background. This segmentation facilitates precise metric extraction for downstream analysis. A time lapse of binary masks is created for each embryo.

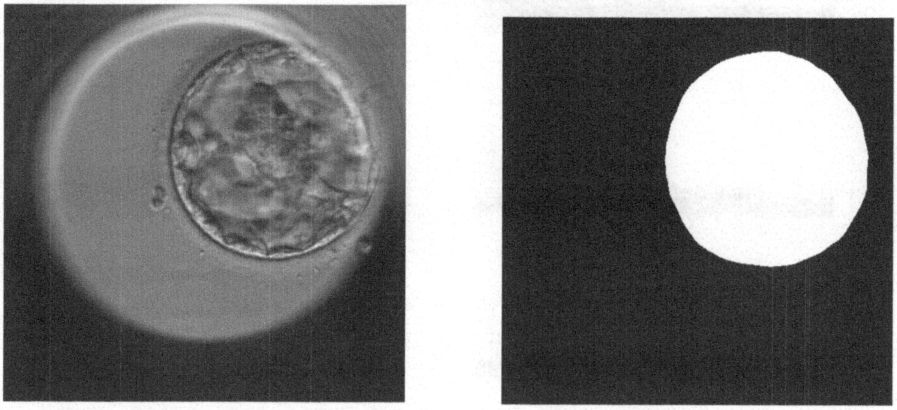

Fig. 2. Raw image(left) and extracted binary mask(right) at 100 h at the beginning of the blastocyst expansion phase.

3.2 Geometric Area Metrics Extraction

By using the extracted binary masks time lapse the surface related metrics are extracted. A comprehensive list of the metrics is presented below. For each frame of the time lapse a value of each metric was extracted and seven time series in total, one for each metric were created.

Area. The sum of all non-zero pixels in the binary mask.

Eccentricity. The ratio of the distance between the foci of an ellipse to its major axis length, representing how elongated the shape is.

Solidity. Ratio of pixels in the region to pixels of th convex hull images.

Major Axis Length. The major axis of an ellipsis fitted on the surface.

Minor Axis Length. The minor axis of an ellipsis fitted on the surface.

Equivalent Diameter. The diameter of a circle with the same Area.

Inertia Tensors. Quantitative measures that describe the distribution of pixel intensities in the binary mask relative to an axis of rotation, used to characterize the shape's resistance to rotational motion and its orientation (Fig. 3).

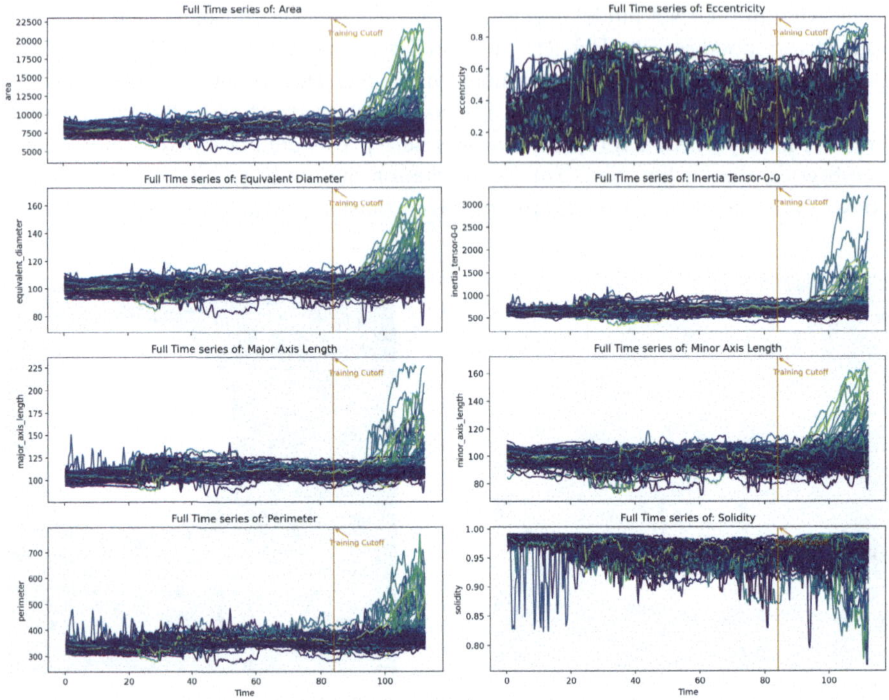

Fig. 3. A sample of the time series spanning from 0 to 120 h.

3.3 Feature Engineering

To perform the classification task, the seven time-series (one for each surface-related metric) need to be condensed into a single feature vector that captures essential information about each embryo's growth. This is achieved by applying a variety of feature extraction methods that summarize statistical, temporal, and structural aspects of the time series data.

To find patterns and connections in the data, the feature engineering approach makes use of an extensive and exploratory methodology. Using a broad range of characteristics, including Fourier and wavelet transforms, this holistic approach guarantees that no potentially important information is missed. Though some features may not seem physiologically reasonable at first, this wide inclusion shows a conscious attempt to maximise the identification of predictive signals. Robust selection techniques are then used to improve the final feature set, guaranteeing that the model keeps only the most significant and pertinent traits for precise predictions.

The features were extracted using the tsfresh library, a tool designed to extract extensive features of timeseries. A brief description of types of features extracted are as follow:

Statistical Features. mean, median, standard deviation, dispersion and central tendency.

Temporal Dynamic Features. Liner regression coefficients, moving average.

Shape and Form Features. Autocorrelation and Partial Autocorrelation, Fourier and Wavelet transforms.

Chage Point Detection Features. First and second derivatives to quantify rate of change, peak and valley detection.

Complexity and Nonlinear Features. Entropy measures.

Each time series is summarized by these extracted features, which also capture subtle changes and high-level trends that are important for forecasting the advancement of the blastocyst stage. We obtain a more manageable dimensionality while maintaining crucial growth properties by reducing them to a single feature vector.

3.4 Feature Selection

The feature selection methodology is as follows: the top 1000 features were chosen from each technique. The final feature set was reduced to 247 features by including features that were kept by at least two of these approaches. Nine of them were clinical features, while the other 238 were signal processing-based features based on time-series data. The use of 247 features strikes a compromise between reducing dimensionality and keeping enough information to make insightful predictions. Ablation investigations were used to validate feature relevance.

We applied a collection of feature selection methods to the training dataset in order to lower the dimensionality of the input. The primary goal was to improve model performance and interpretability by keeping only the most pertinent features.

In particular, the three feature selection techniques listed below were used:

- Iteratively fitting a model and removing the least significant features at each stage until the required number of features remain is known as recursive feature elimination (RFE).
- Logistic Regression: Features with the strongest predictive influence were found using the model's coefficients.
- Gradient Boosting Trees: To rank and choose the most influential features, the gradient boosting model's feature significance scores were used.

The feature selection process was as follows: from each technique, the top 1000 features were selected. By including features that were retained by at least two of these methods, the final feature set was whittled down to 247 features. 238 of them were signal processing-based features derived from time-series data, and nine of them were clinical features. A balance between lowering dimensionality and retaining sufficient data to provide meaningful predictions is achieved by using 247 features. Feature significance was verified through ablation investigations.

3.5 Clinical Features

The clinical features included in the final model were as follows:

- **tPNf:** The duration from fertilization to the fading of pronuclei.
- **t6-tPNf, t6-t5, t5-t4, t3-t2:** Time intervals between key developmental milestones.
- **Multi-nucleation Time:** The observed duration of multi-nucleation in early cell divisions.
- **Blastomere Indication:** Whether cell divisions were even or uneven
- **Morphological Grade (Day 3):** A morphological assessment of embryos on the third day post insemination.
- **Fragmentation Level:** The degree of cytoplasmic fragmentation observed.

4 Methods and Models Used

To establish a baseline performance, a range of models were employed, confirming the data's high predictive capability. The model with the best performance was a gradient boosting tree. Because of its effectiveness, precision, and interpretability, gradient boosting decision trees (GBDT) are a popular machine learning approach. In numerous machine learning tasks, GBDT achieves state-of-the-art results. The implementation of the LightGBM algorithm was specifically adopted since it was more flexible and performed better than other GBDT methods.

4.1 Performance Evaluation

The receiver operating characteristic (ROC) curve, which shows the true positive rate (sensitivity) against the false positive rate (1-specificity) across all potential thresholds, was used to assess the model's performance during training. By choosing particular threshold values, sensitivity and specificity were ascertained. While higher thresholds had the reverse impact, lower thresholds boosted sensitivity while decreasing specificity. The area under the curve (AUC) of the ROC curve, which measures a binary classifier's prediction ability, encapsulates this trade-off. Perfect discrimination is represented by an AUC value of 1.0, whereas random performance is indicated by a value of 0.5. A better balance between sensitivity and specificity is indicated by a greater AUC. Additionally, the AUC can be interpreted as the probability that the classifier assigns a higher score to a randomly chosen positive embryo than to a negative one. For the blind test evaluation, metrics such as accuracy, precision, F1 score, and recall were also employed.

4.2 Embryologist Blind Test

The embryos on the test set were shared with an independent embryologist where the embryologist predicted if an embryo would reach blastocyst stage by examining image data up to 84 h. The prediction was annotated with Yes(Y) and No(N).

5 Results

The results on the blind test are given in the following table. The machine learning model scored AUC 0.852, accuracy score of 0.791 and F1 score of 0.808 The results are being compared directly with the results achieved by the embryologist. The machine learning model outperformed the embryologist in both metrics with a 8.5% increase in the F1 score and 12.01% increase in accuracy.

Metric	ML Model	Embryologist	% Increase
Accuracy	0.791	0.696	12.01
F1 Score	0.811	0.742	8.5
AUC	0.852	N/A	N/A

The two-proportion Z-test was conducted to demonstrate that the predictions made by the machine learning model and those of the embryologist exhibited a statistically significant difference (Fig. 4).

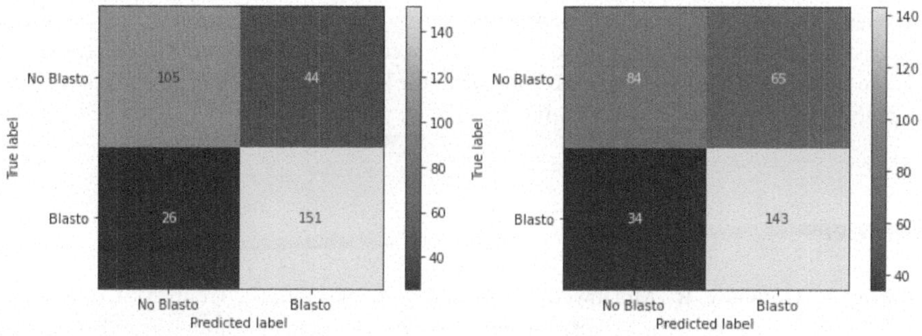

Fig. 4. Confusion Matrices of Machine Learning Model (left) and Embryologist (right) showing the increased performance on blastocyst stage prediction.

6 Discussion

Based on the morphological grades of embryos at an early embryonic stage (keep/freeze or discard), the results demonstrate that machine learning offers substantial prediction skills. Based on the mean AUC throughout the five validation folds, Simundic (2009) claims that the model offers high diagnostic test quality.

Certain developmental events, such as early cleavage and blastulation, have been assessed by the algorithms used in traditional morphokinetic analysis approaches (Meseguer et al. 2012; Conaghan et al. 2013; Basile et al. 2014; Milewski et al. 2016; Fishel et al. 2018; Liu et al. 2018), and scores have been assigned based on these recognised milestones. Numerous studies have looked into the projected accuracy of these morphokinetic algorithms. Barrie et al. (2017) assessed six embryo selection techniques using a single batch of blastocysts and found that none of them had an AUC greater than 0.65. While Storr et al. (2015) used logistic regression across four techniques to show AUC values ranging from 0.585 to 0.748, Liu et al. (2018) discovered a greater range of predictive capacities, with AUCs ranging from 0.509 to 0.762, depending on the approach.

Similar results were found in studies that examined the predictive power of morphokinetics in Day 2 embryos (Ahlstrom et al. 2016; Milewski et al. 2016). According to Ahlstrom et al. (2016), the AUC for embryologist evaluations was 0.74 and the AUC for morphokinetics was 0.67. Based on early cleavage events, embryos are divided into five groups (KID1–KID5) by the KID-score system, which is included into the Embryoscope software (Petersen et al. 2016). According to the ESHRE Special Interest Group of Embryology and Alpha Scientists in Reproductive Medicine (2011), this approach passed morphology-based criteria with prediction values over 0.745.

Knowledge in this field has been considerably enhanced by recent research. In their 2019 study, Blank et al. investigated machine learning techniques that combine clinical factors such parental age, anti-mullerian hormone levels, and reproductive history with conventional morphokinetic methods.

Acknowledgements. This research was supported by the NESTOR project: "Cross-sectoral alliance as the key for innovation-driven business success of Estonian and Greek reproductive healthcare Learning" (No: 101120075) belonging to the HORIZON-CSA Type of Actions, Topic: HORIZON-WIDERA-2022 TALENTS-03-01.

References

1. Barrie, A., Homburg, R., McDowell, G., Brown, J., Kingsland, C., Troup, S.: Examining the efficacy of six published time-lapse imaging embryo selection algorithms to predict implantation to demonstrate the need for the development of specific, in-house morphokinetic selection algorithms. Fertil. Steril. **107**(3), 613–621 (2017)
2. Canat, G., Duval, A., Gidel-Dissler, N., Boussommier-Calleja, A.: A novel deep learning approach to identify embryo morphokinetics in multiple time lapse systems. Sci. Rep. **14**(1), 29016 (2024)
3. Illingworth, P.J., et al.: Deep learning versus manual morphology-based embryo selection in ivf: a randomized, double-blind noninferiority trial. Nat. Med. **30**, 3114–3120 (2024)
4. Jiang, V.S., Bormann, C.L.: Artificial intelligence in the in vitro fertilization laboratory: a review of advancements over the last decade. Fertil. Steril. **120**(1), 17–23 (2023)
5. Kalyani, K., Deshpande, P.S.: A deep learning model for predicting blastocyst formation from cleavage-stage human embryos using time-lapse images. Sci. Rep. **14**(1), 28019 (2024)
6. LeCun, Y., Bengio, Y., Hinton, G.: Deep learning. Nature. **521**, 436–444 (2015)
7. Meseguer, M., Herrero, J., Tejera, A., Hilligsøe, K.M., Ramsing, N.B., Remohi, J.: The use of morphokinetics as a predictor of embryo implantation†. Hum. Reprod. **26**(10), 2658–2671 (2011)
8. Oktay, O., et al.: Attention U-net: Learning Where to Look for the Pancreas. arXiv preprint arXiv:1804.03999 (2018)
9. Ronneberger, O., Fischer, P., Brox, T.: U-net: Convolutional Networks for Biomedical Image Segmentation (2015). arXiv preprint arXiv:1505.04597, LNCS Homepage. http://www.springer.com/lncs. Accessed 21 Nov 2016
10. Sayed, S., Reigstad, M.M., Petersen, B.M., Schwennicke, A., Hausken, J.W., Storeng, R.: Time-lapse imaging derived morphokinetic variables reveal association with implantation and live birth following in vitro fertilization: a retrospective study using data from transferred human embryos. PLoS One. **15**(11), e0242377 (2020)
11. Tartia, A.P., Wu, C.Q., Gale, J., Shmorgun, D., Léveillé, M.C.: Time-lapse kidscored5 for prediction of embryo pregnancy potential in fresh and vitrified-warmed single-embryo transfers. Reprod. Biomed. Online. **45**(1), 46–53 (2022)

12. Yang, L., et al.: Development of a dynamic machine learning algorithm to predict clinical pregnancy and live birth rate with embryo morphokinetics. F S Rep. **3**(2), 116–123 (2022)
13. Zaninovic, N., et al.: A comparison of morphokinetic markers predicting blastocyst formation and implantation potential from two large clinical data sets. J. Assist. Reprod. Genet. **36**(4), 637–646 (2019)
14. Simundic, A.M.: Measures of diagnostic accuracy: basic definitions and practical examples. Biochem. Med. **19**(2), 123–130 (2009)

Comparing Different AI Models and Spectral Indexes on Flood Detection Using Satellite Images from Sentinel-2

Ilias Chamatidis(✉), Denis Istrati, and Nikos D. Lagaros

National Technical University of Athens, Athens 15100, Attica, Greece
ichamatidis@gmail.com, distrati@mail.ntua.gr,
nlagaros@central.ntua.gr

Abstract. Floods are catastrophic events that affect nearly every part of the globe, leading to substantial losses of life and causing extensive economic damage. During flood emergencies, one of the primary challenges faced by response teams is accurately identifying the flooded regions to establish access points and determine safe evacuation routes swiftly. As climate change intensifies, the frequency and severity of extreme weather events like floods are rising, making rapid flood mapping and response even more critical in minimizing damage and saving lives. This study compares several artificial intelligence models and spectral indexes for flood detection in Sentinel 2 multispectral images. The proposed methodology combines transfer learning with different models, such as neural networks, convolutional neural networks, and vision transformers. A variation of this model is the vision transformer (ViT), which can be applied to image classification tasks. Multispectral Instrument (MSI) images from Sentinel-2 contain images from different bands. By combining different zones of those bands different spectral indexes can be calculated. This study uses the Normalized Difference Water Index (NDWI), Modified Normalized Difference Water Index (MNDWI), Water Ratio Index (WRI), Normalized Difference Vegetation Index (NDVI), Enhanced Vegetation Index (EVI), Automated Water Extraction Index (AWEI). By comparing different artificial intelligence models and different spectral indexes the best combination is determined, which is NDVI with VGG CNN model, and can be used for real-time flood detection using Sentinel 2 Multispectral Images.

Keywords: Vision Transformer · CNN · Flood Detection · Sentinel 2

1 Introduction

Floods are a devastating physical phenomenon that is becoming increasingly frequent and intense, making them an important global issue. This trend, which has its roots in climate change and urbanization (uncontrollable constructions and deforestation), underscores the need for reliable and real-time flood detection systems to support disaster response and mitigation efforts. Sentinel-2 satellite imagery contains detailed spatial and temporal resolution in multiple spectral bands and offers a valuable resource for

N. D. Lagaros et al. (Eds.): ADDOPTML 2024, LNCE 646, pp. 300–309, 2025.
https://doi.org/10.1007/978-3-031-92029-5_28

tracking and identifying flood-affected areas. In the last decade, deep learning models, such as convolutional neural networks (CNNs) and Vision Transformers (ViTs), have emerged as powerful tools for processing and analyzing these complex image data. This study explores the comparison between different state-of-the-art models and different spectral indexes to determine which combination is the best. Deep learning has improved the way we approach remote sensing tasks, automating feature extraction and achieving remarkable accuracy compared to traditional methods. Popular CNN models like ResNet, VGG16, and DenseNet have been used for image recognition due to the usage of convolution which can combine spatial information from an image. On the other hand, Vision Transformers (ViTs) are the most recent model in vision models. These models use self-attention mechanisms to capture global relationships in data. While CNNs are good at identifying local features, ViTs might offer advantages in scenarios requiring a larger contextual understanding, such as analyzing Sentinel-2 images for flood detection. This study focuses on comparing Vision Transformers alongside well-established CNN models, ResNet, VGG16, DenseNet, MobileNet, EfficientNet, and SqueezeNet for flood detection using Sentinel-2 imagery. The comparison is done in terms of metrics like accuracy, precision, recall, and AUC. This research provides a baseline comparison using the same dataset and different spectral indexes with each model type for real-time on-site applications to determine which pair of models and spectral index is the best. Furthermore, the study examines how using different spectral indices can improve the performance of these models. These findings aim to provide guidance to the development of advanced machine-learning solutions for real-time environmental monitoring and disaster management, offering new models and performance improvements so floods can be monitored more efficiently.

2 Relevant Work

In [1] performs a comparison of Sentinel-1 and Sentinel-2 images. Sentinel-1 captures images with a synthetic aperture radar (SAR) and Sentinel-2 captures Multispectral Images. These 2 satellites are used to monitor flooding in Europe. The dataset used in this study contains data from 10 years spanning and from 2000 sites. The study compares the images from 2 satellites to determine which one is better. The results show that Sentinel-1 images perform better with accuracy $= 58\%$ compared to Sentinel-2 with accuracy $= 28\%$. This study mentions that Sentinel-1 underperformance was due to cloud coverage. And that Sentinel-1 is suitable for real-time flood detection even with cloud coverage.

Similarly, [2] uses a Random Forest to monitor floods and create maps of them. It uses images from Sentinel-1 and Sentintel-2 to detect changes in the land mass and detect floods that will develop. The RF model takes as an input the S gamma band of VV and VH polarization from SAR images, combined with optical classifications from Sentinel-2, achieving an accuracy $= 80.95\%$ and a kappa index $= 74.1\%$. This study showed that the RFC performed better than traditional Otsu thresholding and NDWI methods, particularly for identifying inundated areas beneath vegetation. Furthermore, [3] experiments with a VGG16 Convolutional Neural Network (CNN) that is pre-trained on ImageNet for automatic flood detection. It combined a novel water index derived from

Green/SWIR and Blue/NIR bands with CNN to enhance flood identification, particularly in cloud-affected regions. The proposed approach achieved an F1 score of 0.96 and a Cohen's Kappa of 0.92, outperforming other water index-based methods. This hybrid method develops a real-time flood mapping method. Also [4] develops an Artificial Neural Networks (ANN) and Support Vector Machines (SVM) to classify and map flooded areas that were caused by a dam that collapsed. The models were applied to pre- and post-flood satellite images from Landsat-8 and Sentinel-2 to distinguish affected areas from other land cover types. The ANN and SVM results for Sentinel-2 indicated affected areas of 2,640,900 m^2 and 2,671,700 m^2, respectively, highlighting the utility of AI in accurate flood damage assessment. Finally [5] develops the OmbriaNet deep neural network architecture, leveraging bitemporal and multimodal input data. It utilized Sentinel-1 SAR and Sentinel-2 multispectral imagery to create high-quality flood maps through supervised segmentation. The Multimodal OmbriaNet achieved an IoU of 70.93% and outperformed baseline methods such as U-Net and traditional thresholding techniques, particularly in cloud-covered and complex scenarios. This novel approach demonstrates the efficiency of fusing data from multiple sensors for flood mapping.

3 Methods

In this section, the methods and models used are described. The first subsection consists of a dataset description and is followed by the AI models used in the experiments.

3.1 Data Preprocess

The dataset used in this study is the SEN12-FLOOD dataset [6], which is a dataset that has Sentinel 2 images. Sentinel 2 are Multispectral images, which contain images from different spectral bands. To create the training dataset first the images are processed to create the images with spectral indexes presented in the next chapter. Then to create a balanced dataset 444 images containing flood events and another 444 images with no-floods events. These 2 sets are the 2 classes that will be used for classification. Furthermore, 3 separate sets are created: training set, validation set, and test set with ratios 60%, 20%, and 20% respectively. The spectral indexes used are:

- AWEI NS: Automated Water Extraction Index—No Shadow
- AWEI SH: Automated Water Extraction Index—Shadow Removal EVI: Enhanced Vegetation Index
- MNDWI: Modified Normalized Difference Water Index
- NDVI: Normalized Difference Vegetation Index
- NDWI: Normalized Difference Water Index
- WRI: Water Ratio Index

3.2 Experimental Setup

The software and hardware used in the experiments are presented in this subsection. The models were developed using PyTorch [7] which is one of the most popular libraries for developing AI models on CUDA architectures and usage of auto-gradients and back-propagation and is one of the most used libraries when developing neural networks. The models were trained using 2 Nvidia GeForce RTX 3090 graphical processing units (GPU) CPU and an Intel(R) Core(TM) i9-10980XE CPU @ 3.00GHz. There are a total of 7 spectral indexes and 7 models, resulting in a total of 49 different model combinations. Each different model was trained for 100 epochs using the training set, and every 2 epochs the validation set was used to test the model's loss and if it overfits or improves. The best model regarding the validation set is kept and is used for testing independently using the test set. The training time for each model was about 20 minutes. After the training finished the best models for each case were used to test the metrics using the test set. The training set consists of 60% of the total 888 images. The images contain 444 flood events and 444 non-flood events labeled respectively as class 0 and class 1. The experiment is set so that the models are tested on the same setup so the results show a clear comparison baseline between all the experiments and which model is the best paired with the best spectral index.

3.3 Models

This subsection describes the AI models used in the experiments, each model was trained with a different spectral index each time using the whole dataset. They were trained using the training set, and the validation set for monitoring purposes (overfitting). The test set was used for an independent test to get the metrics after the training finished. Each model uses the same training set, validation set, and test set. And trained in the same number of epochs so the comparison between them is objective.

3.3.1 Vision Transformer

The Vision Transformer (ViT) described in [8] is a novel approach to image classification that applies the principles of Transformer models. The original task that transformers were developed was natural language processing tasks. But they have successful applications in vision tasks. Instead of processing the entire image as a grid of pixels, the ViT splits the input image into fixed-size patches, which are linearly embedded into vectors. These patch embeddings are then processed through a standard Transformer encoder. Formally, given an image $x \in R^{H \times W \times C}$, it is divided into N patches of size $P \times P$, where H, W, and C are the height, width, and channels, respectively. The patches are flattened and projected to a latent dimension D using a trainable linear projection:

$$z_0 = \left[x^E{}_1; x^E{}_2; \ldots; x^E{}_N \right] + E_{pos} \tag{1}$$

where E is the embedding matrix and E_{pos} represents positional encodings.

3.3.2 ResNet

ResNet, or Residual Networks [9], introduced the concept of skip connections to mitigate the vanishing gradient problem in deep networks. A residual block enables the network to learn the residual mapping instead of the full mapping. Given an input x, a residual block computes:

$$y = F\left(x_i\{W_i\}\right) + x \tag{2}$$

where F $(x, \{W_i\})$ represents the stacked layers' transformation. This skip connection allows gradients to flow directly through the network, enabling the training of networks with hundreds or even thousands of layers.

3.3.3 VGG16

VGG16 [10] is a convolutional neural network that emphasizes simplicity by using a stack of convolutional layers with a fixed kernel size of 3 × 3, followed by max-pooling layers. The network architecture consists of 13 convolutional layers and 3 fully connected layers. Formally, given an input image x, the convolutional operation at each layer is

$$y_{i,j,k} = \sum\nolimits_{m=0}^{M-1} \sum\nolimits_{n=0}^{N-1} x_{i+mj+n} . w_{m,n,k} \tag{3}$$

where w represents the kernel weights and y is the output feature map.

3.3.4 SqueezeNet

SqueezeNet [11] is a lightweight convolutional neural network that achieves AlexNet-level accuracy with 50 times fewer parameters. The core component is the Fire module, which consists of a squeeze convolution layer (using 1 × 1 filters) followed by an expanded layer (using both 1 × 1 and 3 × 3 filters). Given an input x, the Fire module computes:

$$y = \left[f_{1\times1}(x); f_{3\times3}(x)\right] \tag{4}$$

where $f_{1 \times 1}$ and $f_{3 \times 3}$ are the convolution operations.

3.3.5 DenseNet

DenseNet [12] introduces dense connections, where each layer receives input from all preceding layers. This design ensures maximum feature reuse and reduces the number of parameters. Formally, the output of the lth layer is defined as:

$$x_l = H_l\left(\left[x_0, x_1, \ldots, x_{l-1}\right]\right) \tag{5}$$

Where $[x_0, x_1, .., x_{l-1}]$ denotes the concatenation of feature maps from all previous layers, and H_l is a composite function of operations such as Batch Normalization (BN), ReLU, and Convolution.

3.3.6 EfficientNet

EfficientNet [13] is a family of models that scale width, depth, and resolution in a principled manner using a compound scaling method. The compound scaling formula is:

$$\mathrm{d} = a^{\varphi}, w = \beta^{\varphi}, r = \gamma^{\varphi} \tag{6}$$

subject to the constraint $a.\ \beta^2.\ \gamma^2 \approx 2$. Here, d, w, and r represent depth, width, and resolution, respectively, and ϕ is the scaling factor.

3.3.7 MobileNet

MobileNet [14] is a lightweight model designed for mobile and embedded vision applications. It uses depth wise separable convolutions to reduce computation. A depthwise separable convolution consists of a depthwise convolution followed by a pointwise convolution. Formally, given an input x, the depthwise convolution is:

$$y_{i,j,k} = \sum\nolimits_{m=0}^{M-1} \sum\nolimits_{n=0}^{N-1} x_{i+mj+n} \cdot w_{n.n} \tag{7}$$

The pointwise convolution is:

$$z_{i,j,k} = \sum\nolimits_{m=0}^{M-1} y_{i,j,c} \cdot w_{c,k} \tag{8}$$

where w represents the kernel weights.

4 Results

In this section, the results of the experiments are presented, in Table 1 the performance of the ViT is shown. The best spectral index is NDVI with accuracy = 0.82% and the second best NDWI with accuracy = 0.81%. In Table 2 the best performing CNN model is presented, which is VGG with accuracy = 0.83% in the NDVI and accuracy = 0.81% NDWI index. A detailed report of the results of all the CNN models is presented in the Appendix in Table A1.

5 Discussion

The results of this study demonstrate the effectiveness of AI models, Vision Transformers, and various CNN models, in detecting floods through Sentinel 2 images. Sentinel 2 images contain multiple spectral bands, and each band contains different information. This study contacts a comparison between ViT and various CNNs with different spectral indexes. Each spectral index contains different different bands and thus can enhance the performance of the models. The results show that both the ViT and VGG yield high performance, over 80%. Also, the best-performing indexes for all models seem to be NDVI, NDWI, and WRI. From all the combinations the best one is VGG with NDVI, with accuracy = 0.83%. This study proves that AI can be used successfully to automatically detect floods using remote sensing satellite data from Sentinel.

The aim of this study is to provide an objective comparison baseline of different AI vision models and different spectral indices to provide baseline performance for future studies for future researchers to use similar models and indexes for flood detection.

Table 1. Results of ViT with all the spectral indexes.

Spectral index	Accuracy	Precision	Recall	AUC
AWEI NS	0.63	0.63	0.55	0.63
AWEI SH	0.62	0.61	0.55	0.62
EVI	0.76	0.73	0.8	0.77
MNDWI	0.75	0.79	0.67	0.75
NDVI	**0.82**	**0.81**	**0.80**	**0.81**
NDWI	0.81	0.80	0.80	0.81
WRI	0.80	0.79	0.81	0.80

Table 2. Results of the VGG (best CNN) with all the spectral indexes.

Spec. index	Accuracy	Precision	Recall	AUC
AWEI NS	0.72	0.75	0.62	0.72
AWEI SH	0.76	0.78	0.70	0.76
EVI	0.81	0.80	0.81	0.81
MNDWI	0.73	0.72	0.72	0.73
NDVI	**0.83**	**0.80**	**0.84**	**0.83**
NDWI	0.82	0.79	0.85	0.82
WRI	0.83	0.83	0.81	0.83

6 Conclusion

In conclusion, this study focuses on demonstrating the effectiveness of various deep-learning models in detecting floods through remote-sensing images of Sentinel 2. In total 7 models were compared: ViT, ResNet, VGG, SqueezeNet, DenseNet, EfficientNet, and MobileNet along with 7 spectral indexes: AWEI NS, AWEI SH, EVI, MNDWI, NDVI, NDWI, and WRI. The best combination was proven to be VGG with NDVI yielding the highest performance metrics. This study demonstrates that vision AI models can be used successfully to monitor areas through satellites and automatically detect floods. This study shows an objective comparison of the performance in the same experimental setup of different state-of-the-art vision models and different spectral indexes. Which is a clear indication for future researchers when picking a model and a spectral index for similar tasks. Limitations of the study are that the comparison is task-specific, flood detection, and may vary in different applications (e.g. vegetation, urban, etc), also future efforts should be done with similar state-of-the-art models and Sentinel 1 images to have the same baseline comparison for the similar dataset. Finally, the fast performance and inference time of these models make them suitable for real-time flood detection.

Acknowledgements. The authors acknowledge the support of the Hellenic Foundation for Research and Innovation (H.F.R.I.) under the "2nd Call for H.F.R.I. Research Projects to support Faculty Members and Researchers", AMOSS project: "Additively Manufactured Optimized 3D Printed Steel Structures", (Project Number: 02779).

Declarations

- Funding: Project Number: 02779

- Conflict of interest/Competing interests (check journal-specific guidelines for which heading to use): Not applicable
- Ethics approval and consent to participate: Not applicable
- Consent for publication: Not applicable
- Data availability: https://clmrmb.github.io/SEN12-FLOOD/
- Materials availability: Upon request
- Code availability: Upon request
- Author contribution: All authors contributed to this paper

Appendix A

Table A1. Model Performance Metrics for Flood Detection

Model	Spectral index	Accuracy	Precision	Recall	AUC
VGG	ndvi	0.831	0.808	0.847	0.832
VGG	wri	0.831	0.831	0.811	0.830
MobileNetV2	ndwi	0.831	0.831	0.811	0.830
ResNet	ndvi	0.825	0.806	0.835	0.826
EfficientNet	wri	0.825	0.813	0.823	0.825
EfficientNet	evi	0.825	0.806	0.835	0.826
ResNet	ndwi	0.825	0.821	0.811	0.825
VGG	ndwi	0.825	0.793	0.858	0.827
ResNet	awei sh	0.825	0.846	0.776	0.823
EfficientNet	ndwi	0.820	0.804	0.823	0.820
MobileNetV2	wri	0.820	0.797	0.835	0.820
ResNet	wri	0.814	0.795	0.823	0.814
SqueezeNet	wri	0.814	0.788	0.835	0.815
VGG	evi	0.814	0.802	0.811	0.814
SqueezeNet	evi	0.814	0.817	0.788	0.813
DenseNet	wri	0.814	0.795	0.823	0.814
DenseNet	awei ns	0.814	0.809	0.800	0.813
SqueezeNet	ndvi	0.814	0.795	0.823	0.814
EfficientNet	awei sh	0.814	0.802	0.811	0.814
MobileNetV2	ndvi	0.808	0.774	0.847	0.810

(*continued*)

Table A1. (*continued*)

Model	Spectral index	Accuracy	Precision	Recall	AUC
ResNet	awei ns	0.808	0.780	0.835	0.810
MobileNetV2	mndwi	0.803	0.837	0.729	0.800
EfficientNet	ndvi	0.803	0.790	0.800	0.803
EfficientNet	mndwi	0.803	0.797	0.788	0.802
MobileNetV2	awei ns	0.803	0.804	0.776	0.802
ResNet	evi	0.797	0.781	0.800	0.797
DenseNet	evi	0.797	0.757	0.847	0.799
DenseNet	ndwi	0.797	0.781	0.800	0.797
MobileNetV2	awei sh	0.797	0.795	0.776	0.796
DenseNet	ndvi	0.792	0.766	0.811	0.792
MobileNetV2	evi	0.792	0.750	0.847	0.794
SqueezeNet	ndwi	0.775	0.758	0.776	0.775
DenseNet	awei sh	0.775	0.771	0.752	0.774
VGG	awei sh	0.769	0.789	0.705	0.766
SqueezeNet	mndwi	0.769	0.789	0.705	0.766
ResNet	mndwi	0.769	0.755	0.764	0.769
EfficientNet	awei ns	0.758	0.750	0.741	0.757
DenseNet	mndwi	0.758	0.750	0.741	0.757
SqueezeNet	awei sh	0.747	0.743	0.717	0.745
VGG	mndwi	0.735	0.720	0.729	0.735
VGG	awei ns	0.724	0.757	0.623	0.720
SqueezeNet	awei ns	0.713	0.723	0.647	0.710

References

1. Tarpanelli, A., Mondini, A.C., Camici, S.: Effectiveness of sentinel-1 and sentinel-2 for flood detection assessment in Europe. Nat. Haz. Earth Syst. Sci. **22**(8), 2473–2489 (2022)
2. Huang, M., Jin, S.: Rapid flood mapping and evaluation with a supervised classifier and change detection in shouguang using sentinel-1 sar and sentinel-2 opticadata. Rem. Sens. **12**(13), 2073 (2020)
3. Jain, P., Schoen-Phelan, B., Ross, R.: Automatic flood detection in sentinei-2 images using deep convolutional neural networks. In: Proceedings of the 35th Annual ACM Symposium on Applied Computing, pp. 617–623 (2020)
4. Syifa, M., Park, S.J., Achmad, A.R., Lee, C.-W., Eom, J.: Flood mapping using remote sensing imagery and artificial intelligence techniques: a case study in brumadinho, brazil. J. Coast. Res. **90**(SI), 197–204 (2019)
5. Drakonakis, G.I., Tsagkatakis, G., Fotiadou, K., Tsakalides, P.: Ombrianet—supervised flood mapping via convolutional neural networks using multitemporal sentinel-1 and sentinel-2 data fusion. IEEE J. Sel. Top. Appl. Earth Obs. Rem. Sens. **15**, 2341–2356 (2022)
6. Rambour, C., Audebert, N., Koeniguer, E., Le Saux, B., Crucianu, M., Datcu, M.: SEN12-FLOOD : a SAR and Multispectral Dataset for Flood Detection. IEEE Dataport (2020). https://doi.org/10.21227/w6xz-s898
7. Paszke, A., et al.: Pytorch: an imperative style, high-performance deep learning library. In: Advances in Neural Information Processing Systems, vol. 32, (2019)

8. Dosovitskiy, A.: An Image is Worth 16x16 Words: Transformers for Image Recognition at Scale. arXiv preprint arXiv:2010.11929 (2020)
9. He, K., Zhang, X., Ren, S., Sun, J.: Deep residual learning for image recognition. In: Proceedings of the IEEE Conference on Computer Vision and Pattern Recognition, pp. 770–778 (2016)
10. Simonyan, K., Zisserman, A.: Very Deep Convolutional Networks for Large-scale Image Recognition. arXiv preprint arXiv:1409.1556 (2014)
11. Iandola, F.N.: Squeezenet: Alexnet-level Accuracy with 50x Fewer Parameters and 0.5 mb Model Size. arXiv preprint arXiv:1602.07360 (2016)
12. Huang, G., Liu, Z., Van Der Maaten, L., Weinberger, K.Q.: Densely connected convolutional networks. In: Proceedings of the IEEE Conference on Computer Vision and Pattern Recognition, pp. 4700–4708. IEEE (2017)
13. Tan, M., Le, Q.: Efficientnet: Rethinking model scaling for convolutional neural networks. In: International Conference on Machine Learning, pp. 6105–6114. PMLR (2019)
14. Howard, A.G.: Mobilenets: Efficient convolutional neural networks for mobilevision applications. arXiv preprint arXiv:1704.04861 (2017)

Cost-Effective and Constructible Gridshell Design Through the Fusion of Form-Finding and Evolutionary Algorithms

Jonathan Melchiorre(✉)

Department of Structural, Geotechnical and Building Engineering, Politecnico di Torino, DISEG, Corso Duca Degli Abruzzi, 24, 10128 Turin, Italy
jonathan.melchiorre@polito.it

Abstract. The construction industry, despite its considerable environmental footprint, represents a crucial segment of the global economy. Consequently, even minor advancements in this field can generate significant global benefits. Structural optimization emerges as a promising strategy for addressing these challenges, particularly in the design of gridshell structures. Gridshells, known for their ability to span large spaces without intermediary supports, offer remarkable structural efficiency. However, their construction complexity has limited their widespread use. Optimization techniques can address this by improving material efficiency and simplifying construction processes. This study introduces an optimization framework for free-form gridshell structures, combining evolutionary algorithms with the enhanced Multi-body Rope Approach (i-MRA) for form-finding. The primary objectives minimize structural components, reduce material consumption, decrease production waste, and ensure compliance with structural standards. This approach provides an innovative and efficient solution for the conceptual design of lightweight and structurally efficient gridshells.

Keywords: Structural optimization · Multi-objective optimization · Form finding · Lattice structure · Gridshell

1 Introduction

The construction industry is a major contributor to global environmental challenges, exerting a substantial impact on the world ecological footprint. It accounts for approximately 30% of greenhouse gas (GHG) emissions [1], consumes half of the raw materials, and utilizes 40% of global energy resources [2]. Moreover, the sector is responsible for contaminating 40% of drinking water sources, generating half of all landfill waste, and causing 23% of air pollution. Economically, the construction industry is a cornerstone of the global economy, representing 10% of the global Gross Domestic Product (GDP) and valued at roughly 7.5 trillion USD. With projections indicating a doubling of worldwide building floor area by 2050 - driven by population growth and increasing purchasing power in developing and emerging economies - the industry environmental impact is poised to grow significantly [3]. Given these challenges and the economic importance of

N. D. Lagaros et al. (Eds.): ADDOPTML 2024, LNCE 646, pp. 310–321, 2025.
https://doi.org/10.1007/978-3-031-92029-5_29

the sector, integrating structural optimization techniques into standard construction practices [4] could yield substantial benefits [5]. This approach has the potential to mitigate environmental impacts while enhancing economic efficiency, offering both immediate and long-term advantages for the industry and global sustainability efforts [3]. Structural optimization plays a crucial role in developing efficient and cost-effective designs across various engineering disciplines [6]. By applying advanced optimization and machine learning methods [7, 8], engineers can significantly enhance structural performance and durability [9–11] while simultaneously reducing material consumption and construction costs [12, 13]. This approach is aimed to improve the economic feasibility of projects while also contributing to sustainability efforts by minimizing resource use [14].

Gridshell structures are widely recognized for their architectural adaptability and aesthetic appeal. This structural typology presents specific challenges and opportunities for structural optimization [15]. Aesthetics, functionality, construction feasibility, and resource efficiency all have to be balanced when optimizing complex systems. Technicians can use advanced optimization techniques to construct gridshell structures to consider architectural expression, while also demonstrating sustainable and efficient design concepts.

This paper presents an integrated approach for optimizing gridshell structures. The idea is to combine advanced form-finding techniques with metaheuristic multi-objective optimization methods [16]. The scope of this work is to generate a range of optimal solutions that guide designers during the early stages of project development. The methodology integrates the improved Multibody Rope Approach (i-MRA) for gridshell form-finding [17, 18] with a multiobjective genetic algorithm [19]. The approach employs the i-MRA method to define structurally efficient gridshell shapes considering the specific loading conditions, while a genetic algorithm minimizes material use and construction complexity by reducing the variety of structural element typologies required on-site. A custom MATLAB code [20] performs the entire optimization process and analyses. To enhance accessibility for architects and structural designers, an iMRA component has been developed for Grasshopper, a parametric design tool within Rhinoceros3D [21]. The study focuses on simplifying gridshell construction management by minimizing the variety of structural components and optimizing material use to reduce consumption and waste. Furthermore, the Cutting Stock Problem (CSP) is integrated in the method to optimize cutting pattern and further decreasing material waste during the production of the structural components [22]. These efforts are aimed to simplify the construction process, reduce costs, and promote sustainable design techniques in gridshell projects. Two case examples are offered to help validate the proposed approach. The first case study is a square gridshell subjected to self-weight and vertical nodal forces. The second case study focuses on a more complex gridshell with a free form base shape. The analyses compare results under various loading conditions and investigate element typologies, material utilization, and waste minimization. These case studies demonstrate the versatility and effectiveness of the proposed optimization methodology across different gridshell configurations and loading scenarios.

2 Form Finding with Improved Multi-body Rope Approach

The Multibody Rope Approach (MRA) [23, 24] represents a state-of-the-art technique for designing gridshell structures, particularly those involving freeform base geometries and standardized construction elements. This innovative method simulates a dynamic model of falling bodies in space and time, leveraging D'Alembert's principle to iteratively compute the equilibrium configuration of each node in the structure. In the MRA framework, structural components are conceptualized as ropes connecting masses at the nodes [25]. These rope elements exert forces to maintain a predetermined distance between connected masses, ultimately generating a geometry that is both structurally efficient and composed of a minimal variety of element typologies. The resulting configuration resembles an inverted hanging net, yielding a form that is visually striking and structurally optimized [26].

The reduction and standardization of distinct structural element typologies is critical in developing a construction-efficient gridshell design. The assembly complexity of a gridshell grows with the number of distinct structural parts required. The grouping of parts of equal length into various groups is an essential part of the design process. This categorization both simplifies the construction process and ensures a more manageable and cost-effective build. The improved Multibody Rope Approach (i-MRA) is a method designed to consider both the structural and practical aspects of gridshell design.

The idea behind this approach is to optimize the shape for structural efficiency while also considering the construction aspects. The i-MRA improves upon the previous MRA approach by introducing two different methodologies. The first is the Multiple Orders MRA (MO-MRA) feature that is used to group structural elements of identical lengths together. In this way, it is possible to substantially reduce the variety of structural components required for construction. This strategy simplifies the management of the inventory and the structural assembly on-site. Complementing this, the Repulsive Nodes MRA (RN-MRA) incorporates a repulsive force field into the dynamic model. This innovative addition allows for subtle geometric adjustments that further minimize the number of distinct components needed [27], enhancing construction efficiency without compromising structural integrity [28–30]. Tailored specifically to optimize gridshell geometries, the i-MRA method prioritizes construction simplification alongside structural efficiency. Initially developed as a custom MATLAB [20] code, the method has been made more accessible through its conversion into a Grasshopper component [21],

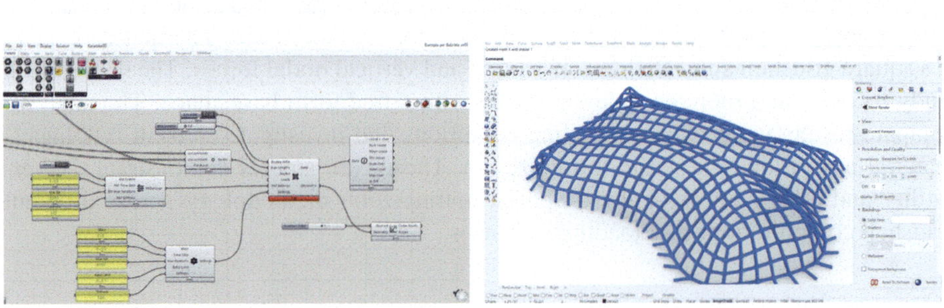

Fig. 1. i-MRA component developed for the Grasshopper3D environment.

as illustrated in Fig. 1. This integration ensures that architects and structural designers can easily incorporate the i-MRA method into their workflow, bridging the gap between advanced optimization techniques and practical design tools.

3 The Optimization Methodology

This paper presents an innovative approach that integrates the improved Multibody Rope Approach (i-MRA) with a multi-objective genetic algorithm (MOGA), utilizing MATLAB [20] "gamultiobj" function. The MOGA operates by generating a population of potential solutions and refining them through successive generations [31], providing a powerful tool for complex optimization problems. The optimization process focuses on two primary objectives: minimizing the number of distinct structural element types required for gridshell construction and reducing total material consumption, accounting for waste generated from cutting industrially produced 12-m steel billets. Considering these two objectives ensures that the resulting gridshells are not only structurally optimized but also efficient in terms of material usage and construction management. In Fig. 2 the optimization workflow is represented. The starting point is a set of design variables that are used to define shape of the gridshell through the formfinding procedure. Starting from a base grid, the i-MRA method is applied to determine the optimal shape under specific loading conditions. After the shape is defined using the form finding process, the Cutting Stock Problem (CSP) [32, 33] is utilised to calculate the number of 12-m billets needed to produce all the structural elements. This critical step has the goal of optimizing cutting patterns and drastically reducing material waste during construction.

Following the optimization process, a finite element model (FEM) of the structure is developed. This model utilizes Circular Hollow Section (CHS) profiles made of S355 steel. To evaluate the structural performance, finite element analyses (FEA) are performed using a custom MATLAB code [34]. The analyses allow the computation nodal displacements and stress distributions within the structural elements. Structural verifications are performed on the basis of the analyses results. The verifications, reported in Eq. 1, consider three structural aspects: limit displacement, yield strength, and Eulerian buckling.

$$\delta_{max} < \delta_{lim} \qquad |\sigma_{max}| < f_y \qquad N_{maxCompression} < N_{Bucking} \tag{1}$$

The limit displacement criterion ensures that the structure deformation under load remains within acceptable bounds. The yield strength check verifies that the stresses in the structural elements do not exceed the material yielding stress. Finally, the Eulerian buckling assessment ensures that the slender elements of the gridshell are not at risk of buckling under compressive loads [35].

In this context, δ_{max} represents the maximum nodal displacement experienced by the structure, while δ_{lim} is the allowed limit, set at $\frac{1}{250}$ of the structure largest dimension, assuming no external constraints are applied. The term $|\sigma_{max}|$ refers to the absolute maximum stress within the structure, which must remain below f_y, the yield stress of the S355 steel used. Finally, $N_{max}{}^{Compression}$ denotes the maximum compressive axial force in the structure, and $N_{Buckling}$ corresponds to the critical axial force for Eulerian buckling. The optimization process incorporates an iterative approach to determine the

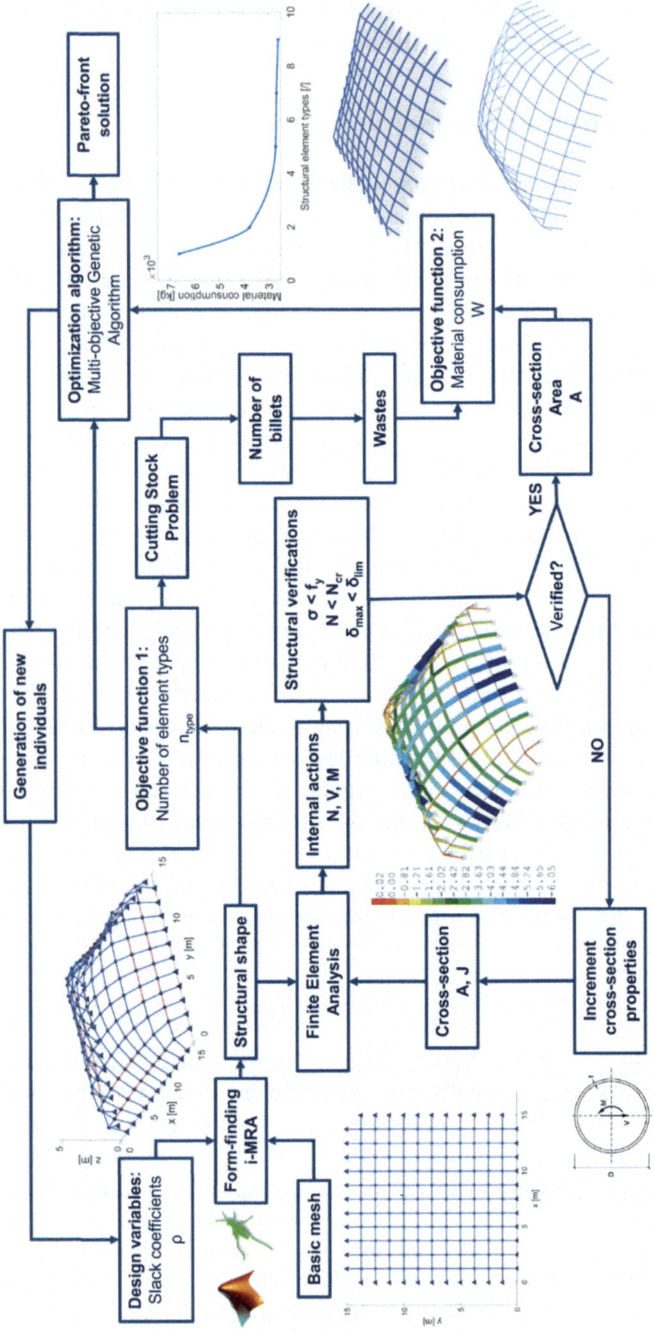

Fig. 2. Workflow of the optimization process.

optimal cross-section dimensions of the CHS profiles. These dimensions are incremented systematically until a configuration is achieved that satisfies all verification criteria. This process ensures the structural integrity of the gridshell while also determining the minimum required cross-sectional area of the structural elements. To calculate the total material necessary for constructing the structure, the optimized crosssectional area is multiplied by the number of billets derived from the Cutting Stock Problem (CSP) solution.

The fitness function evaluated by the multi-objective genetic algorithm (MOGA) is defined by two different quantities: the number of different structural element types (n_type), and the total amount of material used (W). The optimization is an iterative procedure that continues until a set of optimal solutions is generated, forming a Pareto front. The Pareto front can be used as a decision-making tool by the designers, allowing them to select the most appropriate structural configuration for the specific needs of the design. This will help the decision-making process that should be based on a comprehensive set of considerations including architectural vision, structural performance, economic constraints, and project management factors.

4 Discussion and Results

This section presents two case studies that illustrate the application of the proposed optimization process to gridshell form-finding. The first case study focuses on a gridshell structure with a square plan, featuring a side length of 15 m. The initial configuration of this structure is depicted in Fig. 3(a). The optimization outcomes for this square gridshell are comprehensively displayed in Fig. 4. These results showcase the structural responses under various loading conditions, specifically examining the effects of self-weight combined with different magnitudes of nodal overloads. The analysis considers three distinct nodal overloads equal to 1 kN, 5 kN, and 10 kN.

The study compares outcomes with and without the consideration of material waste. Figures 4(a) and (c) illustrate the results where material waste was accounted for in the optimization process. In contrast, Figs. 4(b) and (d) display the outcomes when waste was not factored into the calculations. This comparison allows for a clear assessment of how waste consideration affects the overall design and material efficiency of the

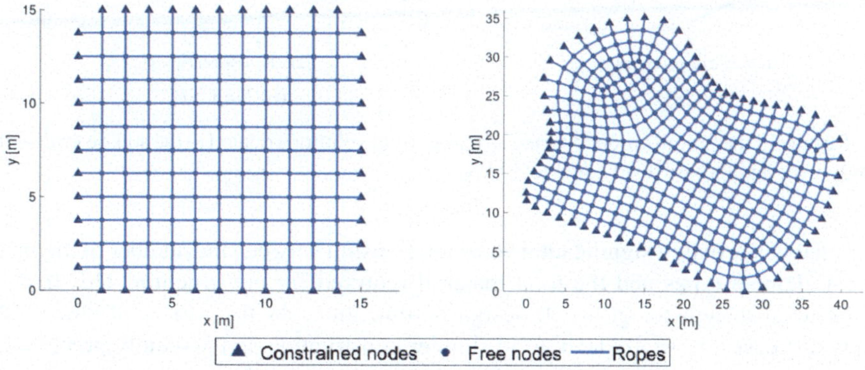

Fig. 3. Base meshes for the two case studies.

gridshell structure. Then, the optimizations were performed using either one or two design variables. Figures 4(a) and (b) show the results obtained using a single design variable, while Figs. 4(c) and (d) depict the outcomes when two design variables were employed. These design variables correspond to the slack coefficients assigned to the ropes during the form-finding process utilizing the improved Multibody Rope Approach (i-MRA) method.

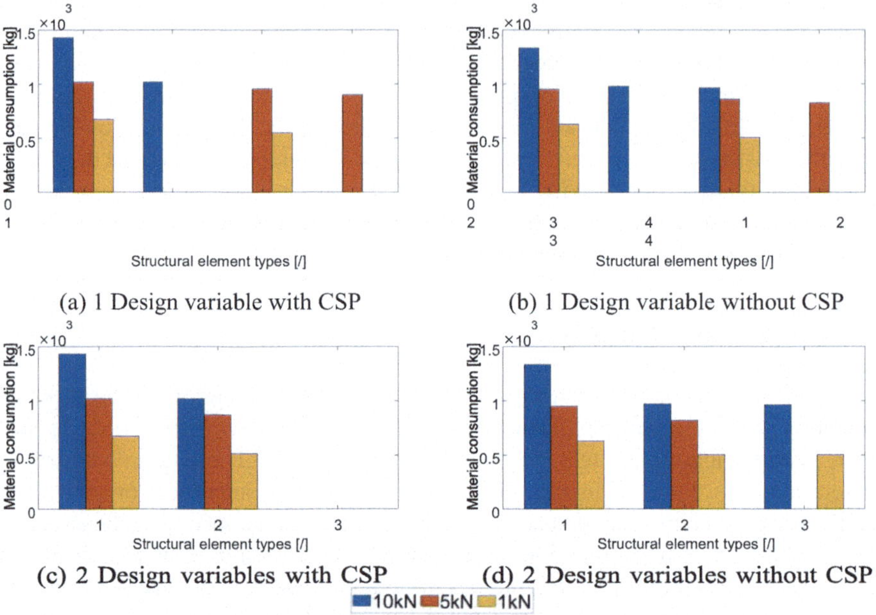

(a) 1 Design variable with CSP (b) 1 Design variable without CSP

(c) 2 Design variables with CSP (d) 2 Design variables without CSP

Fig. 4. Results of the optimization of the square gridshell case study.

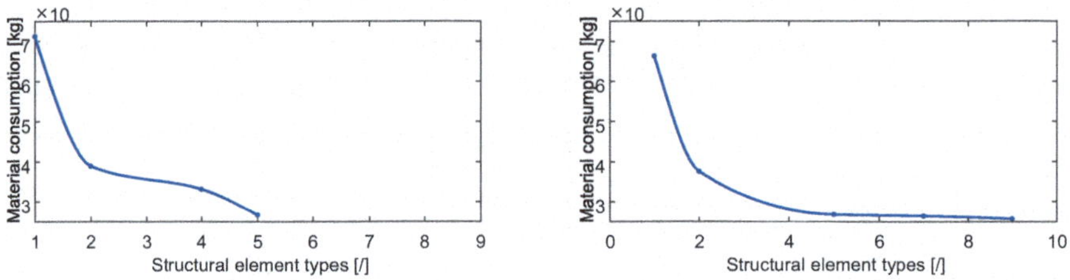

(a) CSP (b) No CSP

Fig. 5. Results of the optimization reported as Pareto front solutions related to the square gridshell case study with a nodal overload equal to 50 kN.

The analysis reveals a significant inverse relationship between the number of distinct structural element types and the total material required for construction. This trade-off is a crucial insight for gridshell design optimization. As the variety of structural elements decreases, typically leading to simpler construction and assembly processes, the overall material consumption tends to increase. Conversely, designs that minimize material usage often require a greater diversity of structural element types, potentially complicating the construction process. This relationship is visually represented in the Pareto front illustrated in Fig. 5.

(a) $n_{type} = 1$ $m = 7.13 ton$ (b) $n_{type} = 2$ $m = 2.55 ton$

Fig. 6. Optimized structural designs for the square gridshell case study, considering the effects of self-weight along with a 50 kN nodal overload.

Furthermore, the analysis reveals a key trade-off in gridshell design between structural profile and element characteristics. Lower-profile structures require thicker elements to withstand higher compression forces but use fewer element types, simplifying construction. Conversely, taller configurations allow for slimmer, more lightweight elements, optimizing material efficiency at the cost of increased element variety. Figure 6 illustrates two optimal configurations, each representing a different balance between material efficiency and construction simplicity.

The study further demonstrates the versatility of the proposed optimization process through a second case study. This example applies the methodology to a more complex gridshell structure, using the initial layout depicted in Fig. 3(b) as its starting point. The optimization outcomes for this case study are illustrated in Fig. 7 as bar graphs. These results reflect various nodal overload conditions and account for production deviations computed using the Cutting Stock Problem (CSP).

Two examples of optimal configurations for the second case study are presented in Fig. 8. As with the previous case, the lower configuration features thicker structural elements and higher material usage, while the elevated configuration is characterized by slimmer structural elements and reduced material consumption.

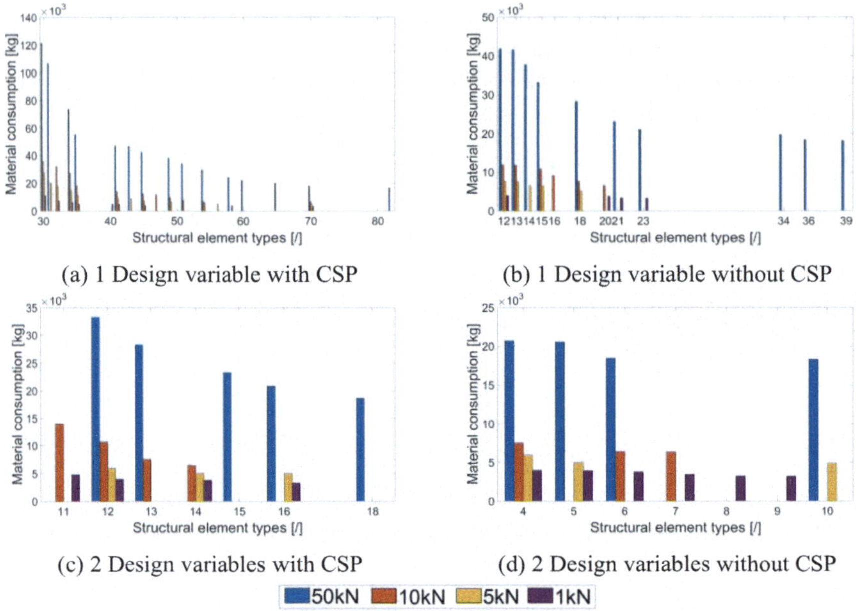

(a) 1 Design variable with CSP (b) 1 Design variable without CSP

(c) 2 Design variables with CSP (d) 2 Design variables without CSP

Fig. 7. Results of the optimization of the free-form base gridshell case study.

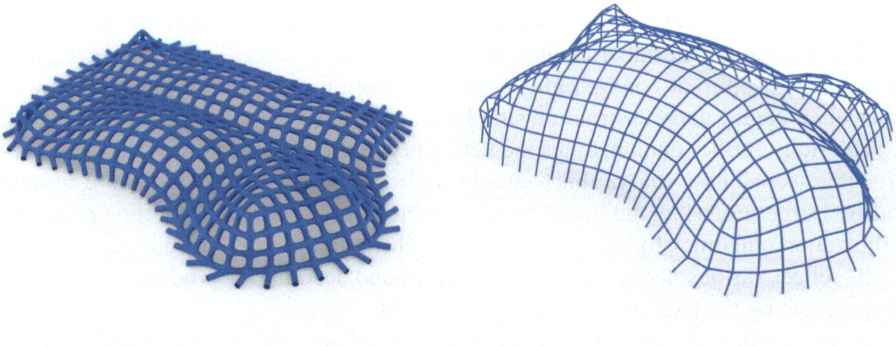

(a) $n_{type} = 30$ $m = 121.5ton$ (b) $n_{type} = 60$ $m = 22.2ton$

Fig. 8. Optimized structural shapes for the free-form base gridshell considering self-weight and a nodal overload of 50 kN.

5 Conclusions

This study presents an innovative optimization procedure for gridshell structures that integrates advanced form-finding techniques with metaheuristic optimization techniques. At the core of this approach is the improved Multibody Rope Approach (i-MRA), which serves as the primary form-finding method. The optimization process focuses on two main objectives: the minimization of the variety of structural element types and the reduction of the overall material consumption. An important novelty of this methodology is the incorporation of the Cutting Stock Problem (CSP) to determine optimal cutting patterns for structural components. This allows for the reduction of material waste and the improvement of the resource efficiency. In this study, the multi-objective genetic algorithm (MOGA) generates a Pareto front, that represents the trade-offs between various design criteria. In this way, the designers are provided with a spectrum of optimal structural configurations that allows them to make informed decisions based on project-specific requirements and constraints. Finally, the practical application of this optimization process is demonstrated through case studies on gridshell structures.

Acknowledgments. This research was supported by project MSCA-RISE-2020 Marie Skłodowska Curie Research and Innovation Staff Exchange (RISE) - http://addoptml.ntua.gr/ - "ADDitively Manufactured OPTimized Structures by means of Machine Learning" (No: 101007595).

References

1. Min, J., et al.: The effect of carbon dioxide emissions on the building energy efficiency. Fuel **326**, 124842 (2022)
2. Klufallah, M.M., Nuruddin, M.F., Khamidi, M.F., Jamaludin, N.: Assessment of carbon emission reduction for buildings projects in Malaysia-a comparative analysis. In: E3S Web of Conferences, EDP Sciences, vol. 3, p. 01016 (2014)
3. Lagaros, N.D.: The environmental and economic impact of structural optimization. Struct. Multidiscip. Optim. **58**(4), 1751–1768 (2018)
4. Cucuzza, R., Aloisio, A., Rad, M.M., Domaneschi, M.: Constructability-based design approach for steel structures: from truss beams to real-world inspired industrial buildings. Autom. Constr. **166**, 105630 (2024)
5. Sgobba, S., Borsa, M., Molfetta, M., Marano, G.C.: Mechanical performance and medium-term degradation of rubberised concrete. Constr. Build. Mater. **98**, 820–831 (2015)
6. Melchiorre, J., Manuello, A., Marmo, F., Adriaenssens, S., Marano, G.: Differential formulation and numerical solution for elastic arches with variable curvature and tapered cross-sections. Eur. J. Mech.-A/Solids **97**, 104757 (2023)
7. Rosso, M.M., Cucuzza, R., Di Trapani, F., Marano, G.C.: Nonpenalty machine learning constraint handling using PSO-SVM for structural optimization. In: Advances in Civil Engineering 2021, pp. 1–17 (2021)
8. Rosso, M.M., Melchiorre, J., Cucuzza, R., Manuello, A., Marano, G.C.: Estimation of distribution algorithm for constrained optimization in structural design. In: WCCMAPCOM 2022, vol. 900 (2022)
9. Marano, G.C., Quaranta, G.: Robust optimum criteria for tuned mass dampers in fuzzy environments. Appl. Soft Comput. **9**(4), 1232–1243 (2009)
10. Olivo, J., Cucuzza, R., Bertagnoli, G., Domaneschi, M.: Optimal design of steel exoskeleton for the retrofitting of rc buildings via genetic algorithm. Comput. Struct. **299**, 107396 (2024)
11. Melchiorre, J., D'Amato, L., Agostini, F., Rizzo, A.M.: Acoustic emission onset time detection for structural monitoring with U-Net neural network architecture. Dev. Built Environ. **18**, 100449 (2024)
12. Marano, G.C., Rosso, M.M., Melchiorre, J.: Optimization as a tool for seismic protection of structures. In: World Conference on Seismic Isolation, pp. 100–113. Springer (2022)

13. Sardone, L., Fiore, A., Manuello, A., Greco, R.: Performative structural design optimization: generative algorithm for a preliminary study of a voided beam. Appl. Sci. **12**(17), 8663 (2022)
14. Sardone, L., Greco, R., Fiore, A., Moccia, C., De Tommasi, D., Lagaros, N.D.: A preliminary study on a variable section beam through algorithm-aided design: a way to connect architectural shape and structural optimization. Procedia Manuf. **44**, 497–504 (2020)
15. Adriaenssens, S., Block, P., Veenendaal, D., Williams, C.: Shell Structures for Architecture: Form Finding and Optimization. Routledge (2014)
16. Marano, G.C., Quaranta, G., Greco, R.: Multi-objective optimization by genetic algorithm of structural systems subject to random vibrations. Struct. Multidiscip. Optim. **39**, 385–399 (2009)
17. Manuello, A., Melchiorre, J., Marano, G.C.: Improved multi-body rope approach for free-form grid shells. In: Italian Workshop on Shell and Spatial Structures (IWSS 2023) (2023)
18. Bertetto, A.M., Melchiorre, J., Marano, G.C.: Improved multi-body rope approach for free-form gridshell structures using equal-length element strategy. Autom. Constr. **161**, 105340 (2024)
19. Melchiorre, J., Soutiropoulos, S., Manuello Bertetto, A., Marano, G.C., Marmo, F.: Gridshell multi-step structural optimization with improved multi-body rope approach and multi-objective genetic algorithm. In: Italian Workshop on Shell and Spatial Structures, pp. 62–72. Springer (2023)
20. MATLAB, version (R2022b), The MathWorks Inc., Natick, Massachusetts (2022). https://www.mathworks.com/products/matlab.html
21. McNeel, R., et al.: Rhinoceros 3D, version 7.0 (https://www.rhino3d.com/). Robert McNeel & Associates, Seattle, WA (2023)
22. Cucuzza, R., Rad, M.M., Domaneschi, M., Marano, G.C.: Sustainable and costeffective optimal design of steel structures by minimizing cutting trim losses. Autom. Constr. **167**, 105724 (2024)
23. Manuello, A.: Multi-body rope approach for grid shells: form-finding and imperfection sensitivity. Eng. Struct. **221**, 111029 (2020)
24. Manuello, A., Melchiorre, J., Sardone, L., Marano, G.C.: Multi-body rope approach for the form-finding of shape optimized grid shell structures. In: Proceedings of the 15th World Congress on Computational Mechanics (2022)
25. Cavaliere, I., Fallacara, G., Manuello Bertetto, A., Melchiorre, J., Marano, G.C.: Multy body rope approach and funicular prototype for a new constructive system for catenary arches. In: Italian Workshop on Shell and Spatial Structures, pp. 259–268. Springer (2023)
26. Fallacara, G., Cavaliere, I., Melchiorre, J., Marano, G.C., Manuello, A.: Reinterpretation of catenary vaulted spaces: construction of a prototype and structural evaluation through multibody rope approach. In: Structures, vol. 66, p. 106746. Elsevier (2024)
27. Melchiorre, J., Invernizzi, S., Manuello Bertetto, A.: Investigating the influence of the improved multibody rope approach on the structural behavior of dakar mosque gridshell structure. Buildings **14**(3),598 (2024)
28. Bazzucchi, F., Manuello, A., Carpinteri, A.: Instability load evaluation of shallow imperfection-sensitive structures by form and interaction parameters. Eur. J. Mech.-A/Solids **66**, 201–211 (2017)
29. Melchiorre, J., Bertetto, A.M., Invernizzi, S., Marano, G.C., Bazzucchi, F.: Dakar mosque gridshell: exploring the benefits of the improved multibody rope approach through postbuckling analysis. In: Structures, vol. 61, p. 106059. Elsevier (2024)
30. Melchiorre, J., Bazzucchi, F., Manuello Bertetto, A., Marano, G.C.: Postbuckling echoes of iMRA introduced variation in gridshells mechanical behaviour. In: Italian Workshop on Shell and Spatial Structures, pp. 379–389. Springer (2023)
31. Deb, K.: Multi-objective optimisation using evolutionary algorithms: an introduction. Springer (2011)

32. Ford, L.R., Jr., Fulkerson, D.R.: A suggested computation for maximal multicommodity network flows. Manage. Sci. **5**(1), 97–101 (1958)
33. Gilmory, P., Gomory, R.: A linear programming approach to the cutting stock problem-part ii. Oper. Res. **11**(6), 863–888 (1963)
34. Paradiso, M., Sessa, S., Vaiana, N., Marmo, F., Rosati, L.: Shear properties of isotropic and homogeneous beam-like solids having arbitrary cross sections. Int. J. Solids Struct. **216**, 231–249 (2021)
35. Bazzucchi, F., Manuello, A., Carpinteri, A.: Interaction between snap-through and eulerian instability in shallow structures. Int. J. Non-Linear Mech. **88**, 11–20 (2017)

Machine-Learning-Based Constraint Handling for Particle Swarm Optimization Within Structural Optimization

Marco Martino Rosso[1,2](✉) and Giuseppe Carlo Marano[1,2]

[1] Department of Structural, Geotechnical and Building Engineering, Politecnico di Torino, Turin, Italy
{marco.rosso,giuseppe.marano}@polito.it
[2] Corso Duca degli Abruzzi, 24, 10129 Turin, Italy

Abstract. Despite their historical roots and formally rigorous mathematical framework, mathematical programming is often not sufficient to effectively deal with real-world complex optimization problems. For instance, focusing on the structural optimization field, heuristic and meta-heuristic computational intelligence methods represented a promising solution for addressing many real-world challenges since their early conception. Several algorithms have been formulated inspired by mimicking natural phenomena, such as Genetic Algorithms or Simulated Annealing, among others. The lack of a strong mathematical basis motivated the idea of always simultaneously implementing different soft-computing techniques both for comparisons and mutual validation, and also because due to the No-Free Lunch, which demonstrated that the perfect algorithm able to solve any optimization problem does not exist. The study of the animal's world behavior, i.e. bird flocking or fish schooling, is the main idea behind one of the nowadays still most widely adopted meta-heuristic algorithms, i.e. the particle swarm optimization (PSO) technique. In the beginning, PSO was formulated to solve unconstrained optimization problems only, and later numerous scholars attempted to introduce some new constraint handling mechanisms in order to exploit the PSO's powerful optimization capabilities even with most likely real-world constrained problems. In this study, the authors focused on the constraint handling problem in PSO for solving structural optimization tasks, by leveraging the nowadays new compelling Machine Learning tools offered by the digital revolution currently in progress. Specifically, the authors formulated a new constraint-handling method based on the support vector machine (SVM) classifier to progressively update the feasible search region while the swarm explores the search space. This novel technique has been tested on a structural optimization conceptual design problem of a Warren truss steel bridge.

Keywords: Structural optimization · Machine Learning · Meta-heuristic Algorithm · Support Vector Machine

N. D. Lagaros et al. (Eds.): ADDOPTML 2024, LNCE 646, pp. 322–330, 2025.
https://doi.org/10.1007/978-3-031-92029-5_30

1 Introduction

Mathematically speaking, optimization problems (OP) are generally referred to as the procedure of finding the absolute minimum (if it exists) of a function called objective function (OF) $f(\boldsymbol{x})$ by varying some governing variables denoted as design parameters $\boldsymbol{x}$ [1]. The same definition encompasses also maximization problems, if considering the opposite of the OF. The optimization problems can be classified in several ways. If there is one OF only, the problem is denoted as single-objective OP (SOOP), whereas if there are different OFs collected in a vectorial notation, the problem is a multi-objective problem (MOOP) [1, 2]. Typically, MOOPs are characterized by opposing OFs, and the optimum is therefore characterized by a series of trade-off solutions among the various considered objectives denoted as Pareto Front. Other possible classification criteria of OPs depend on the presence or not of constraints functions which may limit or not the admissible values assumed by the design variables. This results in characterizing those tuples associated with feasible or infeasible regions of the OF, after projecting these tentative solutions x in the OF space.

In this study, the SOOPs are considered, specifically limited to the structural optimization problems (SOP) field [1–11]. SOPs are generally dealing with solving three main subproblems. The size optimization (SiSOP) task deals with the definition of the optimal cross-section properties associated with every structural element in the domain of analysis, thus governing their resisting and stiffness properties. The shape optimization (ShOP) task instead refers to optimizing those variables governing the shape of the domain of interest, without modifying the interconnection layout of the placed elements. Lastly, the topology optimization task (TSOP) deals with rearranging the structural elements in the domain of analysis to find an optimized configuration under certain loading conditions. This latter is also a more complex task, which often requires special dedicated methods.

In the present study, the authors mainly focused on solving SiSOP and ShOP leveraging the potentials offered by modern approaches based on computational intelligence [1–3]. Indeed, the digital revolution fostered by the progress of artificial intelligence (AI), provided new machine learning (ML) based tools that can be combined with existing approaches for solving complex OP. Indeed, despite the well-established maths branch called operational research, gradient-based algorithms have been demonstrated to suffer from strong limitations with discontinuous and constrained complex real-world SOP. Conversely, meta-heuristic algorithms provided a new powerful framework for addressing those kinds of problems and providing some sub-optimum estimates conceivably close to the actual real optimum. The meta-heuristic algorithms lack of strong mathematical basis, but they were developed by mimicking some natural phenomena and readapted to address numerical OP. Genetic algorithms, Simulated Annealing, and Particle Swarm Optimization can be cited among the others as some of the most popular and nowadays widespread meta-heuristic techniques [12].

In the current study, the ML classification algorithm denoted as Support Vector Machine (SVM), has been integrated with the PSO algorithm as a different constraint handling solution, leveraging the surrogate model provided by SVM for guiding the swarm in the next iterations before updating the model again based on the actual OF evaluation. The idea of hybridizing PSO with SVM is not completely new in scientific literature. Indeed, the first attempts can be found in [13], using PSO and SVM to improve feature selection procedures in data mining and classification tasks. Another example can be found in [14] using PSO-SVM for optimizing the design and control of thermal energy storage systems. Mainly focusing on structural optimization tasks, the adoption of PSO-SVM for reliability-based design optimization can be found in [15] for minimizing a cost function while limiting the failure probability within a fixed threshold. In [16], PSO-SVM has been used for estimating geo-mechanical parameters of rock masses in complex geological conditions.

The current document is organized as follows. In the next section, the main governing principles of PSO and SVM are reviewed, and the proposed PSO-SVM integration is described in detail. Section 3 illustrates a case study example of SOP in which PSO-SVM revealed its potential, i.e. referred to the size and shape optimization of the conceptual design of a truss warren beam structure, which can represent a prototype of an arch bridge structure. The obtained results are in agreement with the existing literature [17, 18], and demonstrated the effectiveness of adopting this PSO-SVM tool for solving real-world structural optimization problems.

2 Particle Swarm Optimization Integrated with Support Vector Machine

The particle Swarm Optimization (PSO) algorithm is a meta-heuristic optimization method formalized by Kennedy and Eberhart [12] inspired by mimicking the biological behavior of bird flocking and fish schooling during the process of getting food in natural environments. It is a population-based method, in which intelligent agents called individuals are exploring the design variable search space, sampling their values, and getting the objective function (OF) evaluation for every position assumed by every agent. In PSO nomenclature, the population of intelligent agents that delivers the tentative solutions is called a swarm. To simulate a real-world swarm, the agents are able to interact with each other transmitting information on the search space to the entire population, according to the neighborhood concept. In this way, at every iteration, it is possible to define a global attractor (global best, Gb) which represents the so far best-found solution in the entire space. Likewise in a real-world biological environment, every individual is attracted not only to the global optimum, but thanks to their own intelligence and cognitive memory, it reminds also their individual so far best-visited position, which is denoted as local attractor (personal best, Pb). Therefore, considering that the particles of the swarm are moving in the search space following a Newtonian dynamics scheme, each particle's movements among the iterations are thus fully characterized by knowing its actual position and its velocity vector, which will determine the next iteration position. Despite the presence of local and global attractors, the movement paths of the individuals in the swarm are not completely deterministic, moving along a trajectory that would be somehow influenced by both an inertial term due to the previous step velocity, and also on a random combination of local and global attractiveness. Therefore, the next-step (k + 1) velocity and displacement updating rule of the Newtonian-dynamics-based PSO version formalized by Kennedy and Eberhart [12] can be thus expressed as follows:

$$ {}^{(k+1)}v_i = {}^{k}v_i + c_1^{(k+1)} r_{1i} \cdot \left[{}^{k}x_i^{Pb} - {}^{k}x_i \right] + c_2^{(k+1)} r_{2i} \cdot \left[{}^{k}x_i^{Gb} - {}^{k}x_i \right] \tag{1} $$

$$^{(k+1)}x_i = {}^{k}x_i + T \cdot {}^{(k+1)}v_i \tag{2}$$

In previous Eqs. (1) and (2), the term xi represents the position of any intelligent agent of the swarm in the search space, i.e. being equal to a specific realization of the design variables vector, whereas vi denotes the velocity term of the *i*-th particle, $T = 1$ is a pseudo-time variable equal to unity merely defined to ensure a physical-based meaning to Eq. (2), r_{1i} and r_{2i} are two random weights sampled by a continuous uniform probability distribution spanning between 0 and 1, c_1 and c_2 are denoted as cognitive and social terms respectively, which are priorly and deterministically weighting the relative importance of the cognitive and social terms related to the local Pb attractor and the global Gb one.

At its origins, the PSO was formalized for working with unconstrained optimization problems, and further constraint-handling methodologies were developed to extend the possibility of a widespread use of PSO, such as penalty-based methods among the others [1, 2]. In the present work, the PSO has been integrated with a machine learning (ML) methodology denoted as support vector machine (SVM) for providing a data-driven constraint handling that constructs a surrogate model of the feasible region in the search space. SVM is a supervised classification ML model based on statistical learning theory and was originally formalized by Vapnik in 1995 [19]. The SVM constructs a sort of Bayesian probability model of the search space while it is attempting to optimally separate the infeasible particle positions from the feasible ones. The ML method is based on the maximization of the margin, i.e. the optimal separation hyperplane, which can be conceived as the constraint position in the current scenario. However, since for structural optimization problems, the inequality constraints' boundaries are often represented by nonlinear functions, the optimal hyperplane can provide a poorly representative separating plane. However, the Kernel trick allows to map of data to a higher dimensional space, in which the apparently non-separable labeled data finally become linearly separable, and the hyperplane margin can be efficiently located with significantly reduced misclassification errors. If the mapping rule to higher dimensional space is invertible, the optimal found hyperplane could be restored in the initial design space, thus becoming a nonlinear boundary which this time better approximates the original constraint boundaries. The powerfulness of the kernel trick allows for the simulation of this process without properly and explicitly mapping data to higher dimensional space, therefore providing a computational effective tool. In this work, the radial basis function (RBF) kernel has been adopted. After a certain number of k of PSO iterations, the SVM can be trained based on all the explored positions of the swarm in the previous iterations, which have been labeled as feasible or infeasible. Once the SVM has been trained, the surrogate model can provide an estimate of the feasible region of the search space, and new PSO positions can be predicted to be feasible or infeasible based on the SVM scores. Periodically, this process can be repeated because the new PSO sampled position represents new data for the SVM algorithm, which can be retrained and improve the estimate of the feasible/infeasible region, especially closer to the real optimum region. It is worth noting that the SVM method can be further tuned to improve the surrogate model estimation, e.g. varying its regularization coefficient C. For a more comprehensive description of the proposed PSO-SVM method, the interested reader can refer to [17].

3 Structural Optimization Example: Conceptual Design of a Warren Truss Beam

The effectiveness of current PSO-SVM implementation on numerical benchmark case studies has been already demonstrated in [17]. The real-world case study herein considered is a planar Warren truss-type beam structure depicted in Fig. 1, carrying a fixed uniform static load of 100 kN/m, acting as point loads in the nodes of the lower chord. A similar geometry has been presented also in [17, 18]. In conceptual design optimization problems, the OF steel truss beams are normally conceived to minimize the material consumption usage, which is expressed directly as the total weight of the structure. Indeed, knowing the steel density $\rho_i = 7850\ \text{kg/m}^3$ of each *i*-th steel member, the weight is the summation of the products of steel density and the relative steel member volume, expressed in turn as the product of its length L_i and its cross-section area A_i. In order to limit the number of design variables, the geometry of the truss has been schematized using four different types of square hollow core (SHC) section beams for the lower chord, the upper chord, the internal diagonals, and the external diagonals. To address also the shape optimization problem, *Hmin* and *Hmax* govern the geometric height and layout of the beam truss. Therefore, the problem statement becomes finding $x = [s_i, B_i, Hmin, Hmax]^T \in \Omega$ such that minimizes the OF $f(x)$ such that:

$$\begin{aligned} \min f(x) = W(x) = \textstyle\sum_{i=1}^{m} \rho_i L_i A_i \ \text{subjected to} \\ \frac{N_{Ed,i}}{N_{tRd,i}} = \frac{N_{Ed,i}}{\frac{f_{yk} \cdot A_i}{\gamma_{M0}}} \le 1 \\ \frac{N_{Ed,i}}{N_{bcRd,i}} = \frac{N_{Ed,i}}{\chi \cdot \frac{f_{yk} \cdot A_i}{\gamma_{M1}}} \le 1 \end{aligned} \tag{3}$$

Moreover, some deformability constraints have been also considered in the optimization process, i.e. limiting the maximum deflection at a limited prescribed value of L/500. The constraints have been considered according to Eurocode design prescriptions EN-1993-2 2006, in terms of Ultimate Limit State (ULS) assessment in terms of maximum tensile strength $N_{tRd,i}$ and maximum buckling compression strength $N_{bcRd,i}$, being $f_{yk} = 275$ MPa the design value of structural steel S275, γ_{M0} is the partial safety factor equal to 1.05, $\chi \le 1$ is the compressive strength reduction factor due to buckling issues, and γ_{M1} is the partial safety factor equal to 1.10 for bridge design. The Warren truss beam total length is L = 20.00 m and the lower chord has been dived into 20 equal length ΔL intervals. The OP in Eq. (3) has been formalized as continuous optimization, therefore PSO explores design variables which are assumed to vary as continuous variables within an admissible range of interest. Indeed, *Hmax* has been limited to (0.05, 20.00) m, and consequently, *Hmin* has been dynamically set to be a positive quantity but lower than *Hmax*. Moreover, an admissible interval of (60;360) mm was set for SHC profiles' width B_i, whilst setting (4;30) mm for SHC profiles' thickness s_i. To evaluate both deflection and stresses in steel members, at every iteration of the PSO-SVM algorithm, the Matlab CALFEM numerical finite element solver has been adopted. The swarm size has been chosen with 100 particles, and the stopping criterion has been set when the threshold number of iterations equal to 200 has been reached. Since the achievement of the actual optimal solution is not guaranteed with meta-heuristic algorithms, it is advisable to perform a series of independent runs of the algorithm, e.g. 50 runs, and then evaluate the statistics of optimal founded solutions.

The optimal results obtained from the 50 independent runs have been reported in Fig. 2. Inspecting the graph, it is worth noting that the dispersion of OF values is quite scattered, demonstrating the meta-heuristic nature of the utilized algorithm, which did not always reach the optimal value of around 3.10 tons (red dashed line). Therefore, to evaluate the significant statistics, it is necessary to consider solely those similar solutions. Thus, the outlier results far from 3.10 tons have been excluded, retaining only 21 successful runs.

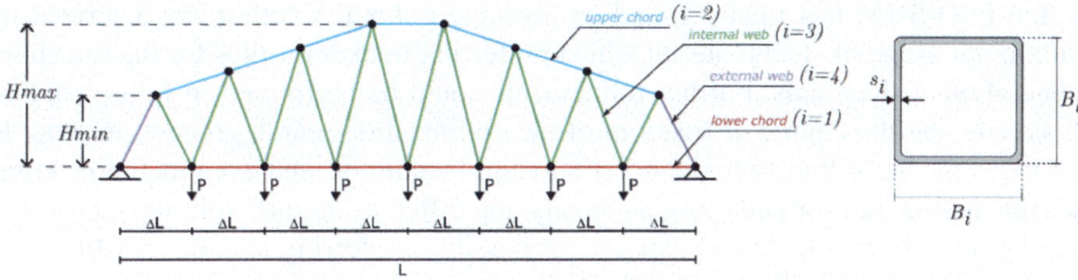

Fig. 1. Warren truss beam structure case study illustration.

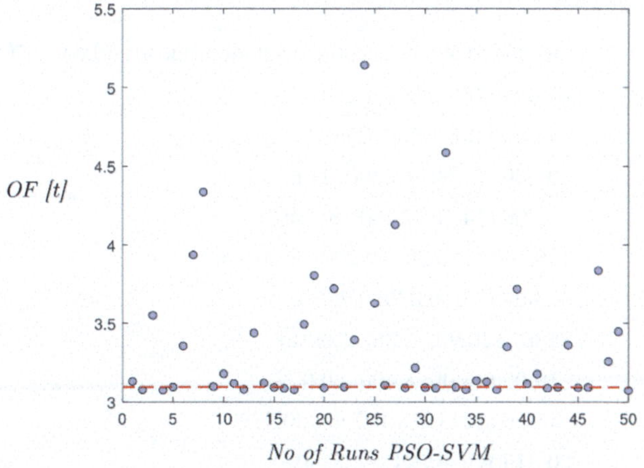

Fig. 2. Best optimal results in terms of OF evaluation for 50 independent runs of the algorithm.

Considering only these 21 latter solutions, the descriptive statistics have been computed and reported in Table 1. The descriptive statistics have been reported in terms of average value, standard deviation, and coefficient of variation (COV), which is the normalized ratio of the standard deviation and the average value. It represents a normalized measure of dispersion. The table also reports the absolute optimal founded solutions among the selected 21 as a reference set of design parameters, which have been rounded considering actual constructability tolerances issues and technical-commercially available steel profiles solutions. It is evident that despite selected optimal results are characterized by the same OF, different combinations of design variables are possible. This fact is evidenced by the relatively high COV of some variables such as the thickness of the lateral outer diagonals s_4, or the thickness of the upper chord s_2, and the corresponding width of the SHC section profiles. Indeed, considering the buckling resistance of steel members, the combination of width and thickness of the profile determines the class of the steel section profiles, permitting the full usage of its plastic resources (classes 1 and 2) or limiting the resistance to the elastic stage (class 3), or even exhibiting local sectional buckling in elastic stage (class 4). Therefore, it is evident that despite the absolute optimal solution is sometimes different from the average result, in this special case, the PSO-SVM has finally found an absolute optimal solution characterized by the minimum usage of steel material whilst preferring thicker profiles for the top chord and lateral outer diagonals. Further refinements could be performed in future studies, such as reducing the degree of freedom of the optimization search process, in order to better focus the algorithm to find similar structural solutions, and even including some clustering approaches for gathering and analyzing different optimal solutions, critically analyzing the different technical aspects for possibly preferring a solution rather than another, despite characterized by similar OF.

Table 1. OF and design variables' optimal results, and their statistics over the 21 selected independent runs characterized by similar OF around 3.10 tons.

Result	Rounded absolute optimal solution (mean; std.dev.; COV)
OF [ton]	3.09 (3.09; 0.007; 0.23%)
s_1 [mm]	4.00 (6.00; 1.500; 25.00%)
B_1 [mm]	95.00 (72.50; 15.700; 21.66%)
s_2 [mm]	26.00 (14.70; 7.400; 50.34%)
B_2 [mm]	105.00 (190.20; 70.200; 36.91%)
s_3 [mm]	4.00 (4.00; 0.030; 0.75%)
B_3 [mm]	130 (128.90; 1.400; 1.09%)
s_4 [mm]	19.00 (14.00; 8.400; 60.00%)
B_4 [mm]	110.00 (211.70; 107.90; 50.97%)
$H\ min$ [m]	0.41 (399.40; 21.50; 5.38%)
$H\ max$ [m]	4.15 (4064.7; 113.500; 2.79%)

4 Conclusions

The nowadays complexity of structural optimization problems (SOP) is demanding new powerful algorithms able to overcome limitations of operational research and mathematical gradient-based existing methods. Indeed, in the current study, a hybrid optimization algorithm has been presented. This was developed by integrating the meta-heuristic particle swarm optimization (PSO) algorithm with the machine learning (ML) support vector machine (SVM) classifier. The method attempts to combine the powerful search capabilities of swarm intelligence with a surrogate model of the search space provided by the SVM technique, acting as a non-penalty constraint handling method. The training labeled data for the SVM directly comes from iterations of the PSO algorithm which explores both feasible and infeasible regions. Periodically, the SVM is retrained, in order to progressively refine the feasible region boundaries. Specifically, when approaching the stopping criterion, the algorithm should migrate from an exploration phase to an exploitation phase to refine the optimal solution. Therefore, the SVM in these latter iterations should accurately represent the constraint boundaries in the optimum nearby. The PSO-SVM algorithm has been effectively employed for studying a real-world structural optimization problem of the conceptual design of a planar Warren truss beam problem. The Warren beam optimal design results are in agreement with the existing literature [18], and provided some interesting insights and remarks for future research improvements.

Acknowledgements. Giuseppe Carlo Marano acknowledges the support he received by the ADDOPTML project: "ADDitively Manufactured OPTi-mized Structures by means of Machine Learning" (No: 101007595, http://addoptml.ntua.gr/) belonging to the Marie Skłodowska-Curie Actions (MSCA) Research and Innovation Staff Exchange (RISE) H2020-MSCA-RISE-2020.

References

1. Rao, S.S.: Engineering Optimization: Theory and Practice. John Wiley & Sons, Hoboken (2019)
2. Spillers, W.R., MacBain, K.M.: Structural optimization. Springer Science & Business Media (2009)
3. Glover, F.W., Kochenberger, G.A. (Eds.): Handbook of metaheuristics, vol. 57. Springer Science & Business Media (2003)
4. Sardone, L., Fiore, A., Manuello, A., Greco, R.: Performative structural design optimization: generative algorithm for a preliminary study of a voided beam. Appl. Sci. **12**(17), 8663 (2022)
5. Sardone, L., Greco, R., Fiore, A., Moccia, C., De Tommasi, D., Lagaros, N.D.: A preliminary study on a variable section beam through Algorithm-Aided Design: a way to connect architectural shape and structural optimization. Procedia Manuf. **44**, 497–504 (2020)
6. Frangedaki, E., Sardone, L., Lagaros, N.D.: Design optimization of tree-shaped structural systems and sustainable architecture using bamboo and earthen materials. J. Archit. Eng. **27**(4), 04021033 (2021)
7. Melchiorre, J., Manuello, A., Marmo, F., Adriaenssens, S., Marano, G.C.: Differential formulation and numerical solution for elastic arches with variable curvature and tapered cross-sections. Eur. J. Mech.-A/Solids **97**, 104757 (2023)

8. Melchiorre, J., Bertetto, A.M., Invernizzi, S., Marano, G.C., Bazzucchi, F.: Dakar Mosque gridshell: exploring the benefits of the improved multibody rope approach through postbuckling analysis. Structures **61**, 106059. Elsevier (2024)
9. Bertetto, A.M., Melchiorre, J., Marano, G.C.: Improved multi-body rope approach for free-form gridshell structures using equal-length element strategy. Autom. Constr. **161**, 105340 (2024)
10. Cucuzza, R., Rad, M.M., Domaneschi, M., Marano, G.C.: Sustainable and cost-effective optimal design of steel structures by minimizing cutting trim losses. Autom. Constr. **167**, 105724 (2024)
11. Olivo, J., Cucuzza, R., Bertagnoli, G., Domaneschi, M.: Optimal design of steel exoskeleton for the retrofitting of RC buildings via genetic algorithm. Comput. Struct. **299**, 107396 (2024)
12. Kennedy, J., Eberhart, R.: Particle swarm optimization. In: Proceedings of ICNN'95-International Conference on Neural Networks, vol. 4. IEEE, 1995
13. Huang, C.L., Dun, J.F.: A distributed PSO–SVM hybrid system with feature selection and parameter optimization. Appl. Soft Comput. **8**(4), 1381–1391 (2008)
14. Yang, I.T., Hsieh, Y.H.: Reliability-based design optimization with cooperation between support vector machine and particle swarm optimization. Eng. Comput. **29**, 151–163 (2013)
15. Olabi, A.G., et al.: Application of artificial intelligence for prediction, optimization, and control of thermal energy storage systems. Therm. Sci. Eng. Progress **39**, 101730 (2023)
16. Li, H., Chen, W., Tan, X., Tan, X.: Back analysis of geomechanical parameters for rock mass under complex geological conditions using a novel algorithm. Tunn. Undergr. Space Technol. **136**, 105099 (2023)
17. Rosso, M.M., et al.: Nonpenalty machine learning constraint handling using PSO-SVM for structural optimization. Adv. Civ. Eng. **1**(2021), 6617750 (2021)
18. Fiore, A., Marano, G.C., Greco, R., Mastromarino, E.: Structural optimization of hollow-section steel trusses by differential evolution algorithm. Int. J. Steel Struct. **16**, 411–423 (2016)
19. Cortes, C., Vapnik, V.: Support-vector networks. Mach. Learn. J. (1995)

Stability Analysis of Vault Prototype Realized by Additive Manufacturing

Amedeo Manuello Bertetto(✉)

Department of Structural Geotechnical and Building Engineering, Politecnico di Torino, Turin, Italy
amedeo.manuellobertetto@polito.it

Abstract. Recent interest in masonry structures has highlighted their enduring potential as good solutions in architecture and engineering. The growing focus on the analysis of ancient masonry vaults, constructed globally throughout history, has revived the need to consider these structures for the roofing of new sustainable and innovative buildings. Concurrently, the stability of both ancient and modern masonry vaults remains a compelling and vital area of research [1–8].

This paper examines funicular structures to facilitate a comparative analysis aimed at identifying the optimal shape for hole configurations. The primary objective is to develop a form-finding design for latticed masonry shells, starting from a defined topology and producing an optimized grid concerning stiffness and stability [8–11]. The vault shapes are achieved by adjusting key parameters, such as the lowering degree and the hole pattern within the vault [1, 2]. Comprehensive structural stability analysis for these structures must be conducted across different loading scenarios and constraint conditions [12–14].

In this study, a scaled prototype of the proposed vault typology is fabricated using additive manufacturing techniques. Experimental and numerical tests are performed to correlate numerical predictions with empirical data, thereby enhancing the understanding of the vault's behavior under load.

Keywords: Masonry vault · Form-finding · multi-body rope approach · Stability

1 Introduction

The growing interest in masonry structures, both ancient and modern, underscores their potential as sustainable architectural solutions. Advances in computer-aided design, form-finding techniques, and digital fabrication have facilitated the development of innovative and unconventional architectures. Research on the stability of masonry vaults, especially in countries like Italy, continues to be a critical area of study. The integration of computational tools and real-time structural data is now crucial for designing these complex structures, enabling the fusion of engineering constraints with visual representation and design flexibility [1–8].

In recent decades, the design of doubly-curved shells and funicular, compression-only vaults has been extensively studied, building on historical techniques like hanging

N. D. Lagaros et al. (Eds.): ADDOPTML 2024, LNCE 646, pp. 331–340, 2025.
https://doi.org/10.1007/978-3-031-92029-5_31

models and graphic statics. Innovators such as Antoni Gaudí, Frei Otto, and Heinz Isler pioneered these approaches using physical models to shape domes and shells. Modern advancements focus on computational methods for optimizing grid shells and spatial structures, employing techniques like genetic algorithms, discrete truss topology, and simulated annealing. Form-finding approaches such as the force density method, dynamic relaxation, and particle-spring models have also been developed, with notable contributions like thrust network analysis for masonry structures. Coupled strategies combining form-finding and grid optimization further enhance the precision and versatility of shell design [1–12].

This paper presents a refined form-finding method based on a multi-body rope algorithm (MRA) to design vault models with freeform shapes composed of standardized elements. The study evaluates the method's effectiveness and identifies optimal configurations for pierced patterns in the vault [8, 12]. Stability analyses explore the best and worst evolution paths as the pierced area increases, providing insights for modern refurbishments of historic roofs where creating new openings is increasingly common. This study explores funicular vault structures to identify optimal shapes and hole patterns, utilizing parameters like lowering degree, slack coefficient, and hole percentage. Thrust Network Analysis (TNA) and global stability evaluations under various loading scenarios informed the designs, which were further validated through additive manufacturing prototypes.

2 MRA the Adopted Form-Finding

2.1 The Method

The study investigates the form-finding of a free-form compressed vault with a 16 × 16 m2 square geometry, focusing on key parameters such as the lowering degree and hole percentage (HP). The Multi-Body Rope Approach (MRA) and I-MRA were used to simulate dynamic, hanging lattice structures, allowing the exploration of different shapes under various loading conditions in both 2D and 3D systems [12]. The approach ensures the final shape consists of only traction forces, which can be reversed to a compression-only vault. The geometry obtained by MRA was validated through Thrust Network Analysis (TNA) using Rhino-vault, showing a small variation (2–5%) between the two methods.

Additionally, the study explored the effect of hole patterns and percentages on vault shapes. Four-hole positioning strategies were considered: central-central, lateral-central, corner-diagonal, and central-corner. The analysis demonstrated that varying the hole percentage (more than 45%) and their distribution did not significantly alter the shape, particularly when comparing extreme conditions. The MRA and TNA methods were found to be interchangeable in this context. The research highlighted the importance of hole percentage and pattern strategy in the design and refurbishment of masonry vaults, focusing on the impact of these factors on structural performance while considering height, thickness, and the slenderness of the vault.

According to the principle of basic MRA the equilibrium of a generic node *i* in the suspended net is [8, 12]:

$$\vec{R}_i = \vec{p}_i + \sum\nolimits_{j=1}^{n_j} \vec{F}_{j,i} + \vec{F}^{I}_{i} + \vec{F}^{II}_{i} = 0 \tag{1}$$

where R_i is the resultant of the equilibrium in the node, p_i represents the vector of the load on the node, $F_{j,i}$ is the force exerted on the rope connecting node i and j. $F^{I}{}_{j,i}$ is the dissipative force and $F^{II}{}_{j,I}$ is the force of inertia defined as follow [8, 12]:

$$\vec{F}^{II}_{i} = -m_i \vec{a}_i \tag{2}$$

$$\vec{F}^{I}_{i} = -c_i \vec{v}_i \tag{3}$$

In particular, the method proposed is characterized by a constraint between the nodes (masses) that is a rope element:

$$F_{rope} = 0 \text{ if } l < l_{rope} \tag{4}$$

$$F_{rope} = \text{k}\left(l - l_{rope}\right) \text{ if } l \geq l_{rope} \tag{5}$$

At the same time the distance between the center of the nodes connected by the ropes are [8, 12]:

$$l(i,j)\left\| \vec{ij} \right\| = \sqrt{(\text{x}_j - \text{x}_i)^2 + \left(\text{y}_j - \text{y}_i\right)^2 + (\text{z}_j - \text{z}_i)^2} \tag{6}$$

When the norm of the vector between the points i and j is ≥ to l_{rope} the rope constraint is active and the condition goes from slack to tension. According to the component of the vectoral equilibrium written in (1) along the *x* direction and assuming the dynamic equilibrium of a generic node in the system:

$$p_{ix} + \sum\nolimits_{j=1}^{n_j} \frac{\left(x_j - x_i\right)}{l(i,j)} F_{j,i} - c_i \dot{x}_i - m_i \ddot{x}_i = 0 \tag{7}$$

where c_i is the dissipative constant and m_i is the mass of the node *i*.

Considering again the *x* component and considering a constant C_3 it is possible to rite the following differential equation:

$$C_3 = 2\omega_n \zeta \dot{x} + \ddot{x} \tag{8}$$

The general integral of the complete solution will be:

$$x(t) = C_1 e^{-2\omega_n \zeta t} + C_2 + \frac{C_3}{2\omega_n \zeta} t \tag{9}$$

Taking into account the particular integral of the complete solution as:

$$x(t) = \frac{C_3}{2\omega_n \zeta} t \tag{10}$$

finally, the solution of Eq. (9) may be rewritten as, obtaining the solutions for the constant C_1 and C_2 assuming the solutions for the previous step for $x(t-\Delta t)$ e $\dot{x}(t-\Delta t)$ [8, 12]:

$$x(t) = -\frac{2\omega_n\zeta x(t-\Delta t) - C_3}{(2\omega_n\zeta)^2} e^{-2\omega_n\zeta t} + \frac{(2\omega_n\zeta)^2\dot{x}(t-\Delta t) - C_3}{(2\omega_n\zeta)^2} + \frac{C_3}{2\omega_n\zeta} t \quad (11)$$

By the equation reported in (11) it is possible to find a solution for the set of bodies connected by ropes. The solutions are given by an iterative process via a step-by-step solution of the dynamic system over the time and the space domains. This result allows to find at the end of the process a condition of equilibrium correspondent to the final or best shape according to the starting conditions. Obviously, these conditions are: (i) the constraints degree of the starting configurations of the net, (ii) the initial geometry (orientation) of the mesh, (iii) the *slack coefficient* imposed at the beginning of the process. In particular, this last parameter is very important in the definition of the lowering degree of the final shape. The slack coefficient mentioned above may be defined as follow [8, 12]:

$$\rho_{i,j} = \frac{l_{rope}}{l_{i,j(0)}} = \frac{l_{rope}}{\sqrt{(x_j - x_i)^2 + (y_j - y_i)^2 + (z_j - z_i)^2}} \quad (12)$$

As the ratio between the imposed length of the final configuration and the length (distance of the vertex) that the lines connecting the bodies assume during the form finding process.

The equilibrium condition of the process is reached and the process is stopped when the velocity and accelerations are $\leq$ to e_1 and e_2 both sufficiently small.

3 Pierced Vault: Pattern Strategy

One of the key factors for the pierced vault is the number and placement of the openings. The authors identified this factor as crucial for both designing new masonry vaults and refurbishing existing ones. The growing demand for more natural light and the constant need to reduce material usage have driven research into this study.

The analysis began by setting the key parameters for the vault range: vault heights of 2.5, 5, and 7.5 m were examined. Simultaneously, various thicknesses of the vaults, specifically 100, 150, 200, 250, and 300 mm, were included.

Several piercing percentages and configurations have been analysed with respect to the holes (see Figs. 2, 3, 4 and 5). Specifically, the analysis considered not only the number of holes but also the geometric patterns used for the growth of the pierced area. These patterns were categorized into four distinct families, each representing different progression paths for the incremental steps, resulting in variations in both the number and placement of the holes. In the figure the four patterns and the growth of the pierced area are reported, along with the hole percentage, the number of holes used at each step, and their positions for each case. Across all the patterns analysed, the maximum pierced percentage is 45.46%, which corresponds to a total of 49 holes (Fig. 1).

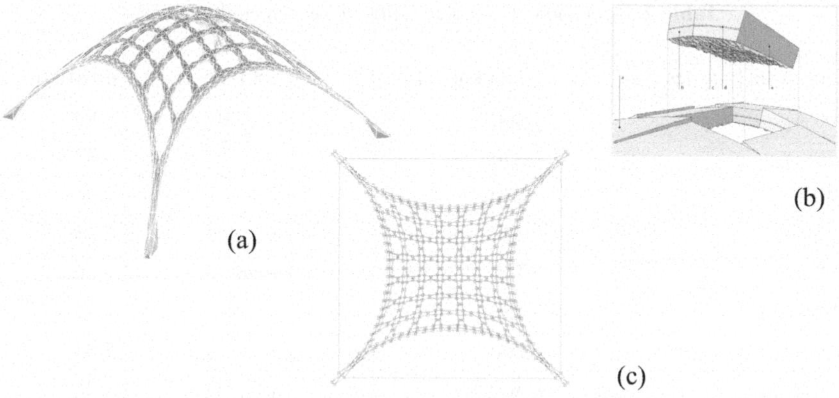

Fig. 1. Configuration of the vault after forming, maximum pierced area percentage: 45.46%. Includes an axonometric view and a top view (a). Diagram illustrating hole creation through the removal of the dowels (b), plan view (c).

In Table 1 mechanical characteristics of the selected masonry for the construction of the vault are reported. In Table 2 the results in term of stress computed in the material varying the piercing area (numbers of holes) the thickness and the progression paths (a, b, c, d).

Table 1. Mechanical characteristics of the selected masonry

Characteristic	value	unit
f_m Normal Stress	6.50	[MPa]
τ_0 Tangential stress	0.28	[MPa]
Γ specific wight	15	[kN/m^3]
E Elstic modulus	4550	[MPa]
G Shear Modulus	1137	[MPa]

In Table 2, the most important results for the four strategies of increasing the hole percentage are summarized. Specifically, the cases are presented varying the thickness, the number of holes, and the strategy used to increase the pierced area. The data for each case corresponds to the final condition where the internal stress values remain below the material limit indicated in Table 1 (6.50 MPa).

Table 2. Mechanical characteristics of the selected masonry

Piercing Area [n holes (%)]	Thickness [mm]	Set type	σ [MPa]
0 (0.00%)	200	a/b/c/d	5.89
1 (0.79%)	150	a/d	4.83
5 (5.40%)	200	a/d	4.94
9 (9.32%)	150	a/d	6.15
13 (13.46%)	150	a/c/d	6.16
33 (35.60%)	150	c	5.98
45 (42.50%)	200	c	5.57
49 (45.46%)	300	a/b/c/d	5.36

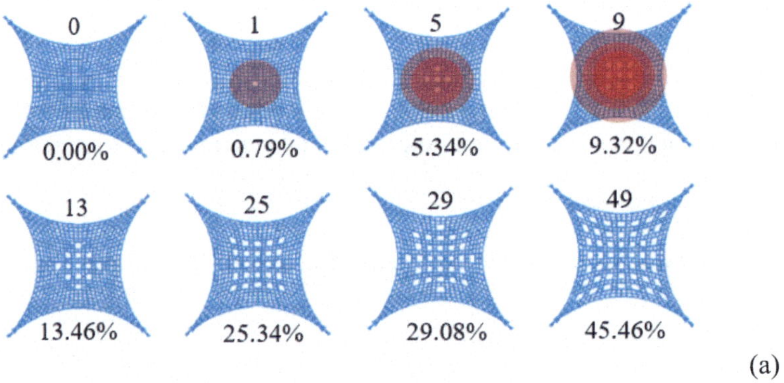

Fig. 2. The various patterns were grouped into four families based on distinct strategies for hole placement, each following a unique path for the progressive growth of the pierced area on the vault. Pattern (a) central-central strategy.

4 The Physical Model

The selected vault exhibits the highest critical load multiplier among structures. This vaulted structure has 49 holes, covering 45.5% of the surface area, stands 7.5 m tall, and has a transverse section thickness of 300 mm. The chosen scaled prototype is reduced by a factor of 1/16 to a side length of 1 m, resulting in a height of 46.9 cm and a transverse section thickness of 18.75 mm. Leveraging the geometry's double symmetry, only a quarter of the vault is considered for designing the voussoir geometry. The primary objective of the preliminary tests is to replicate and assess the overall behavior of the structure under load, with further experiments planned to consider dimensional analysis effects.

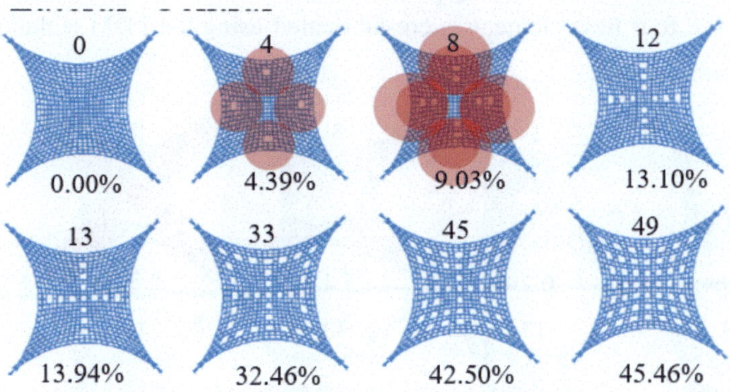

(b)

Fig. 3. The various patterns were grouped into four families based on distinct strategies for hole placement, each following a unique path for the progressive growth of the pierced area on the vault. Pattern (b) lateral-central strategy.

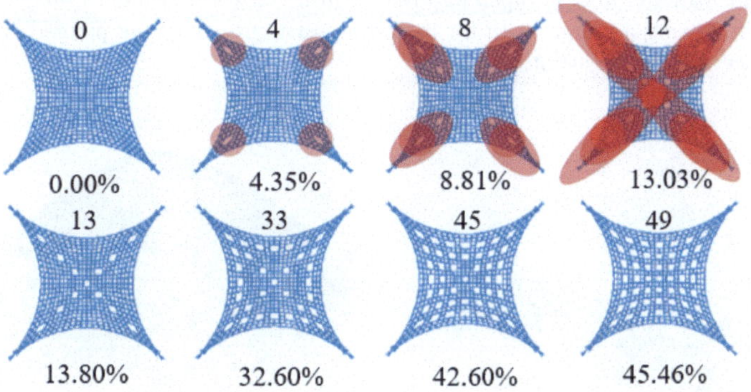

(c)

Fig. 4. The various patterns were grouped into four families based on distinct strategies for hole placement, each following a unique path for the progressive growth of the pierced area on the vault. Pattern (c) diagonal-central strategy.

Using Grasshopper and Rhinoceros, the continuous geometry of the vault was divided into several dowels. The brick design is based on the discretized form derived from the form-finding process, representing the forces acting on the vault when subjected to vertical loads. The bricks are made from cement mortar, requiring the design of a mold system. Given the small scale of the model, ensuring the accuracy of the formwork geometry is crucial. The formwork geometry for each brick was created in Rhinoceros using Grasshopper, and the STL files were exported for 3D printing using a Fuse Deposition Modeling (FDM) technique with PLA material, which is fully eco-friendly. Key 3D printing parameters include a filament diameter of 1.75 mm, a layer height of 0.2 mm, a fill density of 30%, a print speed of 80–150 mm/s, and a printing temperature of 215 °C.

For simplicity, the four base elements were fabricated using the FDM technique. All

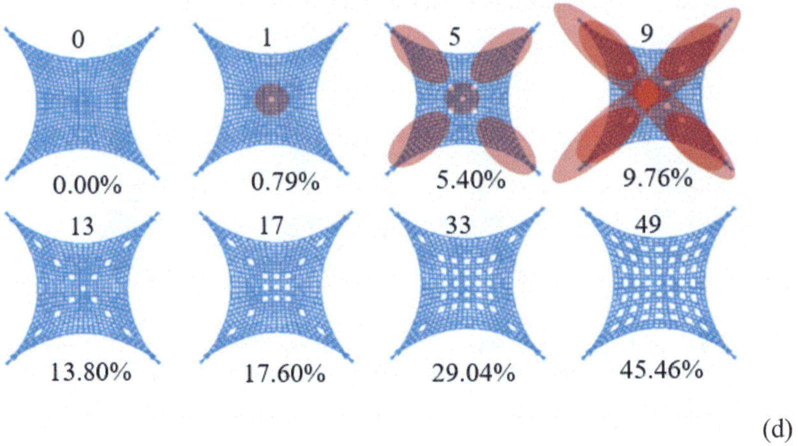

Fig. 5. The various patterns were grouped into four families based on distinct strategies for hole placement, each following a unique path for the progressive growth of the pierced area on the vault. Pattern (d) central-corners strategy.

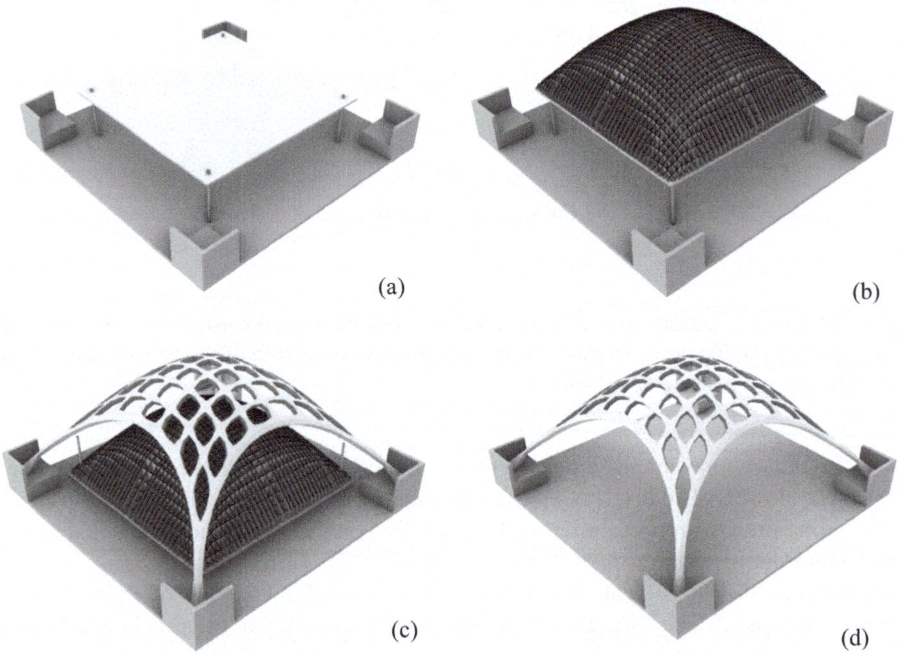

Fig. 6. Construction of the vault: Base (a), Falsework positioning (b), Vault construction and falsework removal (c), Final shape (d).

bricks were cast in the created formworks and made from cement mortar with a 0.4 water-to-cement ratio.

The three-dimensional design of the 100 different brick typologies and their formworks resulted in a total of 400 bricks for the vault. To support the bricks during the prototype construction, a spar was designed parametrically in Grasshopper and cut from cardboard sheets to fit the internal shape of the vault (Fig. 6). The spar was placed on a plywood panel supported by four threaded bars, which allowed height control throughout the construction process. Following the brick placement as per the design, the spar was removed after construction, and during the initial stages of downlifting, the structure's self-weight began to be supported by the vault (Fig. 5d).

5 Conclusions

Masonry structures are experiencing a resurgence as sustainable architectural solutions, with a focus on ancient vaults inspiring innovative designs. This research explores funicular forms to optimize lattice masonry shells, adjusting parameters such as lowering degrees and hole patterns for stability and stiffness. Stability is assessed under varied load and constraint scenarios, supported by a scaled prototype fabricated through additive manufacturing. Experimental and numerical analyses validate the vault's structural performance, advancing understanding of its behavior under load under the light of sustainable point of view and optimization strategies [15, 16]. Table 2 summarizes key results for four strategies of increasing the hole percentage, considering variations in thickness, hole count, and perforation strategies. Each case reflects the final condition where internal stress values stay below the material limit of 6.50 MPa.

Acknowledgments. This research was supported by project MSCA-RISE-2020 Marie Skłodowska Curie Research and Innovation Staff Exchange (RISE) - http://addoptml.ntua.gr/ - "ADDitively Manufactured OPTimized Structures by means of Machine Learning" (No: 101007595).

References

1. Richardson, J.N., Adriaenssens, S., Coelho, R.F., Bouillard, P.: Coupled form finding and grid optimization approach for single layer grid shells. Eng. Struct. **52**, 230–239 (2013)
2. Rippmann, M., Lachauer, L., Block, P.: Interactive vault design. Int. J. Space Struct. **27**(4), 219–230 (2012)
3. Oxman, R.: The New Structuralism: Design, Engineering and Architectural Technologies. John Wiley, New York (2010)
4. Tessmann, O.: Collaborative Design Procedures for Architects and Engineers, PhD thesis, University of Kassel, 2008
5. Fergusson, E.S.: The mind's eye: nonverbal thought in technology. Science **26**, 827–836 (1977)
6. Kilian, A.: Design exploration through bidirectional modeling of constraints, PhD thesis, Massachusetts Institute of Technology, Cambridge, MA, 2006
7. Tomlow, J., Graefe, F., Otto, F., Szeemann, H.: The Model, Institute for Lightweight Structures (IL), no. 34, University of Stuttgart, 1989

8. Manuello, A.: Multi-body rope approach for grid shells: Form-finding and imperfection sensitivity. Eng. Struct. **221**, 111029 (2020)
9. Cucuzza, R., Rosso, M.M., Aloisio, A., Melchiorre, J., Giudice, M.L., Marano, G.C.: Size and shape optimization of a guyed mast structure under wind, ice and seismic loading. Appl. Sci. **12**(10), 4875 (2022)
10. Melchiorre, J., Manuello, A., Marmo, F., Adriaenssens, S., Marano, G.C.: Differential formulation and numerical solution for elastic arches with variable curvature and tapered cross-sections. Eur. J. Mech.-A/Solids **97**, 104757 (2023)
11. Frangedaki, E., Sardone, L., Lagaros, N.D.: Design optimization of tree-shaped structural systems and sustainable architecture using bamboo and earthen materials. J. Archit. Eng. **27**(4) (2021). https://doi.org/10.1061/(asce)ae.1943-5568.0000492
12. Manuello, A., Melchiorre, J., Marano, G.C.: Improved multi-body rope approach for free-form gridshell structures using equal-length element strategy. Autom. Constr. **161**, 105340 (2024)
13. Bazzucchi, F., Manuello, A., Carpinteri, A.: Instability load evaluation of shallow imperfection-sensitive structures by form and interaction parameters. Eur. J. Mech. A. Solids **66**, 201–211 (2017)
14. Bazzucchi, F., Manuello, A., Carpinteri, A.: Interaction between snap-through and Eulerian instability in shallow structures. Int. J. Non-Linear Mech. **88**, 11–20 (2017)
15. Cucuzza, R., Rad, M.M., Domaneschi, M., Marano, G.C.: Sustainable and cost-effective optimal design of steel structures by minimizing cutting trim losses. Autom. Constr. **167**, 105724 (2024)
16. Cucuzza, R., Aloisio, A., Rad, M.M., Domaneschi, M.: Constructability-based design approach for steel structures: from truss beams to real-world inspired industrial buildings. Autom. Constr. **166**, 105630 (2024)

Author Index

N. D. Lagaros et al. (Eds.): ADDOPTML 2024, LNCE 646, pp. 341–342, 2025.
https://doi.org/10.1007/978-3-031-92029-5

The manufacturer's authorised representative in the EU is Springer Nature Customer Service Centre GmbH, Europaplatz 3, 69115 Heidelberg, Germany. If you have any concerns regarding our products, please contact ProductSafety@springernature.com

Printed and bound by CPI Group (UK) Ltd, Croydon, CR0 4YY
30/09/2026
02225590-0010